AGRO-INDUSTRIAL WASTES AS FEEDSTOCK FOR ENZYME PRODUCTION

AGRO-INDUSTRIAL WASTES AS FEEDSTOCK FOR ENZYME PRODUCTION

Apply and Exploit the Emerging and Valuable Use Options of Waste Biomass

Edited by

GURPREET SINGH DHILLON
University of Alberta, Edmonton, AB, Canada

SURINDER KAUR
University of Lethbridge, Lethbridge, AB, Canada

Amsterdam • Boston • Heidelberg • London
New York • Oxford • Paris • San Diego
San Francisco • Singapore • Sydney • Tokyo
Academic Press is an imprint of Elsevier

Academic Press is an imprint of Elsevier
125 London Wall, London EC2Y 5AS, United Kingdom
525 B Street, Suite 1800, San Diego, CA 92101-4495, United States
50 Hampshire Street, 5th Floor, Cambridge, MA 02139, United States
The Boulevard, Langford Lane, Kidlington, Oxford OX5 1GB, United Kingdom

Library of Congress Cataloging-in-Publication Data
A catalog record for this book is available from the Library of Congress

British Library Cataloguing-in-Publication Data
A catalogue record for this book is available from the British Library

ISBN: 978-0-12-802392-1

For information on all Academic Press publications
visit our website at https://www.elsevier.com/

Working together
to grow libraries in
developing countries

www.elsevier.com • www.bookaid.org

Publisher: Nikki Levy
Acquisition Editor: Nina D. Bandeira
Editorial Project Manager: Ana Claudia Garcia
Production Project Manager: Julie-Ann Stansfield
Designer: Greg Harris

Typeset by TNQ Books and Journals

CONTENTS

4. Industrial Enzymes: Recovery and Purification Challenges   95

P. Vijayaraghavan, S.R.F. Raj and S.G.P. Vincent

5. Low-Cost Enzymes and Their Applications in Bioenergy Sector   111

V.L. Queiroz, A.T. Awan and L. Tasic

LIST OF CONTRIBUTORS

R.E. Abraham
Deakin University, Waurn Ponds, VIC, Australia

A. Alberti
State University of Ponta Grossa, Ponta Grossa, Brazil

A.T. Awan
State University of Campinas, Campinas, Brazil

N. Capalash
Panjab University, Chandigarh, India

F. Carvalho
Universidade de Lisboa, Lisbon, Portugal

J.C. de Carvalho
Federal University of Parana, Curitiba, Brazil

M. de la Luz Mora
University of La Frontera, Temuco, Chile

L.P. de Souza Vandenberghe
Federal University of Parana, Curitiba, Brazil

S.K. Deshmukh
The Energy and Resources Institute, New Delhi, India

S. Devatkal
ICAR-Central Institute of Post-Harvest Engineering and Technology, Ludhiana, India;
ICAR-National Research Centre for Meat, Hyderabad, India

G.S. Dhillon
University of Alberta, Edmonton, AB, Canada

P. Fernandes
Universidade Lusófona de Humanidades e Tecnologias, Lisbon, Portugal; Universidade de Lisboa, Lisbon, Portugal

L. Gianfreda
University of Naples Federico II, Portici, Italy

T.A. Gomes
Federal University of Paraná, Curitiba, Brazil

N. Gopalan
CSIR-National Institute for Interdisciplinary Science and Technology (NIIST), Trivandrum, India

G.S. Kaira
CSIR-Central Food Technological Research Institute, Mysuru, India

M. Kapoor
CSIR-Central Food Technological Research Institute, Mysuru, India

K. Kaur
Panjab University, Chandigarh, India

U. Kumar
The Energy and Resources Institute, New Delhi, India

N. Libardi
Federal University of Parana, Curitiba, Brazil

A.U. Muzaddadi
ICAR-Central Institute of Post-Harvest Engineering and Technology, Ludhiana, India

K.M. Nampoothiri
CSIR-National Institute for Interdisciplinary Science and Technology (NIIST), Trivandrum, India

A. Nogueira
State University of Ponta Grossa, Ponta Grossa, Brazil

H.S. Oberoi
Guru Nanak Dev University, Amritsar, India; ICAR-Central Institute of Post-Harvest Engineering and Technology, Ludhiana, India; ICAR-Indian Institute of Horticultural Research, Bengaluru, India

D. Panwar
CSIR-Central Food Technological Research Institute, Mysuru, India

M. Puri
Deakin University, Waurn Ponds, VIC, Australia

S. Puri
Panjab University, Chandigarh, India

V.L. Queiroz
State University of Campinas, Campinas, Brazil

S.R.F. Raj
Nesamony Memorial Christian College, Kanyakumari, India

M.A. Rao
University of Naples Federico II, Portici, Italy

C. Rodrigues
Federal University of Parana, Curitiba, Brazil

R. Scelza
University of Naples Federico II, Portici, Italy

P. Sharma
Panjab University, Chandigarh, India

R. Sharma
Guru Nanak Dev University, Amritsar, India

G. Singh
Panjab University, Chandigarh, India

C.R. Soccol
Federal University of Parana, Curitiba, Brazil

M.R. Spier
Federal University of Paraná, Curitiba, Brazil

K. Sunar
The Energy and Resources Institute, New Delhi, India

L. Tasic
State University of Campinas, Campinas, Brazil

J.O. Ugwuanyi
University of Nigeria, Nsukka, Nigeria

P. Vijayaraghavan
Manonmaniam Sundaranar University, Kanyakumari, India

S.G.P. Vincent
Manonmaniam Sundaranar University, Kanyakumari, India

CHAPTER 1

Microbial Enzyme Factories: Current Trends in Production Processes and Commercial Aspects

L.P. de Souza Vandenberghe, J.C. de Carvalho, N. Libardi, C. Rodrigues, C.R. Soccol

Federal University of Parana, Curitiba, Brazil

INTRODUCTION

Enzymes are special proteins, which catalyze chemical reactions with great specificity and rate enhancements. These reactions are the basis of the metabolism of all living organisms, and provide tremendous and economical biocatalyst conversions. The first half of the past century saw rapid development in enzyme chemistry. The Enzyme Commission, set up by the International Union of Biochemists (1965), has published a system of enzyme classification and more than 4000 enzymes have been recognized. However, 25,000 natural enzymes have now been speculated to exist, which means about 90% of the reservoir of biocatalysts still remains to be discovered and characterized (Menon and Rao, 1999).

When designing a new synthetic process, a suitable catalyst for the reaction must be found, and enzymes are ideal candidates for this role. The industrial use of enzymes has developed rapidly because of their specific functions (Ogawa and Shimizu, 1999). Commercial exploitation of microbial enzymes began much before their natures and properties were worked out. For centuries, extracts of plants have been used to bring about hydrolysis of polymeric materials. However, these sources of enzymes were unreliable and expensive too, hence search for alternative sources began. Largely, this was found in the microbial cultures (Menon and Rao, 1999). The first enzyme produced industrially was the fungal amylase *takadiastase*, employed as a pharmaceutical agent (for digestive disorders) in the United States in 1894. The 1950s saw important growth in the industrial enzyme production and use of microbial enzymes. By 1969, 80% of all laundry detergents contained enzymes, mainly proteases. Additional enzymes, such as lipases, amylases, pectinases, and oxidoreductases were then experimented with in the detergent industry.

Many enzymes are commercially available, and numerous industrial applications have been described. An overview was recently published by Li et al. (2012) about the most

Agro-Industrial Wastes as Feedstock for Enzyme Production
ISBN 978-0-12-802392-1
http://dx.doi.org/10.1016/B978-0-12-802392-1.00001-0

1

important applications of enzymes. The role of enzymes in many processes has been known for a long time. People in ancient Greece used enzymes from microorganisms in baking, brewing, alcohol production, cheese making, etc. With better knowledge and purification of enzymes, the number of applications has increased manyfold, and with the availability of thermostable enzymes a number of new possibilities for industrial processes have emerged (Haki and Rakshit, 2003). Thermostable enzymes, which have been isolated mainly from thermophilic organisms, have found a number of commercial applications because of their overall inherent stability (Demirijan et al., 2001).

The food, feed, agriculture, paper, leather, and textile industries are well suited for enzyme technology because products as well as raw materials consist of biomolecules, which can be produced, degraded, or modified by enzymatic processes (van Beilen and Li, 2002). The majority, almost 75%, of currently used industrial enzymes are hydrolytic in action, being used for the degradation of various natural substances. Approximately 200 microbial original types are used commercially. However, only about 20 enzymes are produced on a truly industrial scale. According to a recent research report from the Austrian Federal Environment Agency, about 158 enzymes were used in the food industry, 64 enzymes in technical application, and 57 enzymes in feedstuff, of which 24 enzymes are used in three industrial sectors (Li et al., 2012).

Enzymes significantly contribute to global annual revenue, and therefore, the emphasis has been on engineering them. Although a great deal of data is accumulating on making alterations in microbial enzymes, there is a lack of definite information on redesigning industrial enzymes. Modification of the existing enzymes has become a trend for fine tuning of biocatalysts in the biotech industry. Tolerance to high or low temperatures, exhibiting activity in alkaline or acidic environments, high performance in nonaqueous media, and increased protease resistance are a few of the requisite properties (Joshi and Satyanarayana, 2015). Proteins are then redesigned in such a way that the industrial processes can be carried out in a more economic and greener way. In protein engineering, various methods are employed in modifying the target proteins. These are mainly rational design, directed evolution, and semirational approaches for designing and constructing novel proteins. The rational methods require prior knowledge of the amino acid sequence, three-dimensional (3D) structure and knowledge of the structure–function aspects of the target proteins. When mutations carried out by computational and random mutagenesis are compared, often the best mutants have been developed by either of the methods.

The use of recombinant DNA technology has made possible the manufacture of novel enzymes. Such enzymes may be discovered by screening microorganisms sampled from diverse environments using modern methods of protein engineering or molecular evolution. As a result, several important new enzymes have been developed. Another important achievement is improvement of microbial production strains. For example, several microbial strains recently developed for enzyme production have been engineered to increase enzyme yield by deleting native genes encoding extracellular proteases. Moreover, certain fungal production strains have been modified to reduce or eliminate their potential for production of toxic secondary metabolites (Olempska–Beer et al., 2006).

In many cases, the substrates in industrial processes are artificial compounds or alternative substrates, such as agro-industrial subproducts. In this way, enzymes known to catalyze suitable reactions for such processes are still unknown. Therefore, screening for novel enzymes that are capable of catalyzing new reactions in such mediums is constantly needed.

MICROORGANISMS AS A POTENTIAL SOURCE FOR HYDROLASES

Over the past few decades, considerable research has been undertaken with the enzymes produced by a wide variety of microorganisms. Enzymes have been derived from several fungi, yeasts, bacteria, and actinomycetes.

One of the most efficient and successful means of finding new enzymes is to screen a large number of microorganisms. The findings of thermophilic organisms, for example, have been possible with the isolation of a large number of beneficial thermophilic microorganisms from different exotic ecological zones of the earth and the subsequent extraction of useful enzymes from them (Demirijan et al., 2001). One extremely valuable advantage of conducting biotechnological processes at elevated temperatures is reducing the risk of contamination by common mesophiles. Elevated process temperatures include higher reaction rates due to a decrease in viscosity and increase in diffusion coefficient of substrates and higher process yield due to increased solubility of substrates and products and favorable equilibrium displacement in endothermic reaction (Kumar and Swati, 2001). Table 1.1 provides some examples of hydrolases produced by different microorganisms.

AGRO-INDUSTRIAL RESIDUES FOR HYDROLASES PRODUCTION

The majority of the commercially available hydrolases are produced by submerged fermentation (SmF), overexpressing selected genes in either native or heterologous microbial hosts. However, there are still many research publications testing and comparing SmF with solid-state fermentation (SSF) processes, as an attempt of improvement of the enzymatic productivity, reducing its costs, and improving the viability for commercial application. According to Hussain et al. (2013), the SSF is the method of preference for α-amylase production, regarding the better quality of production, easy follow-through procedures, lesser production costs, energy savings, and no foam formation. However, xylanases (Ho, 2014) and phytases (Lei et al., 2013) are produced in a better way using the SmF process. Fungal lipases are preferably produced in SSF, while bacteria and yeast lipases are produced in SmF (Treichel et al., 2010).

The definition of a medium composition that can significantly affect the product concentration, yield, and volumetric productivity is of central importance when

Table 1.1 Hydrolases Produced by Different Microorganisms

Enzyme	Microorganisms	References
Amylases	*Aspergillus oryzae, Aspergillus fumigates, Calvatia gigantea, Thermomyces lanuginosus, Streptomyces* sp., *Bacillus* sp., *Halomonas meridian, Aspergillus nidulans, Thermomonospora fusca, Thermococcus profundus, Saccharomyces cerevisiae, Schwanniomyces alluvius, Thermomonospora curvata, Lactobacillus plantarum* A6, *Pseudomonas stutzeri, Streptococcus bovis, Aspergillus usamii, Cryptococcus* S-2, *Fusarium vasinfectum, Paecilomyces* sp., *Micromonospora melanospora*	Gupta et al. (2003)
Cellulases	*Aspergillus niger, A. oryzae, Bacillus subtilis, Trichoderma longibrachiatum*	Li et al. (2012)
	A. niger, Trichoderma reesei, Phanerochaete chrysosporium, Trametes spp., *Pleorotus* spp., *Lentinus edodes, Lentinus tigrinus, Cerrena maxima, Funalia trogii, Coriolopsis polyzona, Pycnoporus coccineus, Pycnoporus sanguineus, Bjerkandera adusta, Fomes fomentarius, Pseudotremella gibbosa, Trichaptum biforme, Irpex lacteus, Ceriporiopsis subvermispora, Laetiporeus sulfurous, Fomitopsis* sp., *Wolfiporia cocos, Piptoporus betulinus, Gloeophyllum trabeum*	Yoon et al. (2014)
	Trichoderma reesei, Penicillium spp., *Acremonium cellulolyticus, Chrysosporium lucknowense, Phanerochaete chrysosporium, Aspergillus terreus, Phanerochaete carnosa, Neurospora crassa*	Zoglowek et al. (2015)
Mannanases	*Bacillus lentus*	Li et al. (2012)
	Trichoderma reesei, Aspergillus spp., *Aureobasidium pullulans, Talaromyces* sp., *Thermoascus* sp., *Thermomyces* sp., *Thielavia* sp., *Rhizopus niveus, Trametes versicolor, Sclerotium rolfsii, Pleurotus ostreatus, Schizophyllum commune, Ceriporiopsis subvermispora, Piptoporus betulinus*	van Zyl et al. (2010)
Lipases	*Bacillus prodigiosus, Bacillus pyocyaneus, Bacillus fluorescens, Serratia marcescens, Pseudomonas aeruginosa, Pseudomonas fluorescens, Bacillus* sp., *Pseudomonas* sp., *Enterococcus faecalis, Lactobacillus plantarum, Staphylococcus* sp., *Penicillium cyclopium Penicillium simplicissimum, A. niger, A. oryzae, Botrytis cinerea, Chromobacterium viscosum, Streptomyces flavogriseus, Trichosporon asteroids, Trichosporon laibacchi, Rhizopus* sp., *Geotrichum candidum, Pichia burtonii, Candida cylidracae, Acinetobacter* sp., *Fusarium solani*	Hasan et al. (2006)
Pectinases	*Aspergillus, Rhizopus, Penicillium, Neurospora, Bacillus, Streptomyces*	Li et al. (2015)
	Aspergillus japonicus, Penicillium glandicola	Bezerra et al. (2012)
	Aspergillus niger CH4, *Penicillium frequentans, Scleroyium rolfsii, Rhizoctonia solani, Mucor pusilus, Clostridium thermosaccharolyticum, Bacillus* sp., *Bacillus polymyxa, Bacillus stearothermophilus, Penicillium italicum, B. subtilis, Pseudomonas syringae pv. Glycinea*	Kashyap et al. (2001)

Table 1.1 Hydrolases Produced by Different Microorganisms—cont'd

Enzyme	Microorganisms	References
Phytases	*A. niger, A. oryzae* *A. niger, Aspergillus ficuum, Aspergillus fumigatus, Saccharomyces cerevisiae, Pichia anomala, Schwanniomyces castellii, Lactobacillus sanfranciscensis, Hanseniaspora uvorum, Yarrowia lipolytica, Kodamaea ohmeri, Candida tropicalis, Candida carpophila, Escherichia coli, B. subtilis, Klebsiella terringa, Lactobacillus sp., Pseudomonas spp.*	Li et al. (2012) Kumar et al. (2010)
Proteases	*Bacillus* sp.	Ladeira et al. (2010)
	B. subtilis	Mukhtar and Haq (2013)
	A. oryzae	Belmessikh et al. (2013)
	A. niger	Mazotto et al. (2013)
	Pseudomonas aeruginosa	Mahanta et al. (2008)
	Synergistes sp.,	Kumar et al. (2008)
	Rhizopus oligosporus	Ikasari and Mitchell (1996)
Xylanases	*A. oryzae, Humicola insolens, B. subtilis* *Aspergillus* sp., *Trichoderma* sp., *Penicillium* sp., *Aureobasidium* sp., *Fusarium* sp., *Chaetomium* sp., *Phanerochaete* sp., *Rhizomucor* sp., *Humicola* sp., *Talaromyces* sp.	Li et al. (2012) Kheng and Omar (2005) and Pandey et al. (1999)
Nitrile Hydrolases	*Gibberella fujikuroi, Rhodococcus* sp., *Bacillus* sp., *Pseudomonas putida*	Kobayashi and Shimizu (2000)
Epoxide Hydrolases	*Rhodotorula glutinis, Mycobacterium* sp., *Nocardia* sp., *Agrobacterium radiobacter, Rhodococcus erythropolis, Aspergillus* spp., *Beauveria* spp.	Weijers and de Bont (1999)
Bile Salt Hydrolases	*Bacteroides* spp., *Enterococcus* spp., *Bifidobacterium* spp., *Lactobacillus* spp., *Lactococcus lactis, Leuconostoc mesenteroides, Streptococcus thermophilus*	Reyes-Nava and Rivera-Espinoza (2014)

developing an industrial bioprocess. The use of agro-industrial residues is an interesting strategy for the reduction of the costs associated with the culture medium formulation. In recent years, there has been an increasing trend toward efficient utilization of agro-industrial residues in fermentative bioprocess, such as coffee pulp and husk, cassava husk, cassava bagasse, sugarcane bagasse, sugar beet pulp, apple pomace, declassified potatoes, citric pulp, sugarcane molasses and vinasse, soybean molasses and vinasse, etc (Soccol and Vandenberghe, 2003).

For α-amylase production, corn steep liquor and chicken feather were already reported as nitrogen sources as well as molasses, sugarcane bagasse, rice husks, corn starch, potato starch, maize starch, wheat starch, and rice starch as carbon sources (Hussain et al., 2013). Several cellulosic residues, such as sugarcane bagasse, wheat bran, paper pulp, corn cob residue, wheat straw, corn stover residue, and many others were reported as carbon sources and inducers of cellulases production in SmF or SSF processes (Singhania et al., 2010). Sahnoun et al. (2015) tested the substitution of yeast extract as the nitrogen source used with agro-industrial residues namely tuna fish powder, wheat gluten waste, and soybean meal, producing α-amylase from *Aspergillus oryzae* in SSF (1L-Erlenmeyer flask, 30°C). After optimization studies, the authors achieved maximum specific enzyme activity of $22,118.34 \, U \, g^{-1}$. In addition, they found a new high-molecular-weight α-amylase, namely Amy C. Singh and Gupta (2014) used Sal (*Shorea robusta*) deoiled cake as substrate for α-amylase production using *Aspergillus flavus* TF-8 in SSF, SmF and mSSF (modified solid-state fermentation), obtaining maximum enzymatic activities of 1.82, 0.36 and $2.51 \, U \, mL^{-1}$, respectively. After optimization of the SmF process, it was reached at $26.38 \, U \, mL^{-1}$.

The most important way to compare different production methods and culture medium compositions, including agro-industrial residues, is the enzyme productivity. Haq et al. (2003) reported the amylase productivity of $102 \, U \, mL^{-1} \, h^{-1}$, for an SmF bioprocess, conducted at 40°C during 48 h, using *Bacillus licheniformis* as the producer strain with wheat bran and pearl-milled starch as substrate. Esterbauer et al. (1991) reported the cellulase production with a volumetric productivity of $100 \, FPU \, L \, h^{-1}$, using wheat straw as substrate. The majority of the published data has been focusing on the maximum enzymatic activity, which makes the comparison between results very difficult.

The great majority of the hydrolases published data are related to bench-scale production processes. Only a few reports deal with the results and problems related to cellulase production in a pilot plant, or to significant parameters for the scaling up of the process from a few to several hundred liters. Oinonen et al. (2004), working in a 700-L fermenter using SmF process, achieved $1160 \, nkat \, mL^{-1}$ of cellulase after 72 h. In 1991, Esterbauer et al. (1991) already reported the cellulase production in a volume of 3000 L, with a yield of $250 \, mg \, cellulase \, mg^{-1}$ substrate and maximum cellulase activity of $18 \, FPU \, mL^{-1}$, using a *Trichoderma reesei* mutant.

In addition, the sustainability of the process should also be taken into the account. Besides the use of agro-industrial residues, strategies such as open culture processes with nonsterile culture conditions could be explored for in situ enzyme production, with mixed cultures of microorganisms (Kleerebezem and Loosdrecht, 2007), taking advantage of ecological selection principles, when high purification degree is not required.

Table 1.2 presents the production of important hydrolases using agro-industrial residues/subproducts.

Table 1.2 Production of Hydrolases Using Different Agro-Industrial Residues

Cellulases

Microorganism	Neurospora crassa	Melanocarpus albomyces	Penicillium janthinellum	Chaetomium thermophilum CT2	Trichoderma reesei Rut C-30	Pleurotus ostreatus	Aspergillus niger NRRL 567	Trichoderma harzianum
Agro-industrial residue	Wheat straw	Cellulose	Sugarcane bagasse	Cellulose	Dried kinnow pulp + wheat bran	Sugarcane bagasse	Apple pomace	Palm oil extraction residues
Bioprocess conditions	SmF; 30°C; pH 6.5; 100h	SmF; 45°C; pH 6.5; 72h	SmF; 30°C; 192h	SmF; 50°C; 216h	SSF; 30°C; 96h	SSF; 192h	SSF; 96h	SSF; 360h
Enzyme Activity	FPAse 1.33 FPU mL⁻¹; CMCase 19.7 U mL⁻¹, BGL 0.58 U mL⁻¹	Cellulase 1160 ECU mL⁻¹ Endo-glucanase 3290 ECU mL⁻¹	Cellulase 0.55 U mL⁻¹ CMCase 21.5 U mL⁻¹ Glucanase 2.3 U mL⁻¹	CMCase 2.7 U mL⁻¹	13.4 FPU gds⁻¹; 13.4 CMCase U gds⁻¹ 18 BGL U gds⁻¹	0.013 U FP gds⁻¹; 0.18 UCMC gds⁻¹	383.7 FPU gds⁻¹; 425 CMCU gds⁻¹; 336 BGL U gds⁻¹	8.2 FPU gds⁻¹ (lab scale) 10.1 FPU gds⁻¹ (pilot scale)
References	Romero et al. (1999)	Oinonen et al. (2004)	Adsul et al. (2004)	Li et al. (2003)	Oberoi et al. (2010)	Membrillo et al. (2008)	Dhillon et al. (2012a,b)	Alam et al. (2009)

Amylases

Microorganism	Aspergillus flavus	Bacillus licheniformis NH1	B. licheniformis	Aspergillus niger UV60	Aspergillus orizae	Aspergillus orizae	B. licheniformis	Bacillus sp. PS-7
Agro-industrial residue	Shorea robusta deoiled cake	Chicken feathers	Wheat bran + pearl milled starch	Food waste	coconut oil cake	Soybean meal	Wheat bran	Wheat bran + soybean meal
Bioprocess conditions	SmF; 200rpm; 30°C; 72h.	SmF; 37°C; pH 7.0; 48h	SmF; 40°C; 48h	SmF; 30°C; pH natural; 96h	SSF; 30°C; 72h;	SSF; 30°C	SSF; 50°C; forced aeration.	SSF; 37°C; 48h
Enzyme Activity	26.38 U mL⁻¹	9.86 U mL⁻¹	436 U mL⁻¹ Qₚ: 102 U mL⁻¹ h⁻¹	137 U mL⁻¹	3388 U gds⁻¹	22,118.34 U g⁻¹	26.350 U g⁻¹ DBB	4,64,000 U g⁻¹ DBB
References	Singh and Gupta (2014)	Erdal and Askin (2010)	Haq et al. (2003)	Wang et al. (2008)	Ramachandran et al. (2004)	Sahnoun et al. (2015)	Babu and Satyanarayana (1995)	Sodhi et al. (2005)

Xylanases

Microorganism	A. nigesr	Lichtheimia ramose	Aspergillus fumigatus	Trichoderma lanuginosus	Penicillium canescens	Thermomyces lanuginosus SD-21	A. niger NRRL - 567	Cochliobolus sativus Cs6
Agro-industrial residue	Palm leaf	Bociaiva fruit residue	Rice straw	Sugarcane bagasse	Soya oil cake + casein	Corn cob + wheat bran	Rice husk	Wheat straw + NaNO₃

Continued

Table 1.2 Production of Hydrolases Using Different Agro-Industrial Residues—cont'd

Bioprocess conditions	SmF	SmF	SmF	SmF	SSF; 30°C; pH 7.0	SSF; 40°C; pH 6.0	SSF; 30°C; pH 5.0; 72h	SSF; 30°C; pH 4.5
Enzyme Activity	1906.5 UL⁻¹	1802 UL⁻¹	1040 UL⁻¹	946 UL⁻¹	18,895 Ug⁻¹	8237 UU g⁻¹	3952 Ugds⁻¹	1,469,4 Ug⁻¹
References	Norazlina et al. (2013)	Silva et al. (2013)	Sarkar and Aikat (2012)	Ali et al. (2013)	Assamoi et al. (2008)	Yang et al. (2011)	Dhillon et al. (2012a,b)	Arabi et al. (2011)

Phytases

Microorganism	Nocardia sp. MB 36	Bacillus subtilis US417	A. niger St-6	Klebsiella sp. DB-3	Rhizomucor pusillus	A. niger NCIM 563	Mucor racemosus	Rhizopus oryzae
Agro-industrial residue	Rice mill waste	Wheat bran	Wheat bran	Orange peels	Wheat bran	Wheat bran	Oils cakes + Wheat bran	Linseed oil cake + Wheat bran
Bioprocess conditions	SmF; 47°C; 40h	SmF; 20h	SmF; 30°C	SmF; 50°C; 72h	SSF; 50°C; pH 6.0; 48h	SSF	SSF;	SSF; 96h
Enzyme Activity	0.254 UmL⁻¹	0.64 UmL⁻¹	85 UmL⁻¹	3.15 UmL⁻¹	9.18 Ug⁻¹	79 Ug⁻¹	44.5 Ug⁻¹	17.68 Ug⁻¹
References	Bajaj and Wani (2011)	Farhat et al. (2008)	Tahir et al. (2010)	Mittal et al. (2013)	Chadha et al. (2004)	Mandviwala and Khire (2000)	Roopesh et al. (2006)	Rani and Ghosh (2011)

Mannanases

Microorganism	Aspergillus sojae	Penicillium chrysogenum QML-2	Aspergillus sydowii	A. niger	Pleurotus ostreatus	Bacillus circulans	A. niger E-30 (mutant)	A. niger
Agro-industrial residue	Sugar beet molasses	Wheat bran + corn stover powder	Banana stem	Defatted copra	Wheat straw	Fibrous soy residue	Wheat straw	Soybean husks
Bioprocess conditions	SmF; 30°C; 96h	SmF; 30°C; pH 6.0; 144h	SmF; 30°C; 168h	Smf; 30°C; 168h	SSF;	SSF	SSF	SSF;
Enzyme Activity		8479.82 Ug⁻¹	1.229 UmL⁻¹	28 UmL⁻¹	35 Ugdm⁻¹	0.54 Umgdm⁻¹	36.67 Ug⁻¹	200 Ugdm⁻¹
References	Ozturk et al. (2010)	Zhang and Sang (2015)	Siqueira et al. (2010)	Lin and Chen (2004)	Valášková and Baldrian (2006)	Heck et al. (2005)	Wu et al. (2011)	Vandenberghe et al. (2012)

Lipases

Microorganism	*Yarrowia lipolytica* M1	*Y. lipolytica* RO12	*Pseudomonas gessardii*	*Candida rugosa*	*P. chrysogenum*	*Penicillium* P74	*Penicillium restrictum*	*Bacillus altitudinis* AP-MSU
Agro-industrial residue	Olive oil	Oil mill wastewater	Slaughter house waste	Oil mill wastewater	Grease waste	Soybean meal	Poultry slaughter house	Fish processing waste
Bioprocess conditions	SmF	SmF	SmF; 37°C; pH 5.0	SmF; pH 5.6	SSF; 30°C; pH 7.0; 168h	SSF; 120h	SSF; 35°C; 20h	SSF; 50°C; pH 7.8
Enzyme Activity	11 U mL^{-1}	1677 U	139 U mL^{-1}	3511 U^{-1}	40.7 U mL^{-1}	203.72 U g^{-1}	21.4 U g^{-1}	2.964 U g^{-1}
References	Mafakher et al. (2010)	Lanciotti et al. (2005)	Ramani et al. (2010)	Gonçalves et al. (2009)	Kumar et al. (2012)	Rigo et al. (2010)	Valladão et al. (2007)	Essakiraj et al. (2010)

Proteases

Microorganism	*Bacillus* sp. SMIA-2	*Bacillus subtilis*	*Aspergillus oryzae*	*A. niger*	*A. oryzae in*	*Pseudomonas aeruginosa* PseA	*Synergistes* sp.	*Rhizopus oligosporus* ACM 145F
Agro-industrial residue	Apple pectin, whey protein and corn step liquor	Soybean meal	Tomato pomace + wheat bran	Chicken feathers	deoiled *Jatropha curcas* seed cake	Jatropha seed cake	Wheat bran	Rice bran
Bioprocess conditions	SmF; 50°C; pH 6.5; 30h	SmF; 35°C; pH 8.5; 48h	SmF; 30°C; pH 6.8; 96h	SmF; 32°C	SSF; 30°C; 84h	SSF; pH 6.0; 72h	SSF; 30°C; 72h	SSF; 37°C; pH 2.0; 72h
Enzyme Activity	70 U mg^{-1}	5.74 U mL^{-1}	2343.5 U g^{-1}	60 U mL^{-1}	14108 U gds^{-1}	1818 U g^{-1}	31.2 U gdm^{-1}	1.6 U mL^{-1}
References	Ladeira et al. (2010)	Mukhtar and Haq (2013)	Belmessikh et al. (2013)	Mazoto et al. (2013)	Thanapimmetha et al. (2012)	Mahanta et al. (2008)	Kumar et al. (2008)	Ikasari and Mitchell (1996)

FPU mL^{-1} = Filter paper unit per milliliter, ECU mL^{-1} = nkatal per milliliter, U g^{-1} = units per gram, U mL^{-1} = units per milliliter, U g^{-1} = units per gram of dry substrate, U g^{-1} DBB = units per gram of dry bacterial bran, Q_p = Volumetric productivity.

RECOVERY AND PURIFICATION OF HYDROLASES

The initial step in all protein purification protocols is the release of the enzyme from the cell or tissue material, or cell extract preparation (Dako et al., 2012). Protein purification varies from a simple one-step purification procedure to large-scale purification processes. The key to obtaining successful and efficient protein-purification strategies is the selection of appropriate techniques that maximize yield and purity and minimize the number of steps required for purification. Each step sequentially enhances the level of purification from a crude extract but also increases purification costs. It is essential to minimize protein losses throughout the process; therefore the use of fewer steps during purification is important since losses can occur in each step. This fact occurs due, for example, to linking to separation matrices, insolubilities, or losses into the fringe fractions during separation procedures (Sá-Pereira et al., 2003).

Preparation of crude samples (preparation, extraction, and clarification, especially for enzymes produced with alternative substrates) to be further purified is a critical step, since it can affect enzyme recovery. Sufficient information can be gathered by complementary techniques that give a profile of the target protein (eg, pI, molecular weight, hydrophobicity) to permit an effective choice of purification strategy.

A pretreatment of the crude enzymatic extract is usually required to remove colored compounds, using dialysis or polyethylenimine (Mishra et al., 1990). Some purification steps involve, for example, ammonium sulfate precipitation, gel filtration, and ion-exchange chromatography and other chromatographic techniques.

Isolation and purification of microbial enzymes has previously been reported. Purification procedure strategies are chosen for each enzyme, including the concentration and purification steps, and the corresponding recovery fields, purification factors, and final specific activities of the different enzymes.

Chromatographic methods are used in a unique or serial steps, when high-level purity enzymes are required, As reported by Sá-Pereira et al. (2003), microbial xylanases are mainly purified by chromatographic methods, using from two to five purification steps (including concentration), and providing recovery yields ranging from 0.2% to 78%. Lower yields are generally obtained when a greater number of purification steps are used. In some cases, higher purification factors correspond to a greater number of purification steps, although no overall correlation was observed. In general, high purification factors correlate with purification schemes for low-molecular-weight xylanases (ranging from 24 to 56 kDa). When two purification steps are used, the purification factors can range from 3 to 36.6, and with three purification-steps can range from 3 to 890.

Ammonium sulfate precipitation generally seems to be an effective method for precipitating high-molecular-weight enzymes. Some reports note that ammonium sulfate promotes flocculation rather than precipitation. In these cases, a decanting funnel can be recommended for visualizing the separation of enzyme phases. Each of the enzyme phases can be analyzed separately in order to balance the enzyme protein concentration and enzyme activity, and to check the purity level of the enzyme by electrophoresis or high-pressure liquid chromatography. Another problem associated with the utilization of ammonium sulfate precipitation is that salt interferes in determinations of activity, for example, leading to overestimations of such activity in crude extracts. Thus, in general a dialysis or desalting step is generally required before the determination of enzyme activity.

Other strategies, such as membrane technology for protein separation, with ultrafiltration, are easily scaled up. Ultrafiltration is based on the principle of separation through a semipermeable membrane filter. Many filter type units are available commercially, either applying a centrifugal or stirred cell type of separation (Bonner, 2007). The principle for all ultracentrifuge methods is the filtration through a membrane that has a selective molecular weight cutoff point, therefore it will separate the extract based on the size of particles, not the charge (Dako et al., 2012).

The cost of the membranes was a serious limitation in the past; however, this barrier is continuously being broken. The effectiveness of membrane separation depends on the enzyme, its molecular weight and conformation, and certainly the adapted porosity and material of the membrane. Ultrafiltration membranes can markedly reduce enzyme recovery yields (MW > 20 kDa), owing to the ability of the enzyme to pass through ultrafiltration membranes with low-molecular-weight cutoff values (5–10 kDa). Another problem can be the material. Since most commercial ultrafiltration membranes are made of cellulose or its derivatives, for example, the presence of cellulases in a xylanase crude extract should be determined before ultrafiltration. Activity losses of about 50–70% after an ultrafiltration step were reported (Sá-Pereira et al. 2003).

Depending on the properties of the enzyme, certain modifications of purification methods must be considered regarding specific problems that can be encountered throughout the process, such as enzyme insolubility and loss of enzyme activity. Although several classic and more modern methods are available to solve these kinds of challenges, the enzyme purification step remains a major challenge for any method of extraction used (Dako et al., 2012).

ENZYME APPLICATIONS AND GLOBAL MARKET

Enzyme technology was presented as the application of free enzymes and whole cell biocatalysts in the production of goods and services. It is obviously an interdisciplinary field, as an important component of sustainable industrial development. Its applications range from straightforward industrial processes, the degradation of various natural substances in the starch processing, detergent and textile industries, to pharmaceutical discovery and the manipulation of DNA/RNA in biotechnology research.

Table 1.3 presents some examples of the multiplicity of enzyme applications in different industry sectors.

Table 1.3 Industrial Applications of Enzymes

Industry	Enzyme Group	Application
Detergent	Protease	Protein stain removal
	Amylase	Starch stain removal
	Lipase	Lipid stain removal
	Cellulase	Cleaning, color clarification, antiredeposition (cotton)
	Mannanase	Mannan stain removal
Starch and fuel	Amylase	Starch liquefaction and saccharification
	Amyloglucosidase	Saccharification
	Pullulanase	Saccharification
	Glucose isomerase	Glucose to fructose conversion
	Cyclodextrin-glycotransferase	Cyclodextrin production
	Xylanase	Viscosity reduction
Food (including dairy)	Protease	Milk clotting, infant formulas, flavor
	Lipase	Cheese flavor
	Lactase	Lactose removal (milk)
	Pectinase	Fruit-based products
Baking	Amylase	Bread softness and volume
	Xylanase	Dough conditioning
	Lipase	Dough stability and conditioning
	Phospholipase	Dough stability and conditioning
	Lipoxygenase	Dough strengthening, bread whitening
	Protease	Biscuits, cookies
Animal feed	Phytase	Phytate digestibility, phosphorus release
	Xylanase	Digestibility
	Beta-glucanase	Digestibility
Beverage	Pectinase	De-pectinization, mashing
	Amylase	Juice treatment, low-calorie beer
	Beta-glucanase	Mashing
	Acetoacetate decarboxylase	Maturation of beer
Textile	Cellulase	Denim finishing
	Amylase	De-sizing
	Catalase	Scouring
Pulp and paper	Xylanase	Bleach boosting
	Lipase	Pitch control, contaminant control
	Protease	Biofilm removal
	Amylase	Starch coating, de-inking
	Cellulase	De-inking, drainage
Fats and oils	Lipase	Transesterification
	Phospholipase	De-gumming, lysolecithin production
Leather	Protease	Unhearing, bating
	Lipase	De-pickling
Personal care	Amyloglucosidase	Antimicrobial
	Glucose oxidase	Bleaching, antimicrobial
	Peroxidase	Antimicrobial

Modified from Soccol, C.R., Vandenberghe, L.P.S., Woiciechowski, A.L., Babitha, S., 2006. Applications Industrial Enzymes. In: Pandey, A., Webb, C., Soccol, C.R., Larroche, C. (Org.), Enzyme Technology. vol. 1. New York, pp. 524–537.

Gross world sales for enzymes are estimated at \$8.0 billion in 2015, with a predicted annual growth rate of 7%. Several case studies, recently published by Li et al. (2012), show how industry has implemented biotechnological processes and assessed benefits in terms of costs and sustainability.

The fast growth over the past decade has also been seen in a diversity of other industries from organic synthesis in pharmaceutical industry to diagnostics enzymes. Contrarily, the detergent industry, once the largest sector in the global enzyme market, experienced a decline due in part to the pricing pressures from the main detergent manufactures after the turn of the century. Demand for cleaning enzymes was accelerated by 2005 as the product lines were reformulated with more-effective new enzymes launched continuously. Bioenergy related enzyme demand was limited by the new legislative mandates for grain based ethanol. While the development of the second generation biofuels derived from cellulosic raw materials will be in favor of demand growth over a long time.

SOME COMMERCIAL HYDROLYSES AND THEIR MANUFACTURE

Since the beginning, the production of enzymes has been relatively concentrated in a few developed nations, including Denmark, Switzerland, Germany, Netherlands, and the United States. In Denmark, Novozymes and Danisco together dominate 70% of the total enzyme market. More recently, in 2010, DSM's sales revenue accounted for 6%. About 100 companies produced enzymes in China, which is estimated less than 1% of the world market share. Some Japanese manufactures are also playing an increasing role in the world enzymes production.

International competition is intense in the enzyme market. Big companies aim to purchase other companies in order to become more efficient and competitive. However, there are still many small- to medium-sized enzyme producers. In China, there are manufacturers that have been gradually washed out from the market. On the other hand, several multinational companies have invested in the enzyme industry in China. Novozymes has three enzyme plants and DSM announced a joint venture with a Chinese company, Yixing Qiancheng Bio-Engineering Company Ltd., to provide α-amylase and xylanase to acquire the food and beverage enzyme markets (Li et al., 2012).

The enzyme market was established in 1832 by Nagase, in Osaka, Japan, producing enzymes for pharmaceuticals, foods, household products, agriculture, and textiles. The leading enzyme manufacturer is Novozymes, with 47% of market share and 902 deposited patents. As the world's biggest enterprise in the production of enzymes, it produces enzymes for household products, foods and beverages, biopharmaceuticals, bioenergy, feed, and other industrial enzymes.

The diversity of commercial enzymes and their manufacturers can be seen in Table 1.4.

Table 1.4 Commercial Hydrolases and Their Manufacturers

Enzyme/Product	Manufacturer	Microorganism
Amylases		
Ban 480L	Novozymes	Not cited
Fungamyl 2500 SG		Not cited
Novamyl 3D		Not cited
Termamyl 120L		*Bacillus licheniformis*
Termamyl 3X		*Bacillus licheniformis*
Termamyl Classic		Not cited
PROI–E0005	PROZOMIX	Not cited
Ronozyme RumiStar 600	DSM	Not cited
Cellulases		
Cellubrix (Celluclast)	Novozymes	*Trichoderma longibrachiatum*
Novozymes 188		*Aspergillus niger*
Bio-feed Beta L, Energex L		*Trichoderma longibrachiatum/ Trichoderma reesei*
Ultraflo L		*T. longibrachiatum/T. reesei*
Viscozyme L		*T. longibrachiatum/T. reesei*
Cellulase 2000L	Rhodia–Danisco	*T. longibrachiatum/T. reesei*
Rohament CL	Rohm–AB Enzymes	*T. longibrachiatum/T. reesei*
Viscostar 150L	Dyadic	*T. longibrachiatum/T. reesei*
Multifect CL	Genecor	*T. reesei*
GC 220		*T. longibrachiatum/T. reesei*
GC 440		*T. longibrachiatum/T. reesei*
GC 880		*T. longibrachiatum/T. reesei*
Accelerase 1500		*T. longibrachiatum/T. reesei*
Spezyme CP		*T. longibrachiatum/T. reesei*
Cellulase AP30K		*A. niger*
Cellulase TAP 106	Amano Enzyme	*Trichoderma viridae*
Cellulase TRL	Solvay Enzymes	*T. longibrachiatum/T. reesei*
Econase CE	Alko–EDC	*T. longibrachiatum/T. reesei*
Biocellulase TRI	Quest Intl	*T. longibrachiatum/T. reesei*
Biocellulase A		*A. niger*
Ultra-low microbial (ULM)	Iogen	*T. longibrachiatum/T. reesei*
Lipases		
Lipolase TM	Novo Nordisk	*Humicola lanuginosa*
Piccnate		*Mucor miehei*
SP526	Nova Nordisk	*Candida antarctica A/B*
SP525 or Novozym 435b		*C. antarctica A/B*
SP 524 Lipolase		*Geotrichum candidum*
ChiroCLEC-CR	Atlus Biologics	*Candida cylindracea*
Lipase AY	Amano	*C. cylindracea*
Lipase AY "Amano" 30		*Candida rugosa*
Lipase A "Amano"		*A. niger*

Table 1.4 Commercial Hydrolases and Their Manufacturers—cont'd

Enzyme/Product	Manufacturer	Microorganism
Lipase F-AP15		*Rhizopus oryzae*
Lipase M "Amano" 10		*Mucor javanicus*
Lipase G "Amano"		*Penicillium* sp.
Lipase MY, Lipase OF-360	Meito Sangyo	*C. cylindracea*
Chirazyme L-3	Boehringer Mannheim	*C. cylindracea*
Chirazyme L-5		*C. antarctica A/B*
Chirazyme L-2		*C. antarctica A/B*
Chirazyme L-8		*G. candidum*
LipomodTM 34P-L034P	Biocatalysts	*C. cylindracea*
LipomodTM 627P-L627P		*Rhizopus oryzae*
LipomodTM 338P-L338P		*Penicillium roqueforti*
LipomodTM 621P-L621		*Penicillium* sp.
Resinase	Novozymes	*C. rugosa*
Palatase		*Rhizomucor miehei*
Lipolase, Lipolase Ultra,		*T. lanuginosus*
Lipo PrimeTM, Lipex		
Lypolyve AN	Lyven	*A. niger*
Piccnate	Gist-Brocades	*M. miehei*

Mannanases

Mannaway	Novozymes	Not cited
Hemicell	Sanphar S. A.	Not cited

Pectinases

Panzym	C.H. Boehringer Sohn	Not cited
Ultrazym	Ciba–Geigy, A.G.	Not cited
Sclase	Kikkoman Shoyu Co.	Not cited
Pectinex	Schweizerische Ferment A.G.	Not cited
Klerzyme	Clarizyme Wallerstein Co.	Not cited
Pectinase	Biocon Pvt. Ltd.	Not cited
Pectinex Mash	Novozymes	Not cited
Ly Peclyve PR	Lyven	Not cited
Pektozyme POWERClear		Not cited
MaxLiq	Danisco	Not cited
Rohapect MA plus	AB Enzymes	Not cited
Rohapect DA6L		Not cited
Rapidase C80Max	Danisco	Not cited
Pectolase	Grinsteelvaeket	Not cited
Pctinol, Rohament	Rohm GmbH	Not cited
Pectinase	Biocon Pvt. Ltd.	Not cited

Continued

Table 1.4 Commercial Hydrolases and Their Manufacturers—cont'd

Enzyme/Product	Manufacturer	Microorganism
Proteases		
Alcalase	Novo Nordisk	*Bacillus licheniformis*
Savinase, Esperase		*Alkalophilic bacillus* sp.
Durazym		*Bacillus amyloliquefaciens* (*savinase*)
Maxacal, Maxatase	Gist-brocades	*A. Bacillus* sp.
Opticlean, Optimase	Solvay Enzymes GmbH	*A. Bacillus* sp.
Maxapem		*A. Bacillus* sp.
Proleather	Amano Pharmaceuticals Ltd.,	*A. Bacillus* sp.
Protease P		*Aspergillus* sp.
Purafect	Genencor International, Inc.	Variant of *B. lentus*
Phytases		
Ronozyme, Biofeed, ZY Phytase	Novozymes and DSM	*Aspergillus oryzae* DSM 14223
ROVABIO	Genecor International	*Penicillium funiculosum*
Phyzyme	Fermic	*A. oryzae*
SP, TP, SF	Alko Biotechnology	*A. oryzae*
Avizyme	Finnfeed	*Trichoderma reesei*
AMAFERM	BioZymes	*A. oryzae*
Allzyme SSF	Alltech	*A. niger*
Finase EC, Finase P/L	AB Vista	*T. reesei*
Quantum™/	AB Vista	*Pichia pastoris* DSM 15927
Natuphos	BASF	*A. niger*
OptiPhos	Enzyvia (JBS United)	*Schizosaccharomyces pombe*
Phyzyme XP	Dupont International	*Pichia pastoris*
Xylanases		
Allzym PT	Alltech	*A. niger*
Fibrozyme		*A. niger*
Amano 90	Amano Pharmaceutical Co.	*A. niger*
Resinase	A/S	Not cited
Bleachzyme F	Biocon	Not cited
Ecosane	Biotec	*T. reesei*
Cartazyme	Clariant	*Termomonospora fusca*
Irgazyme	Ciba–Geiby Ltd.	*T. longibrachiatum*
Grindazym PF	Danisco Ingredients	*A. niger*
Grindazym GP 5000		*A. niger*
Gammafeed X	Gamma Chemie GmbH	*T. longibrachiatum*
Gammazym X4000L		*T. reesei*
Multifect XL	Genecor International Europe	*T. longibrachiatum*
Xylanase 250	Hankyo Bioindustry Co. Ltd.	*T. viridae*
Hemicellulase 100		*A. niger*

Table 1.4 Commercial Hydrolases and Their Manufacturers—cont'd

Enzyme/Product	Manufacturer	Microorganism
Xylanase GS35	Iogen Corp.	*T. reesei*
Bio-feed-plus	Novozymes	*Humicola insolens*
Novozym 431		*T. longibrachiatum*
Pulpzyme		*Bacillus* sp.
Ecopulp X-200	Primalco Ltd. Biotec	*T. reesei*
Bioxylanase	Quest International Ireland	*T. reesei*
Rohalasa 7118	Rohm GmbH	*Aspergillus / Trichoderma*
Vernon 191		*Aspergillus / Trichoderma*
No commercial name	Seikagaku Corporation	*Trichoderma* sp.
Sumizyme X	Shi Nihon Chemical	*Trichoderma koningii*
Solvay pentosanasa	Solvay Enzymes GmbH & Co.	*T. reesei*
Sternzym HC46	SternEnzym GmbH & Co.	*T. reesei*
Sternzym HC40		*A. niger*

Modified from Mittal, A., Singh, G., Goyal, V., Yadav, A., Aggarwal, N.K., 2013. Production of phytase by acido-thermophilic strain of *Klebsiella* sp. DB-3FJ711774.1 using orange peel flour under submerged fermentation. Innovative Romanian Food Biotechnology 10, 18–27; Lei, X., Weaver, J., Mullaney, E., 2013. Phytase, a new life for an "old" enzyme. Annual Review of Animal Biosciences 1, 283–309; Singhania, R., Sukumaram, R., Patel, A., 2010. Advancement and comparative profiles in the production technologies using solid-state and submerged fermentation for microbial cellulases. Enzyme and Microbial Technology 46, 541–549; Harris, A.D., Ramalingam, C., 2010. Xylanases and its application in food industry: a review. Journal of Experimental Sciences 1, 01–11; Singh, A.K., Mukhopadhyay, M., 2012. Overview of fungal lipase: a review. Applied Biochemistry and Biotechnology 166, 486–520; Jisha, V.N., Smitha, B.B., Pradeep, S., Sreedevi, S., Unni, K.N., Sajith, S., Josh, M.S., Benjamin, S., 2013. Versatility of microbial proteases. Advances in Enzyme Research 1, 39–51 and Kuhad, R.C., Singh, A., 2013. Biotechnology for Environmental Management and Resource Recovery, Springer Science & Business.

CONCLUSIONS

Microorganisms are the main sources of enzymes, the efficient and specific biocatalysts, which are applied in different industrial sectors, such as food, animal feed, personal care, detergents, textile and paper, biofuels, and others. The high specificity, fast action, and biodegradability allow the fast bioconversions, with low energy and ecofriendly processes.

There are several techniques and substrates that can be used in enzyme production, either using synthetic or agro-industrial subproducts, which on the one hand bring an alternative way to lower enzyme production costs and on the other hand create supplementary steps in separation and purification operations. The success of the process depends on the source capacity to synthesize the protein of interest, some conditions of the bioprocess, and, certainly, the strategies of enzyme separation and purification. All these steps will depend on the conditions and composition of the medium and are crucial for the establishment of a new highly efficient enzyme production process. Even higher productivities may be reached with the help of the advances of genomics, metagenomics, proteomics, expression systems, and recombinant DNA techniques.

Since new biocatalysts are constantly appearing, the enzyme market is strongly growing with the appearance of new enzyme products. Novel functions, applications, and characteristics of these enzymes are being found and studied with infinite possibilities.

REFERENCES

Adsul, M., Ghule, J., Singh, R., 2004. Polysaccharides from bagasse: applications in cellulase and xylanase production. Carbohydrate Polymers 57, 67–72.

Alam, M.Z., Mamun, A.A., Qudsieh, I.Y., Muyibi, S.A., Salleh, H.M., Omar, N.M., 2009. Solid state bioconversion of oil palm empty fruit bunches for cellulase enzyme production using a rotary drum bioreactor. Biochemical Engineering Journal 46, 61–64.

Ali, U.F., Ibrahim, Z.M., Isaac, G.S., 2013. Ethanol and xylitol production from xylanase broth of *Thermomyces lanuginosus* grown on some lignocellulosic wastes using *Candida tropicalis* EMCC2. Life Sciences Journal 10, 968–978.

Arabi, M., Jawhar, M., Bakri, Y., 2011. Effect of additional carbon source and moisture level on xylanase production by *Cochliobolus sativus* in solid fermentation. Mikrobiologiia 80, 162–165.

Assamoi, A.A., Destain, J., Delvigne, F., Lognay, G., Thonart, P., 2008. Solid-state fermentation of xylanase from *Penicillium canescens* 10-10c in a multi-layer-packed bed reactor. Applied Biochemistry and Biotechnology 145, 87–97.

Babu, K., Satyanarayana, T., 1995. α-amylase production by *Thermophilic coagdans* in solid state fermentation. Process Biochemistry 30, 305–309.

Bajaj, B.K., Wani, M.A., 2011. Enhanced phytase production from *Nocardia* sp. MB 36 using agro-residues as substrates: potential application for animal feed production. Engineering in Life Sciences 11, 620–628.

Belmessikh, A., Boukhalfa, H., Mechakramaza, A., 2013. Statistical optimization of culture medium for neutral protease production by *Aspergillus oryzae*. Comparative study between solid and submerged fermentations on tomato pomace. Journal of the Taiwan Institute of Chemical Engineers 44, 377–385.

Bezerra, J.D.P., Santos, M.G.S., Svedese, V.M., Lima, D.M.M., Fernandes, M.J.S., 2012. Richness of endophytic fungi isolated from *Opuntia ficus-indica* Mill. (Cactaceae) and preliminary screening for enzyme production. World Journal of Microbiology and Biotechnology 28, 1989–1995.

Bonner, P.L.R., 2007. Protein Purification. Taylor and Francis Group, New York, NY. 190 p.

Chadha, B.S., Harmeet, G., Mandeep, M., Saini, H.S., Singh, N., 2004. Phytase production by the thermophilic fungus *Rhizomucor pusillus*. World Journal of Microbiology & Biotechnology 20, 105–109.

Dako, E., Bernier, A.M., Dadie, A.T., Jankowski, K., 2012. The problems associated with enzyme purification. In: Ekinci, D. (Ed.), Chemical Biology 444 p.

Demirijan, D., Moris-Varas, F., Cassidy, C., 2001. Enzymes from extremophiles. Current Opinion in Chemistry and Biology 5, 144–151.

Dhillon, G.S., Brar, S.K., Kaur, S., Metahni, S., M'hamdi, N., 2012a. Lactoserum as a moistening medium and crude inducer for fungal cellulase and hemicellulose induction through solid-state fermentation of apple pomace. Biomass and Bioenergy 41, 165–174.

Dhillon, G.S., Kaur, S., Brar, S.K., Kaur, S., Gassara, F., Verma, M., 2012b. Improved xylanase production using apple pomace waste by *Asperillus niger* in koji fermentation. Engineering of Life Sciences 2, 198–208.

Erdal, S., Aaskin, M., 2010. Production of α-amylase by Penicillium expansum MT-1 in solid-state fermentation using waste Loquat (*Eriobotrya japonica* Lindley) kernels as substrate. Romanian Biotechnological Letters 15, 5342–5350.

Esakkiraj, P., Sandoval, G., Sankaralingam, S., Immanuel, G., Palavesam, A., 2010. Preliminary optimization of solid-state phytase production by moderately halophilic *Pseudomonas* AP-MSU 2 isolated from fish intestine. Annals Microbiology 60, 461–468.

Esterbauer, H., Steiner, W., Labudova, I., 1991. Production of *Trichoderma* cellulase in laboratory and pilot scale. Bioresource Technology 36, 51–65.

Farhat, A., Chouayekh, H., Ben Farhat, M., Bouchaala, K., Bejar, S., 2008. Gene cloning and characterization of a thermostable phytase from *Bacillus subtilis* US417 and assessment of its potential as a feed additive in comparison with a commercial enzyme. Molecular Biotechnology 40, 127–135.

Gonçalves, C., Lopes, M., Ferreira, J.P., Belo, I., 2009. Biological treatment of olive mill wastewater by non-conventional yeasts. Bioresource Technology 100, 3759–3763.

Gupta, R., Gigras, P., Mohapatra, H., Goswami, V.K., Chauhan, B., 2003. Microbial α-amylases: a biotechnological perspective. Process Biochemistry 38, 1599–1616.

Haki, G.D., Rakshit, S.K., 2003. Developments in industrially important thermostable enzymes: a review. Bioresource Technology 89, 17–34.

Haq, I., Ashraf, H., Iqbal, J., 2003. Production of alpha amylase by *Bacillus licheniformis* using an economical medium. Bioresource Technology 87, 57–61.

Harris, A.D., Ramalingam, C., 2010. Xylanases and its application in food industry: a review. Journal of Experimental Sciences 1, 01–11.

Hasan, F., Shah, A.A., Hameed, A., 2006. Industrial applications of microbial lipases. Enzyme and Microbial Technology 39, 235–251.

Heck, J.X., Soares, L.H.B., Ayub, M.A.Z., 2005. Optimization of xylanase and mannanase production by *Bacillus circulans* strain BL53 on solid-state cultivation. Enzyme and Microbial Technology 37, 417–423.

Ho, H.L., 2014. Effects of medium formulation and culture conditions on microbial xylanase production: a review. International Journal of Food Fermentation and Technology 4, 1–11.

Hussain, I., Siddique, F., Mahmood, M., 2013. A review of the microbiological aspect of α-amylase production. International Journal of Agriculture and Biology 15, 1029–1034.

Ikasari, L., Mitchell, D.A., 1996. Leaching and characterization of Rhizopus oligosporus acid protease from solid-state fermentation. Enzyme Microbial Technology 19, 171–175.

Jisha, V.N., Smitha, B.B., Pradeep, S., Sreedevi, S., Unni, K.N., Sajith, S., Josh, M.S., Benjamin, S., 2013. Versatility of microbial proteases. Advances in Enzyme Research 1, 39–51.

Joshi, S., Satyanarayana, T., 2015. *In vitro* engineering of microbial enzymes with multifarious applications: prospects and perspectives. Bioresource Technology 176, 273–283.

Kashyap, D.R., Vohra, P.K., Chopra, S., Tewari, R., 2001. Applications of pectinases in the commercial sector: a review. Bioresource Technology 77, 215–227.

Kheng, P.P., Omar, I.C., 2005. Xylanase production by a local fungal isolate, *Aspergillus niger* USM AI 1 via solid state fermentation using palm kernel cake (PKC) as substrate. Songklanakarin Journal of Science and Technology 27, 325–336.

Kleerebezem, R., Loosdrecht, M., 2007. Mixed culture biotechnology for bioenergy production. Current Opinion in Biotechnology 18, 207–212.

Kobayashi, M., Shimizu, S., 2000. Nitrile hydrolases. Current Opinion in Chemical Biology 4, 95–102.

Kuhad, R.C., Singh, A., 2013. Biotechnology for Environmental Management and Resource Recovery. Springer Science & Business.

Kumar, H.D., Swati, S., 2001. Modern Concepts of Microbiology, second revised ed. Vikas Publishing House Pvt Ltd., New Delhi.

Kumar, A.G., Nagesh, N., Prabhakar, T.G., Sekaran, G., 2008. Purification of extracellular acid protease and analysis of fermentation metabolites by *Synergistes sp.* utilizing proteinaceous solid waste from tanneries. Bioresource Technology 99, 2364–2372.

Kumar, V., Sinha, A.K., Makkar, H.P.S., Becker, K., 2010. Dietary roles of phytate and phytase in human nutrition: a review. Food Chemistry 120, 945–959.

Kumar, S., Mathur, A., Singh, V., Nandy, S., Khare, S.K., Negi, S., 2012. Bioremediation of waste cooking oil using a novel lipase produced by *Penicillium chrysogenum* SNP5 grown in solid medium containing waste grease. Bioresource Technology 120, 300–304.

Ladeira, S., Andrade, M., Delatorre, A., 2010. Utilização de resíduos agroindustriais para a produção de proteases pelo termofílico *Bacillus* sp. em fermentação submersa: Otimização do meio de cultura usando a técnica de planejamento experimental. Química Nova 33, 324–328.

Lanciotti, R., Gianotti, A., Baldi, D., Angrisani, R., Suzzi, G., Mastrocola, D., Guerzoni, M.E., 2005. Use of *Yarrowia lipolytica* strains for the treatment of olive mill wastewater. Bioresource Technology 96, 317–322.

Lei, X., Weaver, J., Mullaney, E., 2013. Phytase, a new life for an "old" enzyme. Annual Review of Animal Biosciences 1, 283–309.

Li, D., Lu, M., Li, Y., 2003. Purification and characterization of an endocellulase from the thermophilic fungus *Chaetomium thermophilum* CT2. Enzyme and Microbial Technology 33, 932–937.

Li, S., Yang, X., Yang, S., Zhu, M., Wang, X., 2012. Technology prospecting on enzymes: application, marketing and engineering. Computational and Structural Biotechnology Journal 2, 1–11.

Li, Q., Coffman, A.M., Ju, L.K., 2015. Development of reproducible assays for polygalacturonase and pectinase. Enzyme and Microbial Technology. http://dx.doi.org/10.1016/j.enzmictec.2015.02.006.

Lin, T., Chen, C., 2004. Enhanced mannanase production by submerged culture of *Aspergillus niger* NCH-189 using defatted copra based media. Process Biochemistry 39, 1103–1109.

Mafakher, L., Mirbagheri, M., Darvishi, F., Nahvi, I., Zarkesh-Esfahani, H., Emtiazi, G., 2010. Isolation of lipase and citric acid producing yeasts from agro-industrial wastewater. New Biotechnology 27, 337–340.

Mahanta, N., Gupta, A., Khare, S.K., 2008. Production of protease and lipase by solvent tolerant *Pseudomonas aeruginosa* PseA in solid-state fermentation using *Jatropha curcas* seed cake as substrate. Bioresource Technology 99, 1729–1735.

Mandviwala, T.N., Khire, J.M., 2000. Journal of Industrial Microbiology & Biotechnology 24, 237–243.

Mazotto, A., Couri, S., Damaso, M., 2013. Degradation of feather waste by *Aspergillus niger* keratinases: comparison of submerged and solid-state fermentation. International Biodeterioration and Biodegradation 85, 189–195.

Membrillo, I., Sánchez, C., Meneses, M., Favela, E., Loera, O., 2008. Effect of substrate particle size and additional nitrogen source on production of lignocellulolytic enzymes by *Pleurotus ostreatus* strains. Bioresource Technology 99, 7842–7847.

Menon, K., Rao, K.K., 1999. Production and application of microbial enzymes. In: Joshi, V.K., Pandey, A. (Eds.), Biotechnology: Food Fermentation. Educational Publishers & Distributors, New Delhi, pp. 1079–1111.

Mishra, C., Forrester, I.T., Kelley, B.D., Burgess, R.R., Leatham, G.F., 1990. Characterization of a major xylanase purified from *Lentinula edodes* cultures grown on a commercial solid lignocellulosic substrate. Applied Microbiology and Biotechnology 33, 226–232.

Mittal, A., Singh, G., Goyal, V., Yadav, A., Aggarwal, N.K., 2013. Production of phytase by acido-thermophilic strain of *Klebsiella* sp. DB-3FJ711774.1 using orange peel flour under submerged fermentation. Innovative Romanian Food Biotechnology 10, 18–27.

Mukhtar, H., Haq, I., 2013. Comparative evaluation of agroindustrial byproducts for the production of alkaline protease by wild and mutant strains of *Bacillus subtilis* in submerged and solid state fermentation. The Scientific World Journal 1–7.

Norazlina, M., Tanes, P.M., Halim, K.H., 2013. Comparison study of xylanolytic enzyme activity produced by *Aspergillus niger* under solid state fermentation using rice straw and oil palm leaf. International Journal of Agricultural Science and Research 3, 245–252.

Oberoi, H.S., Chavan, Y., Bansal, S., Dhillon, G.S., 2010. Production of cellulases through solid state fermentation using kinnow pulp as a major substrate. Food Bioprocess Technology 3, 528–536.

Ogawa, J., Shimizu, S., 1999. Microbial enzymes: new industrial applications from traditional screening methods. Trends Biotechnology 17, 13–21.

Oinonen, A., Londesborough, J., Joutsjoki, V., 2004. Three cellulases from *Melanocarpus albomyces* for textile treatment at neutral pH. Enzyme and Microbial Technology 34, 332–341.

Olempska-Beer, Z.S., Merker, R.I., Ditto, M.D., DiNovi, M.J., 2006. Food-processing enzymes from recombinant microorganisms - a review. Regulatory Toxicology and Pharmacology 45, 144–158.

Ozturk, B., Cekmecelioglu, D., Ogel, Z.B., 2010. Optimal conditions for enhanced β-mannanase production by recombinant *Aspergillus sojae*. Journal of Molecular Catalysis B: Enzymatic 64, 135–139.

Pandey, A., Selvakumar, P., Soccol, C.R., Nigam, P., 1999. Solid-state fermentation for the production of industrial enzymes. Bioresource Technology 77, 149–162.

Ramachandran, S., Patel, A., Nampoothiri, K., 2004. Coconut oil cake - a potential raw material for the production of α-amylase. Bioresource Technology 93, 169–174.

Ramani, K., Kennedy, L.J., Ramakrishnan, M., Sekarana, G., 2010. Purification, characterization and application of acidic lipase from *Pseudomonas gessardii* using beef tallow as a substrate for fats and oil hydrolysis. Process Biochemistry 45, 1683–1691.

Rani, R., Ghosh, S., 2011. Production of phytase under solid-state fermentation using *Rhizopus oryzae*: novel strain improvement approach and studies on purification and characterization. Bioresource Technology 102, 10641–10649.

Reyes-Nava, L.A., Rivera-Espinoza, Y., 2014. Isolation sources of bile salt hydrolase-microorganisms. Herald Journal of Agriculture and Food Science Research 3, 49–54.

Rigo, E., Ninow, J.L., Di Luccio, M., Oliveira, J.V., Polloni, A.E., Remonatto, D., Arbter, F., Vardanega, R., Oliveira, D., Treichel, H., 2010. Lipase production by solid fermentation of soybean meal with different supplements. LWT - Food Science and Technology 43, 1132–1137.

Romero, M., Aguardo, J., González, L., 1999. Cellulase production by *Neurospora crassa* on wheat straw. Enzyme and Microbial Technology 25, 244–250.

Roopesh, K., Ramachandran, S., Nampoothiri, K.M., Szakacs, G., Pandey, A., 2006. Comparison of phytase production on wheat bran and oilcakes in solid-state fermentation by *Mucor racemosus*. Bioresource Technology 97, 506–511.

Sanhoun, M., Kriaa, M., Elgharbi, F., 2015. *Aspergillus oryzae* S2 alpha-amylase production under solid state fermentation: optimization of culture conditions. International Journal of Biological Macromolecules 75, 1–8.

Sá-Pereira, P., Paveia, H., Costa-Ferreira, M., Aires-Barros, M.R., 2003. A new look at xylanases an overview of purification strategies. Molecular Biotechnology 24, 257–281.

Sarkar, N., Aikat, K., 2012. Cellulase and xylanase production from rice straw by a locally isolated fungus *Aspergillus fumigatus* NITDGPKA3 under solid state and fermentation – statistical optimization by response surface methodology. Journal of Technology and Innovative Renewable Energies 1, 54–62.

Silva, C., Lacerda, M., Leite, R., 2013. Production of enzymes from *Lichtheimia ramosa* using Brazilian savannah fruit wastes as substrate on solid state bioprocesses. Electronic Journal of Biotechnology 16, 1–9.

Singh, S., Gupta, A., 2014. Comparative fermentation studies on amylase production by *Aspergillus flavus* TF-8 using Sal (*Shorea robusta*) deoiled cake as natural substrate: characterization for potential application in detergency. Industrial Crops and Products 57, 158–165.

Singh, A.K., Mukhopadhyay, M., 2012. Overview of fungal lipase: a review. Applied Biochemistry and Biotechnology 166, 486–520.

Singhania, R., Sukumaram, R., Patel, A., 2010. Advancement and comparative profiles in the production technologies using solid-state and submerged fermentation for microbial cellulases. Enzyme and Microbial Technology 46, 541–549.

Siqueira, F.G., Siqueira, E.G., Jaramillo, P.M.D., Silveira, M.H.L., Jürgen, A., Couto, F.A., Batista, L.R., Filho, E.X.F., 2010. The potential of agro-industrial residues for production of holocellulase from filamentous fungi. International Biodeterioration & Biodegradation 64, 20–26.

Soccol, C.R., Vandenberghe, L.P.S., 2003. Overview of applied solid-state fermentation in Brazil. Biochemical Engeneering Journal 13, 205–218.

Soccol, C.R., Vandenberghe, L.P.S., Woiciechowski, A.L., Babitha, S., 2006. In: Pandey, A., Webb, C., Soccol, C.R., Larroche, C. (Eds.), Applications Industrial EnzymesEnzyme Technology, vol. 1, pp. 524–537 New York.

Sodhi, H., Sharma, J., Gupta, J., Soni, S., 2005. Production of a thermostable α-amylase from *Bacillus* sp. PS-7 by solid state fermentation and its synergistic use in the hydrolysis of malt starch for alcohol production. Process Biochemistry 40, 525–534.

Tahir, A., Mateen, B., Saeed, S., Uslu, H., 2010. Studies on the production of commercially important phytase from *Aspergillus niger* st-6 isolated from decaying organic soil. Micología Aplicada Internacional 22, 51–57.

Thanapimmetha, A., Luadsongkram, A., Titapiwatanakun, B., Srinophakun, P., 2012. Value added waste of *Jatropha curcas* residue: optimization of protease production in solid state fermentation by Taguchi DOE methodology. Industrial Crops and Products 37, 1–5.

Treichel, H., Oliveira, D., Mazzuti, M.A., Luccio, M.D., Oliveira, J.V., 2010. A review on microbial lipases production. Food Bioprocess Technology 3, 182–196.

Valášková, V., Baldrian, P., 2006. Estimation of bound and free fractions of lignocellulose-degrading enzymes of wood-rotting fungi *Pleurotus ostreatus, Trametes versicolor* and *Piptoporus betulinus*. Research in Microbiology 157, 119–124.

Valladão, A.B.G., Freire, D.M.G., Cammarota, M.C., 2007. Enzymatic pre-hydrolysis applied to the anaerobic treatment of effluents from poultry slaughterhouses. International Biodeterioration & Biodegradation 60, 219–225.

van Beilen, J.B., Li, Z., 2002. Enzyme technology: an overview. Current Opinion in Biotechnology 13, 338–344.

van Zyl, W.H., Rose, S.H., Trollope, K., Görgens, J.F., 2010. Fungal β-mannanases: mannan hydrolysis, heterologous production and biotechnological applications. Process Biochemistry 45, 1203–1213.

Vandenberghe, L.P.S., Soccol, C.R., Spier, M.R., Weingartner, V., Scheuer, T., 2012. Processo para a produção microbiana de mananases utilizando resíduos/subprodutos (Process for microbial production of mannanase using subproducts/residues). Brazilian Patent, PI06879354.

Wang, Q., Wang, X., Wang, X., 2008. Glucoamylase production from food waste by *Aspergillus niger* under submerged fermentation. Process Biochemistry 43, 280–286.

Weijers, C.A.G.M., de Bont, J.A.M., 1999. Epoxide hydrolases from yeasts and other sources: versatile tools in biocatalysis. Journal of Molecular Catalysis B: Enzymatic 6, 199–214.

Wu, M., Tang, C., Li, J., Zhang, H., Guo, J., 2011. Bimutation breeding of *Aspergillus niger* strain for enhancing β-mannanase production by solid-state fermentation. Carbohydrate Research 346, 2149–2155.

Yang, G., Lili, H., Fuliang, Z., 2011. Study on the solid-state fermentation conditions for producing thermostable xylanase feed in a pressure pulsation bioreactor. Advanced Materials Research 236, 72–76.

Yoon, L.W., Ang, T.N., Ngoh, G.C., Chua, A.S.M., 2014. Fungal solid-state fermentation and various methods of enhancement in cellulase production. Biomass and Bioenergy 67, 319–338.

Zhang, H., Sang, Q., 2015. Production and extraction optimization of xylanase and β-mannanase by *Penicillium chrysogenum* QML-2 and primary application in saccharification of corn cob. Biochemical Engineering Journal 97, 101–110.

Zoglowek, M., Lübeck, P.S., Ahringa, B.K., Lübeck, M., 2015. Heterologous expression of cellobiohydrolases in filamentous fungi – an update on the current challenges, achievements and perspectives. Process Biochemistry 50, 211–220.

CHAPTER 2

Fruit and Vegetable Processing Waste: Renewable Feed Stocks for Enzyme Production

R. Sharma[1], H.S. Oberoi[1,2], G.S. Dhillon[3]

[1]Guru Nanak Dev University, Amritsar, India; [2]ICAR-Indian Institute of Horticultural Research, Bengaluru, India; [3]University of Alberta, Edmonton, AB, Canada

INTRODUCTION

The waste derived during the processing of raw vegetable and fruit materials is of major concern to the food industry. Globally, fruit and vegetable processing industry forms a very important part of the food industry, and in most countries agriculture still remains one of the main sources of income (Askar, 1998; Thassitou and Arvanitoyannis, 2001). The fruit and vegetable canning industry, the frozen vegetable industry, the vegetable dehydration industry, the fruit and vegetable drying industry, fruit pulping, tomato juice concentrate, and fruit concentrate belong to this category. According to the processing stage, different types of waste may be produced and contribute with different percentages to the final and cumulative process waste (Thassitou and Arvanitoyannis, 2001; Schieber et al., 2001). Processing of fruits and vegetables generates a substantial amount of residues in the form of pods, peel, pulp, stones, and seeds (Babbar and Oberoi, 2014). The mass of by-products obtained as a result of processing of tropical crops may approach or even exceed that of the corresponding valuable product affecting the economics of growing tropical exotic crops (Miljkovic and Bignami, 2002).

The fruit and vegetable solid waste contains mainly soluble sugars and other hydrolyzable materials and fiber. These residues are disposed of in the municipal bins or are left to rot because of the nonavailability of proper infrastructure to handle such a huge quantity of biomass and/or an established commercial use for such wastes. Disposal of such wastes may present an added cost to processors, and direct disposal to soil or landfill may cause serious environmental problems (Babbar and Oberoi, 2014). On the other hand, costs of drying, storage, and shipment of by-products are economically limiting factors. The problem of disposal of by-products is further aggravated by legal restrictions. Thus, efficient, inexpensive, and environmentally sound utilization of these materials is becoming more important, especially as it affects profitability, the environment, and jobs

Agro-Industrial Wastes as Feedstock for Enzyme Production
ISBN 978-0-12-802392-1
http://dx.doi.org/10.1016/B978-0-12-802392-1.00002-2

23

(Lowe and Buckmaster, 1995). Conversion of abundant fruit and vegetable processing waste (FVPW) into value-added products presents a viable and sustainable option for reducing environmental pollution, improving energy security, and reducing greenhouse emissions. Hence, it is felt that processing waste utilization is a necessity and a challenge for the food industries and needs first and foremost consideration to make units economically viable through the recovery of the value-added products. Therefore, exploitation of fruit and vegetable processing wastes for the development of value-added products is a promising area of research.

CHARACTERIZATION AND COMPOSITION OF FRUIT AND VEGETABLE PROCESSING WASTE

Fruit and vegetable residues/processing wastes are rich in cellulose, hemicellulose, pectin, and minerals and have very low concentrations of lignin in comparison to the crop residues, such as corn stover, wheat straw, rice straw, etc. Such characteristics render FVPW, ideal substrates for the purpose of enzyme production. There are published reports on production of industrially important enzymes using FVPW as substrates, but commercial plants/pilot plants producing enzymes exclusively using FVPW as substrates are unheard of. In addition, only very few fruit and vegetable residues have been attempted for enzyme production. Utilization of the solid wastes after enzyme extraction/recovery also needs to be addressed in a scientific way.

Fruit and vegetable wastes represent a specific waste produced by all the wholesale markets and the food industry. For example, fruit and vegetable processing, packing, distribution, and consumption in the organized sector in India, Philippines, China, and the United States generate approximately 1.81, 6.53, 32.0, and 15.0 million tons of wastes, respectively (Table 2.1). Table 2.1 shows production of fruits and vegetables by different countries in the world and the extent of loss along the food supply chain from production to consumption in many locations resulting in generation of huge amounts of FVPW. India ranks second in worldwide production of fruits and vegetables and accounts for about 10% and 14% of the total production of fruits and vegetables in the world, respectively. With the annual Indian production of fruits and vegetables alone estimated at 243 million tons (Handbook on Horticultural Statistics, 2014), assuming the processing levels at 2% for the organized sector (though they are much higher for unorganized sector) and 30% residue generation, one could expect generation of about 1.45 million tons of FVPW annually in India alone, which creates lot of environmental pollution problems due to its poor disposal. On the other hand, the production of FVPW collected from the market of Tunisia has been estimated at 180 tons per month (Bouallagui et al., 2003).

Without proper storage and transport facilities, the wastes from fruit and vegetable processing industries generally contain large amounts of solid suspensions and a high

Table 2.1 Fruit and Vegetable Wastes Generated After Processing, Packing, Distribution, and Consumption in Organized Sector

Country	Production (Million Tons)			Fruit and Vegetable Processed (%)	Processed Fruit and Vegetables (Million Tons)	Losses and Wastage (%)			Waste Generated (Million Tonnes)
	Fruit	Vegetable	Total			Processing	Distribution	Consumption	
India	74.88	146.55	221.43	2.2	4.87	25	10	7	1.81
China	122.19	473.06	595.25	23.0	136.91	2	8	15	31.98
Philippines	16.18	6.30	22.48	75.0	17.53	25	10	7	6.53
Malaysia	1.07	1.21	2.28	80.0	1.82	25	10	7	0.68
Thailand	10.27	3.81	14.08	30.0	4.22	25	10	7	1.57
United States	25.38	35.29	60.67	65.0	39.43	2	12	28	14.95

Adapted from Wadhwa, M., Bakshi, M.P.S., 2013. Utilization of fruit and vegetable wastes as livestock feed and as a substrate for generation of other value added products. In: Makkar, H.P.S. (Ed.), Food and Agriculture Organization of United Nations. RAP Publication 2013/04, pp. 56.

biochemical oxygen demand (BOD). Some other parameters usually of interest to the waste treatment are pH, chemical oxygen demand (COD), dissolved oxygen (DO), and total solids (Thassitou and Arvanitoyannis, 2001). Indicative values for BOD, COD, suspended solids (SS), and pH for the processing of some fruit and vegetables are summarized in Table 2.2.

Fruit and vegetable industry processing wastes consist of various by-products having an acidic pH (Riggle, 1989) and a moisture content of 80–90% (Grobe, 1994). The chemical composition of the wastes varies and depends on the processed fruit or vegetable. In general the wastes consist of hydrocarbons and relatively small amounts of proteins and fat. Composition of various fruits and vegetables wastes is presented in Table 2.3. Mango seed kernel is rich in carbohydrates, proteins, fats, and minerals. On the other hand, orange, melon, and pumpkin seeds can provide fats and minerals. The fruit and vegetable industry typically generates large volumes of effluents and solid wastes. The effluents contain high organic loads, cleansing and blanching agents, salt, and suspended solids such as fibers and soil particles. They may also contain pesticide residues washed from the raw materials. The main solid wastes are organic materials, including discarded fruits and vegetables. Reductions in wastewater volumes of up to 95% have been reported through implementation of good practices. As an example, recirculation of process water from onion preparation reduces the organic load by 75% and water consumption by 95%. Similarly, the liquid waste load (in terms of BOD) from apple juice and carrot processing can be reduced by 80% (World Bank Publications, 1998). These differences in the nature of the wastes require separate treatment for different wastes. Mirzaei-Aghsaghali and Maheri-Sis (2008) reviewed nutrients in several fruit and vegetable by-products.

It is evident from Table 2.3 that the majority of the FVPW is rich in carbohydrates, proteins, minerals, and fiber. Such characteristics and nutrients profile are essentially required for the production of enzymes through fermentation. Microorganisms need

Table 2.2 Important Parameters for Treatment of Fruit and Vegetable Processing Waste

Fruit and Vegetable	Biochemical Oxygen Demand (mg mL^{-1})	Chemical Oxygen Demand (mg mL^{-1})	Suspended Solids (mg L^{-1})	pH
Carrots	1,350	2,300	4,120	8.7
Corn	1,550	2,500	210	6.9
Tomatoes	1,025	1,500	950	7.9
Green peas	800	1,650	260	6.9
Cherries	2,550	2,500	400	6.5
Grapefruit	1,000	1,900	250	7.4
Apples	9,600	18,700	450	5.9

Adapted from Thassitou, P.K., Arvanitoyannis, I.S., 2001. Bioremediation: a novel approach to food waste management a review. Trends in Food Science and Technology 12, 185–196.

Table 2.3 Composition of Different Fruit and Vegetable Wastes (per 100 g)

Waste	Moisture (g)	Protein (g)	Fat (g)	Minerals (g)	Fiber (g)	Carbohydrates (g)	References
Carrot pomace	4.61	10.06	1.75	7.29	24.73	27–60	Sharma et al. (2012) and Sharoba et al. (2013)
Green pea peels	4.28	13.27	1.34	7.18	19.82	–	Sharoba et al. (2013)
Apple pomace	3.97	4.45	3.49	2.13	48.70	48	Joshi and Attri (2006)
Olive cake	–	–	–	–	15.5	18.4	Haddadin et al. (1999)
Peach pomace	–	–	–	–	19.1	–	Pagan and Ibarz (1999)
Mosambi peel	5.3	5.4	–	5.8	–	14.2	Bandikari et al. (2014)
Pineapple peel	9.4	8.7	–	3.9	–	29.1	Bandikari et al. (2014)
Mango peel	7.3	9.5	–	6.4	27.95	20.7	Snachez et al. (2014) and Bandikari et al. (2014)
Banana peel	10.5	6.02	–	4.2	–	17.8	Bandikari et al. (2014)
Orange bagasse	4.15	7.59	1.57	10.03	39.19	35.56	Sharoba et al. (2013) and Sanchez et al. (2014)
Orange peel	4.23	5.97	1.21	7.12	28.56	25.93	Sharoba et al. (2013) and Sanchez et al. (2014)
Banana peel	–	5.62	–	–	46.79	36.52	Sanchez et al. (2014)
Potato solid waste	85–87	3–5	2.25	6–12	19.86	27–35	Sharoba et al. (2013), Afifi (2011) and Arapaglou et al. (2010)

Continued

Table 2.3 Composition of Different Fruit and Vegetable Wastes (per 100 g)—cont'd

Waste	Moisture (g)	Protein (g)	Fat (g)	Minerals (g)	Fiber (g)	Carbohydrates (g)	References
Tomato solid waste	85–90	17–22	—	3–5	—	35–45	Barth and Powers (2008), Hills and Nakano (1984) and Saev et al. (2009)
Onion peelings and whole bulbs	82–92	—	—	4.7	—	—	Barth and Powers (2008) and Benitez et al. (2011)
Sugar beet pulp, silage and leaves	85	—	—	3.8–8.8	—	—	Mojtahedi and Mesgaran (2009)
Cauliflower leaves	86	16.1	—	14.0	28.0	24.0	Wadhwa et al. (2006)
Cabbage leaves	82.8	20.4	—	17.2	34.0	22.5	Wadhwa et al. (2006)
Peapods, peel, shell waste	91.5	20.2	—	8.5	57.0	46.5	Fallon et al. (2006) and Wadhwa et al. (2006)
Pea vines	91.2	12.2	—	8.8	60.0	47.5	Wadhwa et al. (2006)
Green oats	90.8	10.9	—	9.2	67.0	57.5	Wadhwa et al. (2006)

carbon, nitrogen, and mineral sources for their growth and production of secondary metabolites. In addition, many studies have reported that cellulose, hemicellulose, pectin, etc. help in induction of cellulases, hemicellulases (xylanases and other associated enzymes), and pectinases (Dhillon et al., 2012a; Dhillon et al., 2012b; Thangaratham and Manimegalai, 2014). Value addition of wastes is nearly impossible without the intervention of enzymes, which play an important role in transformation of agro-industrial waste to some important value-added products (Babbar and Oberoi, 2014). A majority of the biological conversions of agro-industrial processing wastes into value-added products have been carried out employing solid-state bioprocessing. Solid-state bioprocessing refers to the utilization of water insoluble substrates for microbial growth and is usually carried out in solid- or semisolid systems in the absence or near absence of water (Zheng and Shetty, 1998). Enzymes can be obtained from bacteria, fungi, animals, plants, insects, or other biological materials. It is important to understand the catalytic property, mode of action, optimal operational conditions, and the type of enzyme or enzyme combination that could be most appropriate to use them effectively for biotransformation processes. This chapter focuses on the biotechnological potential of using and upgrading renewable resources—"leftovers"/wastes, especially the fruit and vegetable processing industries residues (wastes) for developing high-value products, such as enzymes.

IMPORTANT ENZYMES FOR INDUSTRIAL APPLICATIONS

Value addition of agricultural wastes includes the biotransformation of wastes to a useful product like enzymes. Enzymes are protein biomolecules that serve as catalysts to speed up chemical reactions. Enzymes are useful in various areas of applications like manufacturing of food and feedstuff, cosmetics, medicinal products, and as a tool for research and development. At a commercial level, enzymes are applied in detergents, pulp and paper applications, textile manufacturing, leather industry, for fuel production, and for the production of pharmaceuticals (Binod et al., 2013). Agricultural residues are generally composed of cellulose, hemicellulose, lignin, and pectin, and are referred as "lignocellulosic materials," and thus, can be potentially used for the production of various industrially important enzymes. In the lignocellulosic materials, these three fractions are closely associated with each other to form complex structure of the biomass. Basically, cellulose forms a skeleton that is bounded by hemicelluloses and lignin (Fig. 2.1).

Cellulolytic Enzymes

Cellulolytic enzymes are the most important industrial enzymes due to their versatile applications. Basic and applied studies on cellulolytic enzymes have demonstrated their biotechnological potential in various industrial sectors, such as food, animal feed, brewing and wine making, agriculture, biomass refining, pulp and paper, textiles, and laundry products (Pandey et al., 1999; Kuhad et al., 2011). The resurgence in utilization of

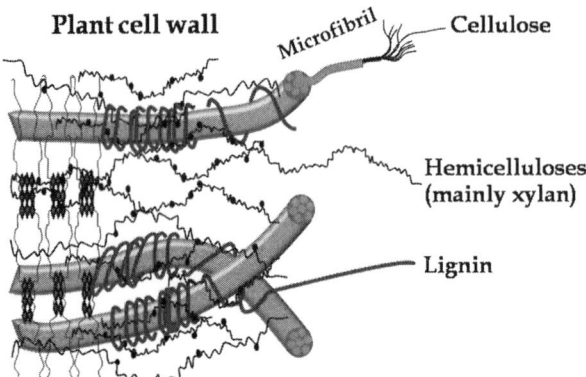

Figure 2.1 Representation of the cellulose, hemicellulose, and lignin fractions in the lignocellulosic complex.

biomass for bioethanol production and other value-added organic compounds has attracted major attention of researchers worldwide toward cellulases. Cellulase is not a single enzyme but is a complex of three major types of cellulases: endoglucanases [endo-1,4-β-glucanase, carboxymethylcellulases (CMCase), endo-1,4-β-D-glucanase], exoglucanases, and cellobiase. Endocellulases (EC 3.2.1.4) randomly cleave internal bonds at amorphous sites that create new chain ends. Exocellulases (EC 3.2.1.91) cleave two to four units from the ends of the exposed chains produced by endocellulase, resulting in tetrasaccharides or disaccharides, such as cellobiose. Cellobiase or β-glucosidase (EC 3.2.1.21) hydrolyzes the exocellulase product into individual monosaccharides. Bioconversion of lignocellulosic waste to soluble sugars relies on the synergistic action of complete cellulase system comprising of CBH, EG, BG, and hemicellulases (Aro et al., 2005).

Hemicellulases

A spectrum of hemicellulase enzymes synergistically act on hemicellulose leading to its complete hydrolysis. These enzymes include endoxylanase (endo-1,4-β-xylanase, EC 3.2.1.8), β-xylosidase (xylan-1,4- β-xylosidase, EC 3.2.1.37), α-glucuronidase (α-glucosiduronase, EC 3.2.1. 139), α-arabinofuranosidase (α-L-arabinofuranosidase, EC 3.2.1.55), and acetylxylan esterase (EC 3.1.1.72). Among them, endoxylanases and β-xylosidases (collectively called xylanases) are the two key enzymes responsible for hydrolysis of xylan, the major component of hemicellulose (Chavez et al., 2006). Endoxylanases act on homopolymeric backbone of 1,4–linked β-D-xylopyranose producing xylooligomers (Ahmed et al., 2009) while β-xylosidases act on these xylooligomers releasing xylose (Knob et al., 2010). In nature, hemicellulases are produced by a wide variety of microorganisms, including bacteria and fungi (Ahmed et al., 2009). Xylanolytic enzymes have attracted a great deal of attention in the past decade, particularly because of their biotechnological potential in various industrial processes, such as food, feed, and pulp and paper industries (Beg et al., 2001).

Proteases

Proteolytic enzymes are capable of hydrolyzing peptide bonds and are also referred to as peptidases, proteases, or proteinases (Barrett and McDonald, 1986). Proteolytic enzymes can be classified based on their origin: microbial (bacterial, fungal, and viral), plant, animal, and human enzymes can be distinguished (Motyan et al., 2013). Proteolytic enzymes belong to the hydrolase class of enzymes (EC 3) and are grouped into the subclass of the peptide hydrolases or peptidases (EC 3.4). Depending on the site of enzyme action, proteases can also be subdivided into exopeptidases or endopeptidases. Exopeptidases catalyze the hydrolysis of the peptide bonds near the N- or C-terminal ends of the substrate. Aminopeptidases can liberate single amino acids (EC 3.4.11), dipeptides (dipeptidyl peptidases, EC 3.4.14), or tripeptides (tripeptidyl peptidases EC 3.4.14) from the N-terminal end of their substrates. Single amino acids can be released from dipeptide substrates by dipeptidases (EC 3.4.13) or from polypeptides by carboxypeptidases (EC 3.4.16–3.4.18), while peptidyl dipeptidases (EC 3.4.15) liberate dipeptides from the C-terminal end of a polypeptide chain. Endopeptidases cleave peptide bonds within and distant from the ends of the polypeptide chain (Rao et al., 1998). Endoproteases are classified according to their catalytic mechanism, which implies specificity in the enzyme active site. Proteases constitute the most important group of industrial enzymes currently in use (Motyan et al., 2013).

Lipases

Lipases have emerged as one of the leading biocatalysts with proven potential for contributing to the multibillion dollar underexploited lipid bioindustry and have been used in in situ lipid metabolism and ex situ multifaceted industrial applications (Sharma and Kanwar, 2012). Lipases are triacylglycerol acylhydrolases (EC 3.1.1.3), which catalyze the hydrolysis of triacylglycerol to glycerol and fatty acids. They often express other activities, such as phospholipase, isophospholipase, cholesterol esterase, cutinase, amidase, and other esterase type of activities (Svendsen, 2000). Lipases are ubiquitous in nature and are produced by several plants, animals, and microorganisms (Thakur, 2012). Some important lipase-producing bacterial genera are *Bacillus*, *Pseudomonas*, and *Burkholderia* (Gupta et al., 2004) and fungal genera include *Aspergillus*, *Penicillium*, *Rhizopus*, and *Candida* (Singh and Mukhopadhyay, 2012). Different species of yeasts belonging to seven different genera including *Zygosaccharomyces*, *Pichia*, *Lachancea*, *Kluyveromyces*, *Saccharomyces*, *Candida*, and *Torulaspora* have also been reported (Romo-Sanchez et al., 2010). Lipases have also been explored for the novel biotechnological applications, such as synthesis of biopolymers and biodiesel, production of enantiopure drugs, and flavor compounds (Dussan et al., 2010; Verma et al., 2012).

Pectinases

Pectinases are responsible for the degradation of long and complex molecules called pectins that occur as structural polysaccharides in the middle lamella and the primary cell walls of young plant cells. Pectinolytic enzymes can be applied in various industrial sectors wherever degradation of pectin is required, and nearly 25% of the global enzyme sales are attributed to pectinase (Kashyap et al., 2000; Kaur et al., 2004). The three major types of pectinases are: (1) pectinesterase, also known as pectinmethyl hydrolase, which catalyzes deesterification of the methoxyl group of pectin forming pectic acid and; (2) depolymerizing enzymes, which include (a) polymethylgalacturonase (PMG); they may be endo-PMGs, which cause random cleavage of α-1,4-glycosidic linkages of pectin, preferentially highly esterified pectin and exo-PMG, which cause sequential cleavage of α-1,4-glycosidic linkage of pectin from nonreducing end of the pectin chain; (b) polygalacturonases (PGs), which catalyze the hydrolysis of α-1,4-glycosidic linkages in pectic acid (polygalacturonic acid). Polygalacturonases are also of two types: endo-PG that catalyzes random hydrolysis of α-1,4-glycosidic linkages in pectic acid and exo-PG that catalyzes hydrolysis of α-1,4-glycosidic in pectic acid in a sequential fashion; and (3) protopectinase, which solubilizes protopectin forming highly polymerized soluble pectin (Favelo-Torres et al., 2006). Pectinolytic enzymes are of significant importance in the current biotechnological era with applications in fruit juice extraction and clarification, scouring of cotton, degumming of plant fibers, wastewater treatment, vegetable oil extraction, tea and coffee fermentations, bleaching of paper, in poultry feed additives, and in alcoholic beverages and food industries (Jayani et al., 2005; Tapre and Jain, 2014).

Amylases

Enzymes responsible for hydrolysis of starch are called amylolytic enzymes or simply amylases. Amylolytic enzymes form a large group of enzymes, among which the most common and the best known are α-amylases, β-amylases, and glucoamylases. Since starch is an essential source of energy, amylolytic enzymes are produced by a variety of living organisms. Alpha-amylases catalyze cleavage of α-1,4-glycosidic bonds in the inner region of the molecule, hence causing a rapid decrease in substrate molecular weight and viscosity. These endoacting enzymes can be divided into liquefying or saccharifying α-amylases, which preferentially degrade substrates containing more than 15 or four glucose units, respectively. Two types of exoacting amylases that are commonly used for starch saccharification are β-amylases and glucoamylases. Both act on glycosidic linkages at the nonreducing ends of amylose, amylopectin, and glycogen molecules, providing low-molecular-weight carbohydrates in the β-anomeric form. The main end product of hydrolysis catalyzed by β-amylases is maltose, while glucoamylase generates glucose (Babbar and Oberoi, 2014). The characterization and potential applications of amylases in different industries have been comprehensively reviewed (Aiyer, 2005).

Ligninolytic Enzymes

Lignin degrading enzymes encompasse mainly three oxidative enzymes: lignin peroxidase (LiP), manganese peroxidase (MnP), and laccase. Laccase (EC 1.10.3.2) is a multicopper protein belonging to the family of the blue oxidase enzymes. This enzyme generally contains four copper ions, grouped in three groups: T1, formed by one ion and is responsible for substrate oxidation and for electron transfer; T2, also formed by one ion and, together with the T3 group, which contains two ions, constitute the copper trinuclear center, involved in oxygen reduction and water release (Torres et al., 2003). Lignin peroxidase (LiP) (EC 1.11.1.14) is a glycoprotein that has a prosthetic group made up of iron protoporphyrin IX, with catalytic activity dependent on H_2O_2. The H_2O_2 required for LiP activity originates from different biochemical pathways, expressed according to the nutritional factors and growth conditions of the microorganism (Pointing, 2001). LiP is an enzyme capable of oxidizing various nonphenolic aromatic compounds, such as benzyl alcohol, cleaving side chains of these compounds, catalyzing aromatic ring opening reactions, demetoxilation, and oxidative dechlorination. Manganese peroxidase (MnP, EC 1.11.1.13) is a glycosylated extracellular enzyme that possesses heme prosthetic group, found only in basidiomycetes. In general, it does not unleash direct transformations on its substrates. Manganese peroxidase reaction mechanism is similar to that of other peroxidases, such as lignin peroxidase, but in this case compounds I and II of MnP oxidize Mn^{2+} (Conessa et al., 2002; Wong, 2009). Demand for these enzymes has increased in recent years due to their potential applications in diverse biotechnological areas. Lignin-degrading enzymes are widely used in pollution abatement, especially in the treatment of industrial effluents that contain hazardous compounds, such as dyes, phenols, and other xenobiotics (Mester and Tien, 2000). The industrial preparation of paper requires separation and degradation of lignin in wood pulp. Pretreatment of wood pulp with lignolytic enzymes might provide milder and cleaner strategies of delignification. Ligninolytic enzymes are commonly used for biobleaching of kraft pulp. The ever-increasing demand for the just-mentioned enzymes in the industrial sector necessitates the production of enzymes from inexpensive raw materials as well as utilization of these enzymes for production of value-added products is discussed in the following sections.

ENZYME PRODUCTION THROUGH FERMENTATION

Enzymes are among the most important products obtained for human needs through microbial sources. Enzyme production is achieved using both solid-state fermentation (SSF) and submerged fermentation (SmF).

Solid-State Fermentation

Solid-state fermentation is defined as any fermentation process performed on a nonsoluble material that acts both as physical support and source of nutrients in absence of free-flowing liquid (Pandey, 1992). SSF involves microbial growth and product formation on solid particles in the absence (or near absence) of water; however, the substrate contains sufficient moisture to allow microbial growth and metabolism. Due to low freely available moisture, SSF can only be carried out by a limited number of microorganisms, mainly yeasts and fungi, although some bacteria have also been employed (Pandey et al., 2000a). The nature of the solid substrate employed is the most important factor affecting SSF processes, and its selection depends upon several factors, mainly related to cost and availability and, thus may involve the screening of several agro-industrial residues. In a SSF process, the solid substrate not only supplies the nutrients to the culture but also serves as an anchor for the microbial cells. Among the several factors that are important for microbial growth and activity in a particular substrate, particle size and moisture level/water activity are the most critical (Auria et al., 1992; Renge et al., 2012).

Agro-industrial residues have been widely used due to their potential advantages for filamentous fungi, which are capable of penetrating into these solid substrates, aided by the presence of turgor pressure at the tip of the mycelium (Ramachandran et al., 2004). In addition, the utilization of these agro-industrial wastes provides alternative substrates and helps in solving pollution problems caused by their disposal (Pandey et al., 1999). SSF is an interesting process for lignocellulose-degrading enzyme production at low cost, due to the possibility of using agricultural and agro-industrial residues as substrates for microbial growth (Subramaniyam and Vimala, 2012). The use of SSF for cultivation of fungi presents many advantages, such as simulation of their natural environment, which results in better adaptation to the medium and higher enzyme production; reduction of bacterial contamination due to absence of free water; the possibility of getting concentrated enzymes as the enzymes may be extracted with a small amount of water (Bianchi et al., 2011). Other advantages of fungal enzyme production by SSF are the use of a smaller amount of water and consequently a decrease in effluent generation, lower requirement of space and energy, more stability of the products obtained, more biomass and enzyme production, less catabolic repression of enzymes, and little protein degradation (Pandey et al., 2000a,b; Viniegra-González et al., 2003). Several authors have reviewed different applications of SSF (Raimbault, 1998; Pandey et al., 2000b; Perez-Guerra et al., 2003; Holker and Lenz, 2005).

Submerged Fermentation

Submerged fermentation involves inoculation of the microbial culture into the liquid medium for production of the desired product. Most of the commercial products are produced through the SmF processes (Subramaniyam and Vimala, 2012). Fermentation processes are broadly classified into aerobic and anaerobic processes. Antibiotics and enzymes are produced through aerobic fermentation, which involves the incorporation of oxygen into the liquid medium, while butanol production proceeds through the anaerobic mode, wherein addition of oxygen is detrimental. Certain fermentation processes, like ethanol production, involve facultative anaerobic organisms, such as *Saccharomyces cerevisiae*, which can grow in the presence of oxygen leading to cell biomass production but switch over to the anaerobic route during the ethanol fermentation.

However, enzymes, such as amylases and proteases generally produced through SmF process employ the aerobic fermentation route. SmF processes can be easily scaled up with ease of automation and do not suffer from the heat mass transfer limitations, which is the main disadvantage of SSF. However, the major drawbacks associated with the SmF processes are the low productivity, high production cost, and complexity of the medium (Babbar and Oberoi, 2014). Some of the enzymes produced using different microbial strains via SSF and SmF on FVPW are presented in Table 2.4. It is clear from this table that excepting xylanase production in some cases, enzyme titers have been significantly higher in case of SSF as compared to SmF.

POTENTIAL OF FRUITS AND VEGETABLE SOLID WASTE FOR THE PRODUCTION OF ENZYMES

In the horticultural sector, there has been significant growth in both acreage and production to fulfill the requirements of global food demand leading to generation of large amounts of by-products. The wastes/by-products obtained from fruits and vegetables present a major disposal problem for the concerned industry. Processing of fruits and vegetables results in different kinds of wastes, such as citrus peel and pomace, apple pomace, banana peel, grape pomace and skin, apricot kernel, mango seed and pulp, litchi peel and pericarp, potato peel, tomato skin and pomace, carrot peel, peapods, cauliflower florets and curds, etc. Nowadays biological conversion processes of fruit and vegetable processing wastes into various value-added products, like enzymes through SSF, have been of major interest (Zheng and Shetty, 2000; Anuradha et al., 2010; Namasivayam and Nirmala, 2011; Rathan and Ambili, 2011; Sharanappa et al., 2011; Padma et al., 2012; de Andrade Silva et al., 2013; Sandhya and Kurup, 2013).

Citrus Waste

Citrus is one of the most important fruit crops of the world. The annual worldwide production of citrus fruits is about 136 million tons (FAO, 2014) and one-third of the fruits are processed. Citrus fruits are processed, mainly to obtain juice, jams, and squashes (Izquirdo and Sendra, 2003). Worldwide industrial citrus wastes were estimated to be more than 15 million tons (Marin et al., 2007). Availability of such a processing waste would have increased significantly owing to increase in the processing levels across the globe. Citrus waste consisting of peel, pulp, and seeds corresponds to half of the fresh fruit weight. Residues of citrus juice production are a source of dried pulp and molasses, fiber-pectin, cold-pressed oils, essences, D-limonene, juice pulps and pulp wash, ethanol, seed oil, pectin, limonoids, and flavonoids (Ozaki et al., 2000; Siliha et al., 2000). Due to the large amount of fruits being processed into juice, there has been an emergence of a considerable industry for utilization of the residual peels, membranes, seeds, and other compounds (Schieber et al., 2001). Dry citrus peels are rich in pectin, cellulose, and

Table 2.4 Enzyme Production Using Fruit and Vegetable Processing Waste via Solid-State Fermentation and Submerged Fermentation

Serial No.	Enzyme	Microorganism	Substrate		Productivity		References
			SSF	SmF	SSF (U g^{-1})	SmF (U mL^{-1})	
1	Amylase	*Bacillus* spp.	Oil cakes, wheat bran	Starch broth	50,000	400	Singh et al. (2010)
2	Xylanase	*Bacillus* spp.	Corn cob, wheat bran	Corn cob and yeast extract	6.18	16.13	Gupta and Kaur (2009)
3	Cellulase	*Bacillus* spp.	Banana waste	Carboxymethylcellulose, cellulose	9.6	1.2	Sukumaran et al. (2005)
4	Cellulase	*Trichoderma viridae* ATCC 13631	Wheat bran	Mandel's liquid medium	60.5	28	Vintila et al. (2009)
5	Lipase	*Aspergillus niger* NCIM 1207	Wheat bran, olive oil	Synthetic oil-based medium	630	18	Mahadik et al. (2002)
6	Pectinase	*A. niger*	Wheat bran, coffee pulp	Pectin-based production medium	2.28	0.48	Maldonado et al. (1998)
7	Pectinase	*A. niger*	Orange peels	Orange peels	1,224	2.29	Mrudula and Anitharaj (2011) and Sharma et al. (2011)
8	Pectinase	*Penicillium citrinum*	Orange waste	Orange waste	34.5	0.067	Sandhya and Kurup (2013)
9	Pectinase	*Bacillus* spp.	Orange peel waste	Orange bagasse	200	0.30	Bayoumi et al. (2008) and Osuigwe et al. (2014)
10	Xylanase	*Trichoderma koeningi*	Pineapple peel powder, corn cobs	Maize straw	2.868	4.62	Bandikari et al. (2014) and Goyal et al. (2008)
11	Laccase	*Pleurotus* spp.	Orange waste	Mandarin peel	12.2	2.37	Inacio et al. (2015)

12	Xylanase	*Aspergillus awamori*	Tomato waste	Oat straw	100	820	Umsza-Guez et al. (2011) and Smith and Wood (1991)
13	Xylanase	*A. niger*	Apple pomace	Palm leaf	5,662	1,906.5	Liu et al. (2008) and Norazlina et al. (2012)
14	Xylanase	*Trichoderma* spp.	Watermelon rinds	Melon peels	70	26.5	Mohamed et al. (2013) and Isil and Nulifer (2005)
15	Xylanase	*Trichoderma* spp.	Sunflower sludge	Pineapple peel	8.75	73.09	Fortkamp and Knob (2014) and Sakthiselvan et al. (2012)

hemicellulose, which make them interesting sources for the production of a variety of bioproducts (Ismail, 1996). Orange waste contains 11.8% fiber, 6.4% protein, 6.7% ash, 19% total sugars, and 0.1% pectin (Martins et al., 2002). Kinnow belongs to the citrus family of fruits, and is known for its high juice content and characteristic juice flavor. During processing, Kinnow also generates a substantial quantity of wastes in the form of peels, pulp, and seeds. Production of pectinases and cellulases using citrus waste is an alternative practical use of waste for the development of value-added products, such as enzymes (Oberoi et al., 2010; De-Gregorio et al., 2002). Mamma et al. (2008) used dry citrus peels as a substrate for the production of multienzyme preparations containing pectinolytic, cellulolytic, and xylanolytic enzymes by the mesophilic fungi *Aspergillus niger* BTL, *Fusarium oxysporum* F3, *Neurospora crassa* DSM 1129, and *Penicillium decumbens* under SSF conditions. In a recent study, Adesina et al. (2013) isolated two fungal strains, *Rhodotorula* spp. and *Mucor mucorales*, and used them in SSF for the production of pectinase. Highest pectinase activity of 82.95 IU per dry weight of substrate (dw) was recorded for *Rhodotorula* spp. using orange peelings as substrate after eight days while the least value of $16.12\,\mathrm{IU\,dw^{-1}}$ was observed on watermelon pomace as substrate after two days. *Mucor* showed the highest pectinase activity of $46.05\,\mathrm{IU\,dw^{-1}}$ on orange peelings as substrate after 12 days. Inacio et al. (2015) studied the production of hydrolytic and oxidative enzymes by *Pleurotus pulmonarius* developed in SSF using orange waste as a substrate. Among the hydrolytic enzymes, pectinase was the main enzyme produced by the fungus presenting the highest enzyme activity of $9.4\,\mathrm{IU\,mL^{-1}}$ after 35 days of cultivation. Among the oxidative enzymes, laccase was the main enzyme produced with maximal activity of $12.2\,\mathrm{IU\,mL^{-1}}$ after 20 days of cultivation. Low enzyme levels of manganese peroxidase ($0.07\pm0.00\,\mathrm{U\,mL^{-1}}$), β-glucosidase ($0.7\pm0.01\,\mathrm{U\,mL^{-1}}$), and β-xylosidase ($0.3\pm0.03\,\mathrm{U\,mL^{-1}}$) were detected at the end of the cultivation after 45 days. The enzymatic levels of amylase, carboxymethyl cellulase, and xylanase were comparable and found to be less than $1.5\,\mathrm{IU\,mL^{-1}}$ after 45 days. Oberoi et al. (2010) studied the potential of using kinnow pulp for production of cellulases by *Trichoderma reesei* Rut C-30. Out of the different combinations tried out, dried kinnow pulp supplemented with wheat bran in the ratio of 4:1 resulted in the highest filter paper cellulase (FPase) activity of 13.4 IU per gds, whereas endo-1,4-β-glucanase (CMCase) activity was found to be best when kinnow pulp was supplemented with wheat bran in the ratio 3:2 using Mandel Weber (MW) medium. The β-glucosidase activity of 18 IU per gds was again found to be maximum in treatment involving 3:2 ratio of kinnow pulp to wheat bran using MW medium.

Apple Residues

Apple (*Malus domestica* Borkh.) is the most favored fruit of millions of people and is a widely grown fruit in temperate regions of the globe (Kaushal and Joshi, 1995; Kaushal et al., 2002; Agrahari and Khurdiya, 2003). The global annual production of apples is about 80.8 million tons (FAO, 2014). Presently, India is the ninth largest producer of apples in the world contributing about one-third of total apple production of the world with an annual production of 1.91 million tons from an area of 0.31 million ha (FAO, 2014). About 71% of apples in India are consumed as fresh, while about 20% is processed into value-added products, of which 65% are processed into apple juice concentrate and the balance quantity into other products that include packed natural ready-to-serve apple juice, apple cider, wine and vermouth, apple purees and jams, and dried apple products (Downing, 1989; Joshi et al., 1991; Joshi, 1997; Kaushal et al., 2002). Apple pomace is the solid residue that remains after the extraction of juice from apple. In India, total production of apple pomace was about 1 million tons per annum approximately of which only 1% was being effectively utilized (Manimehalai, 2007). A huge quantum of apple pomace is not being utilized at present but is dumped in the fields, creating pollution problems because of fermentation and high COD of 250–300 g kg^{-1} (Shalini and Gupta, 2010).

Apple pomace is the main by-product of apple cider and juice processing industries and accounts for about 25% of the original fruit mass. Apple pomace typically contains 7.2–43.6% cellulose, 4.26–24.40% hemicellulose, 15.2–23.5% lignin, 3.5–14.32% pectin, 48.0–83.8% total carbohydrates, 4.7–51.10% fiber, and 2.9–5.7% protein (dry weight basis) (Dhillon et al., 2013). Being a rich source of carbohydrate, pectin, crude fiber, and minerals, it is a good source of nutrients; apple pomace from apple processing industries can be utilized for the development of various enzymes. Numerous enzymatic transformations of apple pomace have been proposed for obtaining valuable products. The highly variable composition of apple pomace and possible strategies for utilization have been extensively reviewed (Kennedy et al., 1999a). The potential of apple pomace as a substrate for the production of β-fructofuranosidase by *Aspergillus fumigatus*, *Aspergillus foetidus*, and *Aspergillus niger* was studied by Hang and Woodams (1994). *Aspergillus. foetidus* produced more β-glucosidase on pomace than the other two species yielding β-glucosidase activity of more than 900 IU kg^{-1} of apple pomace fermented, whereas *A. fumigatus* and *A. niger* produced only 48 and 73 IU kg^{-1}, respectively. Villas-Boas et al. (2002) utilized *Candida utilis* to produce lignolytic enzymes and cellulolytic enzymes using apple pomace as a substrate under solid-state conditions. Apple pomace has been exploited for the production of polygalacturonase by *Lentinus edodoes* through solid-state fermentation (Zheng and Shetty, 2000). Apple pomace has been used for the production of various industrially important enzymes (Villas-Boas et al., 2002; Joshi et al., 2006, 2011; Dhillon et al., 2011a,b,c).

Grapes

Grapes (*Vitis* sp., Vitaceae) are the world's largest fruit crop with an annual production of more than 60 million tons (Schieber et al., 2001). As per FAO (2014), the worldwide grape production was about 71.55 million tons. About 80% of the total crop is used for wine making (Mazza and Miniati, 1993), and pomace represents approximately 20% of the weight of the processed grapes (Meyer et al., 1998). The composition of grape pomace varies considerably, depending on grape variety and technology of wine making. A wide range of products, such as ethanol, tartrates, citric acid, grape seed oil, hydrocolloids, and dietary fiber are recovered from grape pomace (Igartuburu et al., 1997; Bravo and Saura-Calixto, 1998; Girdhar and Satyanarayana, 2000). The by-products obtained from grapes are seeds and pomace left after extraction of juice. Grape seeds are the major waste from the grape processing units. Utilization of grape seeds as support substrate for laccase production in laboratory-scale SSF bioreactors has been suggested (Rodriguez Couto et al., 2006). The lignin content of grape seeds is much higher (around 44%) than many of the other substrates and hence they can be utilized for the production of other ligninolytic enzymes, like LiP and MnP (Niladevi, 2009). Venkatesh et al. (2009) studied production of pectinases utilizing fruit wastes of cashew, banana, pineapple, and grape under controlled fermentation with *A. foetidus* strain. Enzyme production was found to be maximum ($0.35\,IU\,mg^{-1}$) in the medium with grape waste at 40°C after 8 days. Dedavid-e-Silva et al. (2009) studied the production of cellulolytic enzymes by the fungus *Aspergillus phoenicis* using grape waste in SmF at 30°C and 120 rpm for 120 h. Maximum activity of endoglucanase ($0.66\,IU\,mL^{-1}$) and β-glucosidase ($50.36\,IU\,mL^{-1}$) was observed. The potential of grape waste as a growth substrate for the production of enzymes has been reviewed (Botella et al., 2005; Dariot et al., 2007; Diaz et al., 2012; Levin et al., 2012).

Mango

Mango (*Mangifera indica* L., Anacardiaceae) is one of the most important tropical fruits (Ramteke et al., 1999). Major waste of mango processing is peels and stones, amounting to 35–60% of the total fruit weight (Larrauri et al., 1996). Mango seed kernel fat is a promising source of edible oil and has attracted attention since its fatty acid and triglyceride profile is similar to that of cocoa butter (Puravankara et al., 2000). Mango peels were also reported to be a good source of dietary fiber containing high amounts of extractable polyphenolics (Larrauri et al., 1996). Padma et al. (2012) studied polygalacturonase production by *Aspergillus awamori* MTCC 9166 in SSF using different pectin-rich fruit wastes, like apple peel, banana peel, citrus (orange) peel, jackfruit rind, mango peel, and pineapple peel. Highest enzyme production was observed with jackfruit rind ($420\,IU\,g^{-1}$) and mango peel ($250\,IU\,g^{-1}$) at 65% moisture content, 28°C, pH 5.2, 10^6 spores per g inoculum size for jackfruit rind and 10^8 spores per g for mango peel after 96 h. Kumar et al. (2013) investigated α-amylase production using mango kernel as a substrate with *Fusarium solani*. Maximum amylase production ($0.889\,U\,g^{-1}$) was recorded using a substrate concentration of 5% (w/v), pH 4, and temperature of 30°C on day 9 of incubation. Saravanan et al. (2012) optimized culture medium for cellulase production by *T. reesei* using mango peel as substrate with statistical methodology based on experimental designs. Under optimum conditions validated experimentally, an enhanced cellulase activity of $7.8\,IU\,mL^{-1}$ was obtained.

Pineapple

The worldwide production of pineapple has been estimated at approximately 24.78 million tons (FAO, 2014). The pulpy waste material resulting after juice production still contains substantial amounts of sucrose, starch, and hemicellulose and may therefore be used for ethanol production (Tanaka et al., 1999; Nigam, 2000). Folakemi et al. (2008) studied cellulase production from cellulosic pineapple waste using *Trichoderma longibrachiatum*, *A. niger*, and *Saccharomyces cerevisiae*. The results obtained from the fermentation showed that *T. longibrachiatum* produced the highest amount of glucose among the cultures tested (0.92 mg per 0.5 mL). This was produced from pineapple pulp at pH 4.5 and temperature of 45°C after seven days of fermentation. Fortkamp and Knob (2014) studied production of xylanases by *Trichoderma viridae* using pineapple peel as substrate. Highest xylanase production $(35.12\,U\,mL^{-1})$ was obtained with 2% substrate concentration within seven days at 28°C. Hariharan and Nambisan (2013) studied optimization of lignin peroxidase, manganese peroxidase, and laccase production from *Ganoderma lucidum* under SSF of pineapple leaf. Pineapple residues composed of peel, core, crown, and stem are rich in holocellulose (80%) and can thus be extensively exploited for the production of cellulases and hemicellulases (Sarvanan et al., 2013). The same authors reported cellulase production of about $8.61\,IU\,mL^{-1}$ after process optimization through SmF using pineapple waste as substrate. One of the most important envisaged uses of pineapple waste is in the production of bromelain.

Banana

Banana (*Musa paradisiaca* L., Musaceae) represents one of the most important fruit crops, with a global annual production of about 50 million tons (FAO, 2014). Peels constitute up to 30% of the ripe fruit. About 1000 banana plants are estimated to yield 20–25 tons of pseudostems providing about 5% edible starch (Anand and Maini, 1997). Attempts at utilization of banana waste include the biotechnological production of protein (Chung and Meyers, 1979), ethanol (Tewari et al., 1986), α-amylase (Krishna and Chandrasekaran, 1996; Krishna et al., 2012; Paul and Sumathy, 2013), hemicellulases (Medeiros et al., 2000; Rehman et al., 2014), and cellulases (Krishna, 1999; Sun et al., 2011; Dabhi et al., 2014). Krishna and Chandrasekaran (1996) used banana fruit stalk as a substrate for α-amylase production by *Bacillus subtilis*. Oberoi et al. (2011) reported $28.2\,g\,L^{-1}$ ethanol from banana peel after its enzymatic hydrolysis. Commercial cellulase, β-glucosidase, and pectinase at $8\,FPU\,g^{-1}$ cellulose, $15\,IU\,g^{-1}$ cellulose, and $66\,IU\,g^{-1}$ pectin hydrolyzed banana peel resulting in glucose and reducing sugars concentrations of 28.2 and $48\,g\,L^{-1}$, respectively (Oberoi et al., 2012). Unakal et al. (2012) used banana waste as a substrate for the production of α-amylase by *B. subtilis* under SSF conditions yielding maximum enzyme activity up to $7.26\,IU\,mL^{-1}$ per minute after 24 h. Banana skin has been used as a support substrate for the production of extracellular laccase by the white-rot fungus *Trametes pubescens* (Osma et al., 2007). Production of cellulolytic and ligninolytic enzymes from banana waste under SSF conditions has been investigated, and it has been reported that the level of ligninolytic enzymes produced from banana waste was higher than the cellulolytic enzymes (Shah et al., 2005). The results of this study indicated the suitability of banana waste for ligninolytic enzymes production under SSF.

Passion Fruit

Yellow passion fruit (YPFW; *Passiflora edulis* flavicarpa) is the edible fruit widely grown in many tropical and subtropical areas and is a predominant fruit of South America. Other common names for passion fruit are maracuya, marcha (Spanish), and maracujá (Portuguese). It has a soft to firm juicy interior, filled with numerous seeds. Brazil is the largest yellow passion fruit producer with an annual production of 0.49 million tons in 2004 (Agrianual, 2007). The country has since long had a well-established yellow passion fruit juice industry with large-scale extraction plants for export of passion fruit to several parts of the world (Dhawan et al., 2004). During industrial production of the passion fruit juice, several thousands of fruit barks are discharged, being considered as organic waste. This material is particularly rich in pectin, represents around 52% of the total weight of the residue, and contains a moisture content of 88% (Cordova et al., 2005). Because of the significant juice production, the peels, as a major waste, have become a substantial burden on the environment. Hence, it is necessary to find a feasible way to transform the peels into useful products or to dispose them suitably, seeking a positive environmental impact. Zilly et al. (2012) investigated the potential of YPFW peel alone and with wheat bran (WB) as a support substrate for the production of hydrolytic and oxidative enzymes by several food-grade white-rot fungi *Pleurotus ostreatus*, *Pleurotus pulmonarius*, *Macrocybe titans*, *Ganoderma lucidum*, and *Grifola frondosa*, under SSF conditions. Among oxidative enzymes, all cultures presented high-laccase activity in both substrates, YPFW (ranging from 6040 to $10,200\,U\,L^{-1}$) and WB (ranging from 5900 to $10,900\,U\,L^{-1}$) with the exception of the *G. frondosa* cultures ($1500\,U\,L^{-1}$ in YPFW and $1830\,U\,L^{-1}$ in WB). Among the polysaccharide depolymerase enzymes, pectinase appears to be the main enzyme produced by all species, especially in YPFW cultures, with *M. titans* as the best producer ($1720\,U\,L^{-1}$). Low levels of endoxylanase and no exo- and endocellulase activities were observed in all YPFW extracts. In WB cultures, the production of endoxylanase was higher, especially by *P. ostreatus*, *P. pulmonarius*, and *G. lucidum*. All extracts (from YPFW and WB cultures) presented high levels of arylglycosidases (β-glucosidase, β-xylosidase, and β-galactosidase).

Jackfruit

Jackfruit (*Artocarpus heterophyllus* Lam.) is a monoecious evergreen tree that is popular in several tropical countries (Babitha et al., 2006). There may be 100 or up to 300 seeds in a single fruit. Seeds make up around 10–15% of the total fruit mass and have high carbohydrate and protein content (Kumar et al., 1988). Pectinase enzyme was produced from *A. niger* NCIM 548 under SSF using jackfruit waste as substrate (Rao et al., 2014). The maximum yield of pectinase at 39.836 IU per gds was obtained using optimized conditions that included jackfruit waste 10 g, moisture content 70% (v/w), pH 5.0, temperature 30°C, glucose 3.5% (w/w), $(NH)_4SO_4$ 1.0% (w/w), and fermentation time of 72 h. In another study, Anuradha et al. (2010) after screening various isolates found *A. awamori* MTCC 9166 as an isolate with highest polygalacturonase activity. Isolate was tested for utilization of raw pectins for enzyme production. Polygalacturonase production was high in raw pectin sources, such as orange peel ($16.8\,U\,mL^{-1}$), jackfruit rind ($38\,IU\,mL^{-1}$), carrot peel ($36\,IU\,mL^{-1}$), and beetroot peel ($24\,IU\,mL^{-1}$).

Kiwifruit

Kiwifruit (*Actinidia chinensis* Planch, Actinidiaceae) waste results from rejected kiwifruits, which comprise up to 30% of the total kiwifruit crop. In addition, kiwifruit pomace after juice production is also considered as a waste. In 2013 the total production of kiwifruit in the world was around 3.26 million tons (FAO, 2014). Part of this production is rejected because the kiwifruit does not have the right shape. A comprehensive review of the components and potential uses of kiwifruit waste has been published (Kennedy et al., 1999b), inferring that very little work has so far been done in finding uses for kiwifruit pomace. The total dietary fiber content of kiwifruit pomace amounts for approximately 25% on a dry weight basis (Martin-Cabrejas et al., 1995). The possibility of utilizing rejected kiwi fruits and peelings for laccase production has only been investigated (Rosales et al., 2005). Since, kiwifruit pomace and kiwifruit are rich in dietary fiber, they can be exploited for production of cellulases and hemicellulases.

Date Palm

The date palm (*Phoenix dactylifera*), a tropical and subtropical tree, belongs to the family Palmae (Arecaceae), and today its worldwide production, utilization, and industrialization is continuously increasing. The production of date fruits is on the increase as recorded for some of the major date producing countries like Egypt (1,352,950 metric tons), Saudi Arabia (1,078,300 metric tons), Iran (1,023,130 metric tons), UAE (775,000 metric tons), and Algeria (710,000 metric tons). Date fruit is marketed all over the world as a high-value confectionery, and as a fresh fruit it remains an important subsistence crop in most of the desert areas (Chandrasekharan and Bahkali, 2013). Date trees produce large quantities of agricultural waste, and according to one estimate, each date tree produces about 20 kg of dry leaves yearly. Other wastes such as date pits represent an average of 10% of the date fruits. Enormous wastes mainly in the form of date fruits that fall from the tree before maturity, date fruit seeds (pits), and date press cakes are being generated by the date palm agro-industry and date processing industries. These wastes pose serious environmental problems besides contributing to a great loss of raw materials (Chandrasekharan and Bahkali, 2013). Production of endopectinase from date pomace by *A. Niger* PC5 was studied using a two-level fractional factorial design for screening of the most important factors among concentrations of ammonium sulphate, potassium dihydrogen phosphate, and date pomace, pH, total spore amount, aeration rate, and fermentation time. Second-stage results exhibited maximum amounts of endopectinase ($10.88 \, IU \, mL^{-1}$) with pH 5, 0.3% ammonium sulphate, and 76.05 h of incubation (Bari et al., 2010). In another study, yeast *Candida guilliermondii* CGLA10 was reported to produce α-amylase using date wastes (Acourene and Ammouche, 2012). The results of the study showed that the starch components in the date wastes at $5.0 \, g \, L^{-1}$ strongly induced the production of α-amylase. Among the various nitrogen sources tested, urea at $5.0 \, g \, L^{-1}$ supported maximum biomass ($5.76 \pm 0.56 \, g \, L^{-1}$) and α-amylase ($2.304.19 \pm 31.08 \, mol \, L^{-1}$), after 72 h incubation at 30°C, with an initial pH of 6.0 and potassium phosphate concentration of $6.0 \, g \, L^{-1}$ (Acourene and Ammouche, 2012).

By-Products of Vegetable Processing

Tomato Waste

Tomato (*Lycopersicon esculentum* Solanaceae) juice is the most important vegetable juice with respect to per capita consumption, followed by carrot juice. According to statistics from the World Processing Tomato Council, over 30 million tons of tomatoes are processed annually worldwide to produce tomato juice, ketchup, canned tomatoes, and many other products (WPTC, 2013). During tomato processing, a waste known as tomato pomace is generated. This material represents about 5% by weight of the processed tomatoes and consists mainly of tomato peels, pulp residues, and seeds. Tomato pomace has no commercial value and is currently disposed of as a solid waste or used to a limited extent for animal feeding. However, a careful examination of the characteristics of tomato pomace reveals that it is a rich source of nutrients and valuable phytochemicals. In particular, important phenolics and carotenoids are present in the peel fraction of the waste (Zuorro et al., 2014). Tomato pomace consists of the dried and crushed skins and seeds of the fruit. The seeds account for approximately 10% of the fruit and 60% of the total waste, respectively, and are a source of protein (35%) and fat (25%) (Bhullar and Sogi, 2000). Among tomato components, lycopene has attracted the greatest attention in recent years for its potential health benefits (Papaioannou et al., 2012). Umsza-Guez et al. (2011) studied SSF of tomato pomace for the production of some hydrolytic enzymes, including xylanase, exo-polygalacturonase (exo-PG), cellulase (CMCase), and α-amylase by *A. awamori*. Maximum xylanase and exo-PG activities were measured during the first days of culture, reaching values around 100 and 80 IU per gds, respectively. For CMCase and α-amylase, production remained almost constant throughout fermentation, with average values of 19 and 21.5 IU per gds, respectively.

Carrot

Carrot (*Daucus carota* L., Apiaceae) juices and blends thereof are among the most popular nonalcoholic beverages (Chen and Tang, 1998). Despite considerable improvements in processing techniques, a major part of valuable compounds, such as carotenes, uronic acids, and neutral sugars is still retained in the pomace, which is generally utilized as feed or as fertilizer. Juice yield is reported to be only 60–70%, and up to 80% of carotene may be lost in pomace (Sims et al., 1993; Singh et al., 2006). Total carotene content of pomace may be up to 2 g per kg dry matter, depending on processing conditions (Stoll et al., 2001). Carrot pomace is a by-product obtained during carrot juice processing. Patil et al. (2012) reported maximum polygalacturonase production of $49.58 \, IU \, mL^{-1}$ from carrot waste using *Bacillus* sp. RCPTM1. Madhanasundareswari and Jeyachitra (2015) studied production and optimization of growth conditions for invertase enzyme by *Aspergillus* in SSF using carrot peel as substrate and reported maximum activity of $6.2 \, U \, mL^{-1}$.

Onion

The amount of onion (*Allium cepa* L., Alliaceae) waste produced annually in the European Union is estimated at approximately 450,000 tons, while the annual global production of dry onion has been reported at 86 million tons (FAO, 2014). The major by-products resulting from industrial peeling of onion bulbs are brown skin, the outer two fleshy leaves, and the top and bottom bulbs. Owing to their strong characteristic aroma and their susceptibility to phytopathogens, onion wastes are not suitable as fodder. However, they are a source of flavor components and fiber compounds and particularly rich in quercetin glycosides (Hertog et al., 1992; Waldron, 2001). With respect to the recovery of fructans and fructooligosaccharides, the outer two fleshy leaves have been demonstrated to be the most suitable sources (Jaime et al., 2000). The production of alcohol and snacks from onion pomace has also been reported (Horiuch et al., 2000; Kee et al., 2000, 2001). Ananthi et al. (2014) studied solid-state lipase production from *Bacillus cereus* MSU AS on seven agricultural waste substrates (coconut oil cake, neem oil cake, onion skin waste, wheat bran, jackfruit waste, banana skin peels, and pomegranate skin peel). Among the tested substrates, onion skin waste supplemented medium yielded maximum lipase activity $219 \, IU \, mL^{-1}$ under optimized conditions, and hence it was selected as a major substrate for lipase production.

Olive

The olive oil production industry generates large amounts of waste, of which olive mill wastewater (OMW) and crude olive oil cake (solid waste—OOC) have the highest organic load and therefore, present the largest pollutants of the oil processing industry (Moftah et al., 2013). The quality and quantity of the constituents of these wastes are dependent on many factors, such as type and maturity of the olives, climatic conditions and region of origin, cultivation methods, and the technology used for oil extraction

(Roig et al., 2006). The by-products resulting from olive oil extraction are the vegetation water, also called black water or vegetable water, and the olive husk including skins and stones. Depending on the processing conditions, 50–110 kg of water results from 100 kg of olives (Vitolo et al., 1999). The husk can be reprocessed for the recovery of olive oil, or extracted with an organic solvent to yield husk oil. Dried husk is utilized as fuel or animal feed (Gasparrini, 1999). In a recent study, Moftah et al. (2013), the solid and liquid wastes from the olive oil processing industry were evaluated as substrates for lipase production by *Yarrowia lipolytica*. The highest lipolytic activity of 850 IU dm^{-1} was achieved after four days of submerged cultivation in supplemented olive mill wastewater. In addition, olive oil cake appeared to be a convenient substrate for lipase production under the SSF mode. However, the most significant improvement in lipase production under SSF was achieved by an alkaline treatment of the substrate (more than 10-fold), when the amount of lipase produced reached up to ≈40 IU gL^{-1} of substrate. De Felice et al. (1997) investigated lipase production as well as OMW degradation with batch cultures of *Y. lipolytica* W29 (ATCC 20255), showing that the yeast was capable of metabolizing the waste and under optimum conditions, a lipase activity of 770 IU dm^{-1} was obtained. Gonçalves et al. (2012) also cultivated *Y. lipolytica* W29 on OMW for lipase production under batch and fed-batch culture conditions, reporting that the enzyme yields were higher in the former.

Red Beet

More than 200,000 tons of red beet (*Beta vulgaris* L. ssp. vulgaris, Chenopodiaceae) are produced in Western Europe annually, most of which (90%) is consumed as vegetable. The remainder is processed into juice, coloring foodstuff and food colorant, the latter commonly known as beetroot red (Henry, 1996). Though still rich in betalains, the pomace from the juice industry accounting for 15–30% of the raw material is disposed as feed or manure (Otto and Sulc, 2001). Alternatively, high content of pectins (20–25%) and its availability and low cost makes sugar beet pulp a potential source of pectins. Sugar beet pulp is mainly composed of (% on dry basis) pectin, 28.7; cellulose, 20; hemicellulose, 17.5; protein, 9.0; and lignin 4.4 (Jacob, 2009). Since the pectin content of beet pulp is high, it can be used for the production of pectinolytic enzymes without adding pectinaceous material as inducer. Azzaz et al. (2013) studied the production of pectinase using sugar beet pulp as main carbon source by *A. niger*. Maximum production of 3.2 U mL^{-1} pectinase by *A. niger* was achieved at16% beet pulp concentration, inoculum size of 5%, 48 h of incubation period, initial pH of growth medium at 7.0, and yeast extract as a nitrogen source at a concentration of 0.33 gL^{-1}. Heerd et al. (2014) investigated the comparison of pectinase production by *Aspergillus sojae* ATCC 20235 and *A. sojae* CBS 100928 under optimized conditions. Highest enzyme yield (909.5 ± 2.7 U g^{-1}) was obtained by *A. sojae* ATCC 20235 after 8 days at 30°C with 30% sugar beet pulp as substrate in combination with wheat bran with medium wetted at 160% with 0.2 M HCl.

Potato

While consumption of potatoes (*Solanum tuberosum* L., Solanaceae) has decreased, processed products, such as French fries, chips, and puree, have experienced growing popularity (ZMP, 2000). Peels are the major waste of potato processing. In the past decade there was a significant increase in residue production in the potato processing industry, primarily due to the supply to the fast food industry (Pereira et al., 2005). These residues have high organic matter content. Approximately 40% of potatoes are wasted, representing approximately 10 tons/day of residue (Barampouti and Vlyssides, 2005; Misha and Arora, 2004). Much of these residues consist of polysaccharides, such as cellulose, hemicellulose, and lignin. Its use as feedstock for bioprocesses has therefore become feasible due to its low economic cost (Couto and Sanroman, 2006; Holker et al., 2004). Dos-Santos et al. (2012) optimized the process of SSF of potato peel for the production of cellulolytic enzymes. They reported the optimum time for enzyme CMCase production as 82.88 h, with moisture content of 51.48% and temperature of 29.46°C; however, for FPase (filter paper activity), the most optimum time was 80.62 h, moisture content of 50.19% and temperature at 30°C. In the case of xylanase production, the optimized process parameters were time of 81.92 h, moisture content of 50.72%, and temperature of 28.85°C. Elayaraja et al. (2011) studied the production of a thermostable alkaline α-amylase from *Bacillus firmus* CAS 7 strain using potato peels. The maximum production of α-amylase was found to be 676 U mL^{-1} at 35°C, pH 7.5, and 1% substrate concentrations. Results reported by Rosales et al. (2002) indicated that potato peelings have an enormous potential as support for laccase production by the white-rot fungus *Trametes hirsuta* under solid-state conditions with a considerable reduction in production costs. Saez et al. (2013) used potato processing wastes (peels and discarded tubers) as nutritional support for production of lignolytic enzymes by *Anthracophyllum discolour*. The study demonstrated that this fungus has a high ligninolytic activity, mainly MnP in potato peel (193.5 ± 14.5 IU mL^{-1}) than in discarded potato (111.3 ± 23.5 IU mL^{-1}).

It is clear from the previous information that only a few fruit and vegetable residues have been studied for production of different enzymes employing SmF and SSF processes. Residues from jackfruit, pineapple, pomegranate, watermelon, etc., which form a significant part of the fruit, need to be properly analyzed for various important constituents so that enzyme production could be further ameliorated using such substrates. One of the major problems in using the fruit and vegetable residues is the use of these residues in combinations for enhanced enzyme production and also ensuring year-round production of enzymes. In addition, most of the residues after enzyme extraction still contain significant quantities of fiber, minerals, proteins, etc. In fact, SSF further improves the protein content of the residual biomass through impregnation of the fungal mycelium. The residual biomass from the SmF process after enzyme extraction could be used for composting and animal feed, while that obtained from SSF after enzyme extraction could be used as animal feed supplements, soil amendment, compost, and lignin-rich material. However, a lot of work still needs to be done in the area of ascertaining the use of residues after enzyme extraction/recovery.

CONCLUSIONS

It is clear from the information presented in this chapter that fruit and vegetable processing wastes present a great potential for production of enzymes. Most of the FVPW are either rich in holocellulose (cellulose + hemicellulose) or pectin or pectin along with holocellulose in addition to minerals. Since cellulose, hemicellulose, and pectin help in induction of cellulases, hemicellulases, and pectinases and presence of minerals in FVPW further help in ameliorating the enzyme titers. Many studies have reported production of a variety of enzymes using different microorganisms and different substrates. Enzymes with known industrial applications, such as cellulases, hemicellulases, amylases, pectinases, proteases, laccases, etc. have been produced either through SmF or SSF processes using FVPW. However, scale-up studies are important to establish their commercial potential. Studies are further needed to find a use for the waste left out after enzyme extraction especially from an SSF process. In addition, some of the residues, such as jackfruit rind and pineapple waste, which significantly contribute to the fruit weight, should be studied for production of multienzyme preparations.

LIST OF ABBREVIATIONS

BG Cellobiase or β-glucosidase
BOD Biochemical oxygen demand
CBH Cellobiohydrolase
CMCase Carboxymethylcellulases
COD Chemical oxygen demand
DO Dissolved oxygen
EG Endoglucanase
FAO Food and agriculture organization
FVPW Fruit and vegetable processing waste
IU International units
LiP Lignin peroxidase
MnP Manganese peroxidase
OMW Olive mill wastewater
OOC Olive oil cake
PG Polygalacturonase
PMG Polymethylgalacturonase
SmF Submerged fermentation
SSF Solid-state fermentation
TSS Total suspended solids
YPFW Yellow passion fruit waste

ACKNOWLEDGMENTS

The authors thankfully acknowledge the financial assistance received under the AMAAS subproject on "Optimization of parameters for utilization of paddy straw, kinnow pulp and peapods for production of cellulases, ethanol and feed supplements," funded by the Indian Council of Agricultural Research, Government of India, for conducting a study on the basis of which this chapter could be compiled.

REFERENCES

Acourene, S., Ammouche, A., 2012. Optimization of ethanol, citric acid, and a-amylase production from date wastes by strains of *Saccharomyces cerevisiae*, *Aspergillus niger*, and *Candida guilliermondii*. Journal of Industrial Microbiology and Biotechnology 39, 759–766.

Adesina, F.C., Olutola, A.A., Adeyefy, A.O., Habiba, U.O., Shadrach, A.O., 2013. Production of pectinase by fungi isolated from degrading fruits and vegetable. Nature and Science 11 (10).

Afifi, M.M., 2011. Enhancement of lactic acid production by utilizing liquid potato wastes. International Journal of Biological Chemistry 5, 91–102.

Agrahari, P.R., Khurdiya, D.S., 2003. Studies on preparation and storage of RTS beverage from pulp of culled apple pomace. Indian Food Packer 57 (2), 56–61.

Agrianual, 2007. Anuário da agricultura brasileira, tenth ed. FNP Consultoria, São Paulo.

Ahmed, S., Riaz, A., Jamil, D., 2009. Molecular cloning of fungal xylanases: an overview. Applied Microbiology and Biotechnology 84, 19–35.

Aiyer, P.V., 2005. Biotechnological applications of industrially important enzymes. African Journal of Biotechnology 4 (13), 1525–1529.

Anand, J.C., Maini, S.B., 1997. Utilisation of fruit and vegetable wastes. Indian Food Packer 51, 45–63.

Ananthi, S., Ramasubburayan, R., Palavesami, A., Immanuel, G., 2014. Optimization and purification of lipase through solid state fermentation by *Bacillus cereus* MSU as isolated from the gut of a marine fish *Sardinella longiceps*. International Journal of Pharmacy and Pharmaceutical Sciences 6 (5), 291–298. ISSN:0975-1491.

de-Andrade Silva, C.A., Priscila, M., Lacerda, F., Leite, R.S.R., Fonseca, G.G., 2013. Production of enzymes from *Lichtheimia ramosa* using Brazilian savannah fruit wastes as substrate on solid state bioprocesses. Electronic Journal of Biotechnology 6 (5), 1–7. http://dx.doi.org/10.2225/vol16-issue5-fulltext-7. ISSN:0717–3458. http://www.ejbiotechnology.info.

Anuradha, K., Padma, P.N., Venkateshwar, S., Reddy, G., 2010. Fungal isolates from natural pectic substrates for polygalacturonase and multienzyme production. Indian Journal of Microbiology 50, 339–344.

Arapoglou, D., Varzakas, T., Vlyssides, A., Israilides, C., 2010. Ethanol production from potato peel waste (PPW). Waste Management 30, 1898–1902.

Aro, N., Pakula, T., Penttila, M., 2005. Transcriptional regulation of plant cell wall degradation by filamentous fungi. FEMS Microbiology Reviews 29 (4), 719–739.

Askar, A., 1998. Importance and characteristics of tropical fruits. Fruit Processing 8, 273–276.

Auria, R., Palacios, J., Revah, S., 1992. Determination of the inter-particular effective diffusion coefficient for CO_2 and O_2 in solid state fermentation. Biotechnology and Bioengineering 39, 898–902.

Azzaz, H.H., Murad, H.A., Kholif, A.M., Morsy, A.M., Mansour, A.M., El-Sayed, H.M., 2013. Pectinase production optimization and its application in banana fiber degradadtion. Egyptian Journal of Nutritional and Feeds 16 (2), 117–125.

Babbar, N., Oberoi, H.S., 2014. Enzymes in value-addition of agricultural and agro-industrial residues. In: Brar, S.K., Verma, M. (Eds.), Enzymes in Value-Addition of Wastes. Nova Publishers, pp. 29–50 (Chapter 2).

Babitha, S., Soccol, C.R., Pandey, A., 2006. Jackfruit seed – a novel substrate for the production of monascus pigments through solid-state fermentation. Food Technology Biotechnology 44 (4), 465–471.

Bandikari, R., Poondla, V., Sarathi, V., Reddy, O., 2014. Enhanced production of xylanase by solid state fermentation using *Trichoderma koeningi* isolate: effect of pretreated agro-residues. Biotechnology 4, 655–664.

Barampouti, E.M.P.,Vlyssides,A.G., 2005. Dynamic modeling of biogas production in an UASB reactor for potato processing wastewater treatment. Chemical Engineering Journal 106, 53–58.

Bari, M.R., Alizadeh, M., Farbeh, F., 2010. Optimizing endopectinase production from date pomace by *Aspergillus niger* PC5 using response surface methodology. Food and Bioproducts Processing 88, 67–72.

Barrett, A.J., McDonald, J.K., 1986. Nomenclature: protease, proteinase and peptidase. Biochemical Journal 237, 935–944.

Barth, S., Powers, T., 2008. Agricultural Waste Characteristics. Agricultural Waste Management Field Handbook. United States Department of Agriculture, South Carolina, pp. 1–32.

Bayoumi, R.A.,Yassin, H.M., Swelim, M.A., Abdel-All, E.Z., 2008. Production of bacterial pectinases from agro-industrial wastes under solid state fermentation conditions. Journal of Applied Scientific Research 4, 1708–1721.

Beg, Q.K., Kapoor, M., Mahajan, L., Hoonda, G.S., 2001. Microbial xylanases and their industrial applications: a review. Applied Microbiology Biotechnology 56, 326–338.

Benitez,V., Molla, E., Martın-Cabrejas, M.A., Aguilera,Y., Lopez-Andreu, F.J., Cools, K.,Terry, L.A., Esteban, R.M., 2011. Characterization of industrial onion wastes (*Allium cepa* L.): dietary fibre and bioactive compounds. Plant Foods for Human Nutrition 66, 48–57.

Bhullar, J.K., Sogi, D.S., 2000. Shelf life studies and refining of tomato seed oil. Journal of Food Science and Technology India 37, 542–544.

Bianchi, V.L.D., Moraes, I.O., Capalbo, D.M.F., 2011. Fermentação em estado sólido. In: Biotecnologia Industrial. Edgard Blücher, São Paulo, vol. 2(13), pp. 247–276.

Binod, P.B., Palkhiwla, P., Gaikaiwani, R., Nampoothiri, K.M., Duggal, A., Dey, K., Pandey, A., 2013. Industrial enzymes – present status and future perspectives for India. Journal of Scientific and Industrial Research 72, 271–286.

Botella, I., De Ory, C.,Webb, D., 2005. Hydrolytic enzyme production by *Aspergillus awamori* on grape pomace. Biochemical Engineering Journal 26, 100–106.

Bouallagui, H., Cheikh, R.B., Marouani, L., Hamdi, M., 2003. Mesophilic biogas production from fruit and vegetable waste in tubular digester. Bioresource Technology 86, 85–90.

Bravo, L., Saura-Calixto, F., 1998. Characterization of dietary fiber and the in vitro indigestible fraction of grape pomace. American Journal of Enology and Viticulture 49, 135–141.

Chandrasekaran, M., Bahkali, A.H., 2013. Valorization of date palm (*Phoenix dactylifera*) fruit processing by-products and wastes using bioprocess technology. Saudi Journal of Biological Sciences 20, 105–120.

Chávez, R., Bull, P., Eyzaguirre, J., 2006. The xylanolytic enzyme system from the genus *Penicillium*. Journal of Biotechnology 123 (4), 413–433.

Chen, B.H.,Tang,Y.C., 1998. Processing and stability of carotenoid powder from carrot pulp waste. Journal of Agricultural and Food Chemistry 46, 2312–2318.

Chung, S.L., Meyers, S.P., 1979. Bioprotein from banana waste. Developments in Industrial Microbiology 20, 723–731.

Conessa, A., Punt, P.J.,Van Den Hondel, C.A., 2002. Fungal peroxidases: molecular aspects and applications. Journal of Biotechnology 93 (2), 143–158. ISSN:0168-1656.

Cordova, K.R.V., Gama, T.M., Winter, C.M.G., Kaskantzis Neto, G., Freitas, R.J.S., 2005. Características físico-químicas da casca do maracujá-amarelo (*Passiflora edulis* Flavicarpa Degener) obtida por secagem. Boletim do CEPPA 23, 221–230.

Couto, S.R., Sanroman, M.A., 2006. Application of solid-state fermentation to food industry a review. Journal of Food Engineering 76, 291–302.

Dabhi, B.K.,Vyas, R.V., Shelat, H.N., 2014. Use of banana waste for the production of cellulolytic enzymes under solid substrate fermentation using bacterial consortium. International Journal of Current Microbiology and Applied Sciences 3 (1), 337–346.

Daroit, D.J., Silveira, S.T., Hertz, P.F., Brandelli, A., 2007. Production of extracellular b-glucosidase by *Monascus purpureus* on different growth substrates. Process Biochemistry 42, 904–908.

De Felice, B., Pontecorvo, G., Carfagna, M., 1997. Degradation of waste waters from olive oil mills by *Yarrowia lipolytica* ATCC 20255 and *Pseudomonas putida*. Acta Biotechnology 17, 231–239.

Dedavid-e-Silva, L.A., Lopes, F.C., Silveira, S.T., Brandelli, A., 2009. Production of cellulolytic enzymes by *Aspergillus phoenicis* in grape waste using response surface methodology. Applied Biochemistry Biotechnology 152, 295–305.

De-Gregorio, A., Mandalari, G., Arena, G., Nucita, F., Tripodo, M.M., Lo Curto, R.B., 2002. SCP and crude pectinase production by slurry-state fermentation of lemon pulps. Bioresource Technology 83, 89–94.

Dhawan, K., Dhawan, S., Sharma, A., 2004. Passiflora: a review update. Journal of Ethnopharmacology 94, 1–23.

Dhillon, G.S., Brar, S.K., Valero, J.R., 2011a. Bioproduction of hydrolytic enzymes using apple pomace waste by *Aspergillus niger*: applications in biocontrol formulations and for hydrolysis of chitin/chitosan. Bioprocess and Biosystems Engineering. http://dx.doi.org/10.1007/s00449-011-0552-9.

Dhillon, G.S., Brar, S.K., Kaur, S., Valero, J.R., 2011b. Chitinolytic and chitosanolytic activities from crude cellulase extract produced by *A. niger* grown on apple pomace through koji fermentation. Journal of Microbiology and Biotechnology 2, 1312–1321.

Dhillon, G.S., Brar, S.K., Kaur, S., Valero, J.R., 2011c. Optimization of xylanase production using apple pomace waste by *A. niger* through solid-state fermentation process. Engineering in Life Sciences 12, 1–10.

Dhillon, G.S., Brar, S.K., Kaur, S., Sabrine, M., M'hamdi, N., 2012a. Lactoserum as a moistening medium and crude inducer for fungal cellulases and hemicellulase induction through solid-state fermentation of apple pomace. Biomass and Bioenergy 41, 165–174.

Dhillon, G.S., Brar, S.K., Kaur, S., 2012b. Improved xylanase production using apple pomace waste by *Aspergillus niger*. Engineering in Life Sciences 12 (3), 1–10.

Dhillon, G.S., Kaur, S., Brar, S.K., 2013. Perspective of apple processing wastes as low-cost substrates for bioproduction of high value products: a review. Renewable and Sustainable Energy Reviews 27, 789–805.

Díaz, A.B., De-Ory, I., Caro, I., Blandino, A., 2012. Enhance hydrolytic enzymes production by *Aspergillus awamori* on supplemented grape pomace. Food and Bioproducts Processing 90, 72–78.

Dos Santos, T.C., Gomes, D.P.P., Bonomo, R.C.F., Franco, M., 2012. Optimization of solid state fermentation of potato peel for the production of cellulolytic enzymes. Food Chemistry 133, 1299–1304.

Downing, D.L., 1989. Apple cider. In: Downing, D.L. (Ed.), Processed Apple Product. AVI Publ, West Port, Conn, pp. 168–186.

Dussan, K.J., Cardona, C.A., Giraldo, O.H., Gutierrez, L.F., Perez, V.H., 2010. Analysis of a reactive extraction process for biodiesel production using a lipase immobilized on magnetic nanostructures. Bioresource Technology 101, 9542–9549.

Elayaraja, S., Velvizhi, T., Maharani, V., Mayavu, P., Vijayalakshmi, S., Balasubramanian, T., 2011. Thermostable α-amylase production by *Bacillus firmus* CAS 7 using potato peel as a substrate. African Journal of Biotechnology 10 (54), 11235–11238. http://dx.doi.org/10.5897/AJB11.1572. ISSN:1684-5315. http://www.academicjournals.org/AJB.

Fallon, E., Tremblay, N., Desjardins, Y., 2006. Relationships among growing degree-days, tenderness, other harvest attributes and market value of processing pea (*Pisum sativum* L.) cultivars grown in Quebec. Canadian Journal of Plant Sciences 86, 525–537.

FAOSTAT, 2014. Food and Agriculture Organization. Rome, Italy www.fao.org.

Favela-Torres, E., Volke-Sepúlveda, T., Viniegra-González, G., 2006. Production of hydrolytic depolymerising pectinases. Food Technology Biotechnology 44 (2), 221–227.

Folakemi, O.P., Priscilla, J.O., Ibiyemi, S.A., 2008. Cellulase production by some fungi cultured on pineapple waste. Nature and Science 6 (2). ISSN:1545-0740.

Fortkamp, D., Knob, A., 2014. High xylanase production by *Trichoderma viridae* using pineapple peel as substrate and its application in pulp bleaching. African Journal of Biotechnology 13 (22), 2248–2259.

Gasparrini, R., 1999. Treatment of olive oil processing residues. Oils and Fats International 2, 32–33.

Girdhar, N., Satyanarayana, A., 2000. Grape waste as a source of tartrates. Indian Food Packer 54, 59–61.

Gonçalves, C., Oliveira, F., Pereira, C., Belo, I., 2012. Fed-batch fermentation of olive mill wastewaters for lipase production. Journal of Chemical Technology and Biotechnology 87, 1215–1218.

Goyal, M., Kalra, K.L., Sareen, V.K., Sonu, G., 2008. Xylanase production with xylan rich lignocellulosic wastes by a local soil isolate of *Trichoderma viridae*. Brazalian Journal of Microbiology 39 (3), 5335–5541.

Grobe, K., 1994. Composter links up with food processor. BioCycle 34, 42–43.

Gupta, R., Gupta, N., Rathi, P., 2004. Bacterial lipases: an overview of production, purification and biochemical properties. Applied Microbiology and Biotechnology 64, 763–781.

Gupta, U., Kaur, R., 2009. Xylanase production by a thermo-tolerant *Bacillus* species under solid-state and submerged fermentation. Brazilian Archives of Biology and Technology 52 (6), 1363–1371.

Haddadin, M.S., Abdulrahim, S.M., Al-Kawaldeh, G.Y., Robinson, R.K., 1999. Solid state fermentation of waste pomace from olive processing. Journal of Chemical Technology and Biotechnology 74, 613–618.

Handbook on Horticulture Statistics, 2014. Department of Agriculture and Co-operation, Ministry of Agriculture, Government of India.

Hang, Y.D., Woodams, E.E., 1994. Apple pomace, a potential substrate for production of glucosidase by *Aspergillus foetidus*. Lebensmittel-Wissenschaft und -Technologie 27, 587–589.

Hariharan, S., Nambisan, P., 2013. Optimization of lignin peroxiade, managanese peroxide and laccase production from *Ganoderma lucidum* under solid state fermentation of pineapple leaf. BioResources 8 (1), 250–271.

Heerd, D., Diercks-Horn, S., Fernández-Lahore, M., 2014. Efficient polygalacturonase production from agricultural and agro-industrial residues by solid-state culture of *Aspergillus sojae* under optimized conditions. SpringerPlus 3, 742.

Henry, B.S., 1996. Natural food colours. In: Hendry, G.F., Houghton, J.D. (Eds.), Natural Food Colorants, second ed. Blackie Academic & Professional, London, pp. 40–79.

Hertog, M.G.L., Hollman, P.C.H., Katan, M.B., 1992. Content of potentially anticarcinogenic flavonoids of 28 vegetables and 9 fruits commonly consumed in the Netherlands. Journal of Agricultural and Food Chemistry 40, 2379–2381.

Hills, D.J., Nakano, L., 1984. Effects of particle size on anaerobic digestion of tomato solid waste. Agricultural Wastes 10, 285–295.

Holker, U., Lenz, J., 2005. Solid-state fermentation – are there any biotechnological advantages? Current Opinion in Microbiology 8, 301–306.

Holker, U., Hofer, M., Lenz, J., 2004. Biotechnological advantages of laboratory scale solid-state fermentation with fungi. Applied Microbiology and Biotechnology 64, 175–186.

Horiuchi, J.I., Yamauchi, N., Osugi, M., Kanno, T., Kobayashi, M., Kuriyama, H., 2000. Onion alcohol production by repeated batch process using flocculating yeast. Bioresource Technology 75, 153–156.

Igartuburu, J.M., Pando, E., Rodriguez Luis, F., Gil-Serrano, A., 1997. An acidic xyloglucan from grape skins. Phytochemistry 46, 1307–1312.

Inacio, F.D., Ferreira, R.O., de Araujo, C.A.V., Peralta, R.M., de Souza, C.G.M., 2015. Production of enzymes and biotransformation of orange waste by oyster mushroom, *Pleurotus pulmonarius* (Fr.) Quél. Advances in Microbiology 5, 1–8.

Isil, S., Nilufer, A., 2005. Investigation of factors affecting xylanase activity from *Trichoderma harzianum* 1073 D3. Brazilian Archives of Biology and Biotechnology 48 (2), 187–193.

Ismail, A.S., 1996. Utilization of orange peels for the production of multienzyme complexes by some fungal strains. Process Biochemistry 1, 645–650.

Izquirdo, L., Sendra, J.M., 2003. Citrus fruits composition and characterization. In: Caballero, B., Trugo, L., Finglas, P. (Eds.), Encyclopedia of Food Science and Nutrition. Academic Press, Oxford, p. 6000. 1335-1341.

Jacob, N., 2009. Pectinolytic enzymes. In: Nigam, P.S., Pandey, A. (Eds.), Biotechnology for Agro-industrial Residues Utilization. Springer, Netherlands.

Jaime, L., Martinez, F., Martın-Cabrejas, M.A., Molla, E., Lopez-Andreu, F.J., Waldron, K.W., Esteban, R.M., 2000. Study of total fructan and fructooligosaccharide content in different onion tissues. Journal of the Science of Food and Agriculture 81, 177–182.

Jayani, R.S., Saxena, S., Gupta, R., 2005. Microbial pectinolytic enzymes: a review. Process Biochemistry 40, 2931–2944.

Joshi, V.K., Sandhu, D.K., Attri, B.L., Walia, R.K., 1991. Cider preparation from apple juice concentrate and its consumer acceptability. Journal of Horticulture 48, 321–327.

Joshi, V.K., 1997. Fruit Wines. Directorate of Extension Education, second ed. Dr YS Parmar University of Horticulture and Forestry, Solan, India, pp. 1–35.

Joshi, V.K., Attri, D., 2006. Solid state fermentation of apple pomace for the production of value added products. Natural Product Radiance 5, 289–296.

Joshi, V.K., Parmar, M., Rana, N.S., 2006. Pectin esterase production from apple pomace in solid-state and submerged fermentations. Food Technology and Biotechnology 44, 253–256.

Joshi, V.K., Parmar, M., Rana, N., 2011. Purification and characterization of pectinase produced from apple pomace and evaluation of its efficacy in the fruit juice extraction and clarification. Indian Journal of Natural Products and Resources 2, 189–197.

Kashyap, D.R., Chandra, S., Kaul, A., Tewari, R., 2000. Production, purification and characterization of pectinase from a *Bacillus sp.* DT7. World Journal of Microbiology and Biotechnology 16, 277–282.

Kaur, G., Kumar, S., Satyarnarayana, T., 2004. Production, characterization and application of a thermostable polygalactouronase of a thermophilic mould *Sporotrichum thermophile*. Bioresource Technology 94, 234–239.

Kaushal, N.K., Joshi, V.K., 1995. Preparation and evaluation of apple pomace based cookies. Indian Food Packer 49 (5), 17–24.

Kaushal, N.K., Joshi, V.K., Sharma, R.C., 2002. Effect of stage of apple pomace collection and the treatment on the physical-chemical and sensory qualities of pomace papad (fruit cloth). Journal of Food Science and Technology 39, 388–393.

Kee, H.J., Ryu, G.H., Park, Y.K., 2000. Preparation and quality properties of extruded snack using onion pomace and onion. Korean Journal of Food Science and Technology 32, 578–583.

Kee, H.J., Ryu, G.H., Park, Y.K., 2001. Physical properties of extruded snack made of dried onion and onion pomace. Journal of the Korean Society of Food Science and Nutrition 30, 64–69.

Kennedy, M., List, D., Lu, Y., Foo, L.Y., Newman, R.H., Sims, I.M., Bain, P.J.S., Hamilton, B., Fenton, G., 1999a. Apple pomace and products derived from apple pomace: uses, composition and analysis. In: Linskens, H.F., Jackson, J.F. (Eds.), Modern Methods of Plant Analysis, vol. 20. Springer, Berlin, Heidelberg, pp. 76–119.

Kennedy, M., List, D., Lu, Y., Foo, L.Y., Robertson, A., Newman, R.H., Fenton, G., 1999b. Kiwifruit waste and novel products made from kiwifruit waste: uses, composition and analysis. In: Linskens, H.F., Jackson, J.F. (Eds.), Modern Methods of Plant Analysis. vol. 20. Springer, Berlin, Heidelberg, pp. 121–152.

Knob, A., Terrasan, C.R.F., Carmona, E.C., 2010. β-Xylosidases from filamentous fungi: an overview. World Journal of Microbiology and Biotechnology 26, 389–407.

Krishna, C., Chandrasekaran, M., 1996. Banana waste as substrate for alpha-amylase production by *Bacillus subtilis* (CBTK 106) under solid-state fermentation. Applied Microbiology and Biotechnology 46, 106–111.

Krishna, C., 1999. Production of bacterial cellulases by solid state bioprocessing of banana wastes. Bioresource Technology 69, 231–239.

Krishna, P.R., Srivastava, A.K., Ramaswamy, N.K., Suprasanna, P., D'souza, S.F., 2012. Banana peel as substrate for α-amylase production using *Aspergillus niger* NCIM 616 and pocess optimization. Indian Journal of Biotechnology 11, 314–319.

Kuhad, R.C., Gupta, R., Singh, A., 2011. Microbial cellulases and their industrial applications. SAGE-Hindawi Enzyme Research 1–10.:280696. http://dx.doi.org/10.4061/2011/280696.

Kumar, D., Yadav, K.K., Muthukumar, M., Garg, N., 2013. Production and characterization of amylase from mango kernel by *Fusarium solani* NAIMCC-F-02956 using submerged fermentation. Journal of Environmental Biology 34, 1053–1058.

Kumar, S., Singh, A.B., Abidi, A.B., Upadhyay, R.G., Singh, A., 1988. Proximate composition of jackfruit seeds. Journal of Food Science and Technology 25, 141–152.

Larrauri, J.A., Ruperez, P., Borroto, B., Saura-Calixto, F., 1996. Mango peels as a new tropical fibre: preparation and characterization. Lebensmittel-Wissenschaft und -Technologie 29, 729–733.

Levin, L., Diorio, L., Grassi, E., Forchiassin, F., 2012. Grape stalks as substrate for white rot fungi, lignocellulolytic enzyme production and dye decolorization. Revista Argentina de Microbiologia 44 (2). ISSN:0325-7541.

Liu, C., Sun, Z.T., Du, J.H., Wang, J., 2008. Response surface optimization of fermentation conditions for producing xylanase by *Aspergilllus niger* SL-05. Journal of Indian Microbiology and Biotechnology 135 (7), 703–711.

Lowe, E.D., Buckmaster, D.R., 1995. Dewatering makes big difference in compost strategies. Biocycle 36, 78–82.

Madhanasundareswari, K., Jeyachitra, K., 2015. Production and optimization of growth conditions for invertase enzyme by *Aspergillus* in solid state fermentation (SSF) using carrot peel as substrate. Scrutiny International Research Journal of Agriculture, Plant Biotechnology and Bio Products, SIRJ-APBBP 2 (1).

Mahadik, N.D., Puntambekar, U.S., Bastawde, K.B., Khire, J.M., Gokhale, D.V., 2002. Production of acidic lipase by *Aspergillus niger* in solid state fermentation. Process Biochemistry 38 (5), 715–721.

Maldonado, M.C., Strasser de Saad, A.M., 1998. Production of pectinestrase and polygalactouronase by *Aspergillus niger* in submerged and solid state systems. Journal of Industrial Microbiology and Biotechnology 20, 34–38.

Mamma, D., Kourtoglou, E., Christakopoulos, P., 2008. Fungal multienzyme production on industrial by-products of the citrus-processing industry. Bioresource Technology 99, 2373–2383.

Manimehalai, N., 2007. Fruit and waste utilization. Beverage and Food World 34 (11), 53–56.

Marin, F.R., Solr, C.R., Benavente, G.O., Castillo, J., Perz, H., 2007. By-products from different citrus processes as a source of customized functional fibres. Food Chemistry 100, 736–741.

Martin-Cabrejas, M.A., Esteban, R.M., Lopez-Andreu, F.J., Waldren, K., Selvendran, R.R., 1995. Dietary fibre content of pear and kiwi pomaces. Journal of Agricultural and Food Chemistry 43, 662–666.

Martins, E.S., Silva, D., Da Silva, R., Gomes, E., 2002. By-products from different citrus processes as a source of customized functional fibres. Process Biochemistry 37, 949–954.

Mazza, G., Miniati, E., 1993. Grapes. In: Anthocyanins in Fruits, Vegetables, and Grains. CRC Press, Boca Raton, Ann Harbor, London, Tokyo, pp. 149–199.

Medeiros, R.G., Soffner, M.L.A.P., Thome, J.A., Cacais, A.G., Estelles, R.S., Salles, B.C., Ferreira, H.M., Lucena-Neto, S.A., Silva, F.G., Filho, E.X.F., 2000. The production of hemicellulases by aerobic fungi on medium containing residues of banana plant as substrate. Biotechnology Progress 16, 522–524.

Mester, T., Tien, M., 2000. Oxidation mechanism of ligninolytic enzymes involved in the degradation of environmental pollutants. International Biodeterioration and Biodegradation 46, 51–59. http://dx.doi.org/10.1016/S0964-8305(00)00071-8.

Meyer, A.S., Jepsen, S.M., Sorensen, N.S., 1998. Enzymatic release of antioxidants for human low-density lipoprotein from grape pomace. Journal of Agricultural and Food Chemistry 46, 2439–2446.

Miljkovic, D., Bignami, G.S., 2002. Nutraceuticals and Methods of Obtaining Nutraceuticals from Tropical Crops. USA. Application number: 10/992.502 Published in Google Patent No 10/992.502.

Mirzaei-Aghsaghali, A., Maheri-Sis, N., 2008. Nutritive value of some agro-industrial by-products for ruminants – a review. World Journal of Zoology 3 (2), 40–46.

Mishra, B.K., Arora, A., 2004. Optimization of a biological process for treating potato chips industry wastewater using a mixed culture of *Aspergillus foetidus* and *Aspergillus niger*. Bioresource Technology 94, 9–12.

Moftah, O.A.S., Grbavcic, S.Z., Moftah, W.A.S., Lukovic, N.D., Prodanovic, O.L., Jakovetic, S.M., Jugovic, Z.D., 2013. Lipase production by *Yarrowia lipolytica* using olive oil processing wastes as substrates. Journal of Serbian Chemical Society 78 (6), 781–794.

Mohamed, S.A., Al-Malki, A.L., Khan, J.A., Kabli, S.A., Al-Garni, S.M., 2013. Solid state production of polygalacturonase and xylanase by *Trichoderma* species using cantaloupe and watermelon rinds. Journal of Microbiology 51 (5), 605–611.

Mojtahedi, M., Mesgaran, M.D., 2009. Variability in the chemical composition and in situ ruminal degradability of sugar beet pulp produced in north-east Iran. Research Journal of Biological Sciences 4, 1262–1266.

Motyan, J.A., Tóth, F., Tőzsér, J., 2013. Applications of proteolytic enzymes in molecular biology. Biomolecules 3, 923–942. http://dx.doi.org/10.3390/biom3040923.

Mrudula, S., Anitharaj, R., 2011. Pectinase production in solid state fermentation by *Aspergillus niger* using orange peel as substrate. Global Journal of Biotechnology and Biochemistry 6 (2), 64–71.

Namasivayam, S.K.R., Nirmala, D., 2011. Enhanced production of alpha amylase using vegetables waste by *Aspergillus niger* strain SK01 marine isolate. Indian Journal of Geo-Marine Sciences 40 (1), 130–133.

Nigam, J.N., 2000. Continuous ethanol production from pineapple cannery waste using immobilized yeast cells. Journal of Biotechnology 80, 189–193.

Niladevi, K.N., 2009. Ligninolytic enzymes. In: Singh Nigam, P., Pandey, A. (Eds.), Biotechnology for Agro-Industrial Residues Utilization, pp. 387–414.

Norazlina, I., Pushpahvalli, B., Kuhalim, K.H., Norakma, M.N., 2012. Comparable study of xylanase production fron *Aspergillus niger* via solid state culture. Journal of Chemical Engineering 6 (12), 1106–1113.

Oberoi, H.S., Chavan, Y., Bansal, S., Dhillon, G.S., 2010. Production of cellulases through solid state fermentation using kinnow pulp as a major substrate. Food and Bioprocess Technology 4, 528–536.

Oberoi, H.S., Vadlani, P.V., Saida, L., Bansal, S., Highes, J.D., 2011. Ethanol production from banana peels using statistically optimized simultaneous saccharification and fermentation process. Waste Management 31, 1576–1584.

Oberoi, H.S., Sandhu, S.K., Vadlani, P.V.V., 2012. Statistical optimization of hydrolysis process for banana peels using cellulolytic and pectinolytic enzymes. Food and Bioproducts Processing 90, 257–265.

Osma, J.F., Toca-Herrera, J.L., Rodriguez, Couto, S., 2007. Banana skin: a novel waste for laccase production by *Trametes pubescens* under solid-state conditions and application to synthetic dye decolouration. Dyes Pigments 75, 32–37.

Osuigwe, M.J., Nkem, T., Emuebie, O.R., Isreal, E.J., Kayode, A.F., 2014. Physicochemical factors influencing pectinolytic enzyme produced by *Bacillus licheniformis* under submerged fermentation. Nature and Science 12 (8), 110–116.

Otto, K., Sulc, D., 2001. Herstellung von Gemüsesäften. In: Schobinger, U. (Ed.), Frucht- und Gemüsesäften. Ulmer, Stuttgart, pp. 278–297.

Ozaki, Y., Miyake, M., Inaba, N., Ayano, S., Ifuku, Y., Hasegawa, S., 2000. Limonoid glucosides in fruit, juice and processing by-products of satsuma mandarin (*Chus unshiu* Marcov.). Journal of Food Science 60, 186–189.

Padma, P.N., Anuradha, K., Nagaraju, B., Kumar, V.S., Reddy, G., 2012. Use of pectin rich fruit wastes for polygalacturonase production by *Aspergillus awamori* MTCC 9166 in solid state fermentation. Journal of Bioprocessing and Biotechniques 2, 2.

Pagan, J., Ibarz, A., 1999. Extraction and rheological properties of pectin from fresh peach pomace. Journal of Food Engineering 39, 193–201.

Pandey, A., 1992. Recent process developments in solid-state fermentation. Process Biochemistry 27, 109–117.

Pandey, A., Selvakumar, P., Soccol, C.R., Nigam, P., 1999. Solid state fermentation for the production of industrial enzymes. Current Science 77 (1), 149–162.

Pandey, A., Soccol, C.R., Nigam, P., Soccol, V.T., 2000a. Biotechnological potential of agro-industrial residues: sugarcane bagasse. Bioresource Technology 74, 69–80.

Pandey, A., Soccol, C.R., Mitchell, D., 2000b. New developments in solid state fermentation: I – bioprocesses and products. Process Biochemistry 35, 1153–1169.

Papaioannou, E.H., Karabelas, A.J., 2012. Lycopene recovery from tomato peel under mild conditions assisted by enzymatic pre-treatment and non-ionic surfactants. Acta Biochimica Polonica 59, 71–74.

Patil, R.C., Murugkar, T.P., Shaikh, S.A., 2012. Extraction of pectinase from pectinolytic bacteria isolated from carrot waste. International Journal of Pharma and Bio Sciences 3 (1), 262–266.

Paul, M.S., Sumathy, V.J.H., 2013. Production of α-amylase from banana peels with *Bacillus subtilis* using solid state fermentation. International Journal of Current Microbiology and Applied Sciences 2 (10), 195–206.

Pereira, C.A., Carli, L., Beux, S., Santos, M.S., Busato, S.B., Kobelnik, M., 2005. Utilização de farinha obtida a partir de rejeito de batata na elaboração de biscoitos. Publicatio UEPG Ciências Exatas e da Terra, Ciências Agrárias e Engenharia 1, 19–26.

Perez-Guerra, N., Torrado-Agrassar, A., Lopez-Macias, C., Pastrana, L., 2003. Main characteristics and applications of solid state fermentation. Electronic Journal of Environment Agriculture and Food Chemistry 2 (3), 343–350.

Pointing, S.B., 2001. Feasibility of bioremediation by white-rot fungi. Applied Microbiology and Biotechnology 57, 20–33. ISSN:0175-7598.

Puravankara, D., Boghra, V., Sharma, R.S., 2000. Effect of antioxidant principles isolated from mango (*Mangifera indica* L.) see d kernels on oxidative stability of buffalo ghee (butter-fat). Journal of the Science of Food and Agriculture 80, 522–526.

Raimbault, M., 1998. General and microbiological aspects of solid substrate fermentation. Electron Journal of Biotechnology 1 (3), 174–188.

Ramachandran, S., Patel, A.K., Nampoothiri, K.M., Francis, F., Nagy, V., Szakacs, G., 2004. Determination of the interparticular effective diffusion coefficient for CO_2 and O_2 in solid state fermentation. Bioresource Technology 93, 169–174.

Ramteke, R.S., Vijayalakshmi, M.R., Eipeson, W.E., 1999. Processing and value addition to mangoes. Indian Food Industry 18, 155–163.

Rao, M.B., Tanksale, A.M., Ghatge, M.S., Deshpande, V.V., 1998. Molecular and biotechnological aspects of microbial proteases. Microbiology and Molecular Biology Reviews 62, 597–635.

Rao, P.V.V., Satya, P., Reddy, D.S.R., 2014. Jack fruit waste: a potential substrate for pectinase production. Indian Journal of Scientific Research 9 (1), 058–062.

Rathan, R.K., Ambili, M., 2011. Cellulase enzyme production by *Streptomyces* sp. using fruit waste as substrate. Australian Journal of Basic and Applied Sciences 5, 1114–1118.

Renge, V.C., Khedkar, S.V., Nandurkar, N.R., 2012. Enzyme synthesis by fermentation method. Scientific Reviews in Chemical Communications 2 (4), 585–590. ISSN:2277-2669.

Rehman, S., Aslam, H., Ahmad, A., Khan, S.A., Sohail, M., 2014. Production of plant cell wall degrading enzymes by monoculture and coculture of *Aspergillus niger* and *Aspergillus terrus* under solid state fermentation of banana peels. Brazilian Journal of Microbiology 45, 1485–1492.

Riggle, D., 1989. Revival time for composting food industry wastes. BioCycle 29, 35–37.

Rodriguez Couto, S., Lopez, E., Sanroman, M.A., 2006. Utilization of grape seeds for laccase production in solid-state fermentors. Journal of Food Engineering 74, 263–267.

Roig, A., Cayuela, M.L., Sánchez-Monedero, M.A., 2006. Lycopene recovery from tomato peel under mild conditions assisted by enzymatic pre-treatment and non-ionic surfactants. Waste Management 26, 960–969.

Romo-Sanchez, S., Alves-Baffi, M., Arévalo-Villena, M., Úbeda-Iranzo, J., Briones-Pérez, A., 2010. Yeast biodiversity from oleic ecosystems: study of their biotechnological properties. Food Microbiology 27, 487–492.

Rosales, E., Rodríguez Couto, S., Sanromán, A., 2002. New uses of food waste: application to laccase production by *Trametes hirsuta*. Biotechnology Letters 24, 701–704.

Rosales, E., Rodriguez Couto, S., Sanroman, M.A., 2005. Reutilization of food processing wastes for production of relevant metabolites. Application to laccase production by *Trametes hirsuta*. Journal of Food Engineering 66, 419–423.

Saev, M., Koumanova, B., Simeonov, I., 2009. Anaerobic codigestion of wasted tomatoes and cattle dung for biogas production. Journal of University Chemical Technology and Metallurgy 44, 55–60.

Saez, H.S., Araneda, R.S., Uribe, E.H., Jerez, M.C.D., 2013. Use of potato processing wastes as nutritional support production of lignolytic enzymes by *Anthracophyllum discolour*. In: III Symposium on Agricultural and Agroindustrial Waste Management March 12–14, Sao Pedro, SP, Brazil.

Sakthiselvan, P., Naveena, B., Partha, N., 2012. Effect of medium composition and ultrasonication on xylanase production by *Trichoderma harzianum* MTCC 4358 on novel substrate. African Journal of Biotechnology 11 (57), 12067–12077.

Sanchez, R.O., Hernandez, P.B., Morales, G.R., Nunez, F.U., Villafuerte, J.O., Lugo, V.L., Flores, N.R., Diaz, C.E.B., Vazquez, P.C., 2014. Characterization of lignocellulosic fruit waste as an alternative feedstock for bioethanol production. Bioresources 9 (2), 1873–1885.

Sandhya, R., Kurup, G., 2013. Screening and isolation of pectinase from fruit and vegetable wastes and the use of orange waste as a substrate for pectinase production. International Research Journal of Biological Sciences 2 (9), 34–39.

Saravanan, P., Muthuvelayudham, R., Viruthagiri, T., 2012. Application of statistical design for the production of cellulase by *Trichoderma reesei* using mango peel. Enzyme Research 7.:157643. http://dx.doi.org/10.1155/2012/157643.

Sarvanan, P., Muthuvelayudham, R., Viruthagiri, T., 2013. Enhanced production of cellulase from pineapple waste by response surface methodology. Journal of Engineering 8 (Article ID: 979547).

Schieber, A., Stintzing, F.C., Carle, R., 2001. By-products of plant food processing as a source of functional compounds – recent developments. Trends in Food Science and Technology 12, 401–413.

Shah, M.P., Reddy, G.V., Banerjee, R., Ravindra Babu, P., Kothari, I.L., 2005. Microbial degradation of banana waste under solid state bioprocessing using two lignocellulolytic fungi (*Phylosticta* spp. MPS-001 and *Aspergillus* spp. MPS-002). Process Biochemistry 40, 445–451.

Shalini, R., Gupta, D., 2010. Utilization of pomace from apple processing industries: a review. Journal of Food Science and Technology 47 (4), 365–371.

Sharanappa, A., Wani, K.S., Patil, P., 2011. Bioprocessing of food industrial waste for α-amylase production by solid state fermentation. International Journal of Advanced Biotechnology and Research 2 (4), 473–480. ISSN:0976-2612.

Sharma, K.D., Karki, S., Thakur, N.S., Attri, S., 2012. Chemical composition, functional properties and processing of carrot: a review. Journal of Food Science and Technology 49, 22–32.

Sharma, C.K., Kanwar, S.S., 2012. Purification of a novel thermophilic lipase from *B. licheniformis* MTCC-10498. ISCA Journal of Biological Sciences 1 (3), 43–48.

Sharma, N.R., Sasankan, A., Singh, A., Soni, G., 2011. Production of ploygalaturonase and pectim methyl esterase from agrowaste by using various isolates of *Aspergillus niger*. Insight Microbiology 1 (1), 1–7.

Sharoba, A.M., Farrag, M.A., El-Salam, A.M., 2013. Utilization of some fruits and vegetables waste as a source of dietary fiber and its effect on the cake making and its quality attributes. Journal of Agroalimentary Processes and Technologies 19 (4), 429–444.

Siliha, H., El-Sahy, K., Sulieman, A., Carle, R., El-Badawy, A., 2000. Citrus wastes: composition, functional properties and utilization. Obst-, Gemüse- und Kartoffelverarbeitung [Fruit, Vegetable and Potato Processing] 85, 31–36.

Sims, C.A., Balaban, M.O., Matthews, R.F., 1993. Optimization of carrot juice color and cloud stability. Journal of Food Science 58, 1129–1131.

Singh, A.K., Mukhopadhyay, M., 2012. Overview of fungal lipase: a review. Applied Biochemistry and Biotechnology 166, 486–520.

Singh, B., Panesar, P.S., Nanda, V., 2006. Utilization of carrot pomace for the preparation of a value added product. World Journal of Dairy and Food Sciences 1 (1), 22–27. ISSN:1817-308X.

Singh, R.K., Mishra, S.K., Kumar, N., 2010. Optimization of culture conditions for amylase production by thermophilic *Bacillus* sp. in submerged fermentation. Research Journal of Pharmaceutical, Biological and Chemical Sciences 1 (4), 867–876.

Smith, D.C., Wood, T.M., 1991. Xylanase production by *Aspergillus awamori* development of a medium and optimization of the fermentation paramerters for the production of extracellular xylanase and beta-xyloasidase while maintaining low protease production. Biotechnology Bioenergy 38 (8), 883–890.

Stoll, T., Schieber, A., Carle, R., 2001. Carrot pomace—an underestimated by-product? In: Pfannhauser, W., Fenwick, G.R., Khokhar, S. (Eds.), Biologically-Active Phytochemicals in Food: Analysis, Metabolism, Bioavailability and Function. Royal Society of Chemistry, Cambridge, UK, pp. 525–527.

Subramaniyam, R., Vimala, R., 2012. Solid state and submerged fermentation for the production of bioactive substances: a comparative study. International Journal of Science and Nature 3 (3), 480–486. ISSN:2229-6441.

Sukumaran, R.K., Singhania, R.K., Pandey, A., 2005. Microbial cellulases – production, application, and challenges. Journal of Scientific and Industrial Research 64, 832–844.

Sun, H.Y., Li, J., Zhao, P., Peng, M., 2011. Banana peel: a novel substrate for cellulase production under solid-state fermentation. African Journal of Biotechnology 10 (77), 17887–17890.

Svendsen, A., 2000. Review: lipase protein engineering. Biochemia et Biophysica Acta 1543, 223–238.

Tanaka, K., Hilary, Z.D., Ishizaki, A., 1999. Investigation of the utility of pineapple juice and pineapple waste material as lowcost substrate for ethanol fermentation by *Zymomonas mobilis*. Journal of Bioscience and Bioengineering 87, 642–646.

Tapre, A.R., Jain, R.K., 2014. Pectinases: enzymes for fruit processing industry. International Food Research Journal 21 (2), 447–453.

Tewari, H.K., Marwaha, S.S., Rupal, K., 1986. Ethanol from banana peels. Agricultural Wastes 16, 135–146.

Thakur, S., 2012. Lipases, its sources, properties and applications: a review. International Journal of Scientific and Engineering Research 3 (7), 1–29.

Thassitou, P.K., Arvanitoyannis, I.S., 2001. Bioremediation: a novel approach to food waste management a review. Trends in Food Science and Technology 12, 185–196.

Thangaratham, T., Manimegalai, G., 2014. Optimization and production of pectinase using agro waste by solid state and submerged fermentation. International Journal of Current Microbiology and Applied Sciences 3 (9), 357–365.

Torres, E., Bustos-Jaimes, I., Le Borgne, S., 2003. Potential use of oxidative enzymes for the detoxification of organic pollutants. Applied Catalysis Environmental 46, 1–15. ISSN:0926-3373.

Umsza-Guez, M.A., Diaz, A.B., Ory, I., Blandino, A., Gomes, E., Caro, I., 2011. Xylanase production by *Aspergillus awamori* under solid state fermentation conditions on tomato pomace. Brazilian Journal of Microbiology 42 (4), 1585–1597.

Unakal, C., Kallur, R.I., Kaliwal, B.B., 2012. Production of α-amylase using banana waste by *Bacillus subtilis* under solid state fermentation. European Journal of Experimental Biology 2 (4), 1044–1052.

Venkatesh, M., Pushpalatha, P.B., Sheela, K.B., Girija, D., 2009. Microbial pectinase from tropical fruit wastes. Journal of Tropical Agriculture 47 (1–2), 67–69.

Verma, N., Thakur, S., Bhatt, A.K., 2012. Microbial lipases: industrial applications and properties: a review. International Research Journal of Biological Sciences 1 (8), 88–92. ISSN:2278-3202.

Vilas-Boas, S.G., Esposito, E., Mendonca, M.M., 2002. Novel lignocellulolytic ability of *Candida utilis* during solid-substrate cultivation on apple pomace. World Journal of Microbiology and Biotechnology 18, 541–545.

Viniegra-González, G., Favela-Torres, E., Aguiler, C.N., de Jesus Rómero-Gomez, S., Diaz-Godinez, G., Augur, C., 2003. Advantages of fungal enzyme production in solid state over liquid fermentation systems. Biochemical Engineering Journal 13, 157–167.

Vintila, T., Dragomirescu, M., Jurcoane, S., Caprita, R., Maiu, M., 2009. Production of cellulase by submerged and solid-state cultures and yeasts selection for conversion of lignocellulose to ethanol. Romanian Biotechnological Letters 14 (2), 4275–4281.

Vitolo, S., Petarca, L., Bresci, B., 1999. Treatment of olive oil industry wastes. Bioresource Technology 67, 129–137.

Wadhwa, M., Bakshi, M.P.S., 2013. Utilization of fruit and vegetable wastes as livestock feed and as a substrate for generation of other value added products. In: Makkar, H.P.S. (Ed.), Food and Agriculture Organization of United Nations. RAP Publication 2013/04, p. 56.

Wadhwa, M., Kaushal, Bakshi, M.P.S., 2006. Nutritive evaluation of vegetable wastes as complete feed for goat bucks. Small Ruminant Research 64, 279–284.

Waldron, K., 2001. Useful ingredients from onion waste. Food Science and Technology 15, 38–41.

Wong, W.S., 2009. Structure and action mechanism of ligninolytic enzymes dominic. Applied Biochemistry and Biotechnology 157, 174–209.

World Bank Publications, 1998. Pollution prevention and abatement handbook: toward cleaner production. Nature 777, 316–319 ISBN:0-8213-3638-X.

WPTC (The World Processing Tomato Council), 2013. World Production Estimate as of 25 October 2013. www.wptc.to/.

Zheng, Z., Shetty, K., 1998. Cranberry processing waste for solid state fungal inoculant production. Process Biochemistry 33, 323–329.

Zheng, Z., Shetty, K., 2000. Enhancement of pea (*Pisum sativum*) seedling vigour and associated phenolic content by extracts of apple pomace fermented with *Trichoderma* spp. Process Biochemistry 36, 79–84.

Zilly, A., Bazanella, G.C.S., Helm, C.V., Araújo, C.A.V., de Souza, C.G.M., Bracht, A., Peralta, R.M., 2012. Solid-state bioconversion of passion fruit waste by white-rot fungi for production of oxidative and hydrolytic enzymes. Food and Bioprocess Technology 5, 1573–1580.

ZMP (Zentrale Markt- und Preisberichtsstelle GmbH), 2000. ZMP Marktbilanz Kartoffeln. ZMP, Bonn.

Zuorro, A., Lavecchia, R., Medici, F., Piga, L., 2014. Use of cell wall degrading enzymes for the production of high-quality functional products from tomato processing waste. Chemical Engineering Transactions 38, 355–360. http://dx.doi.org/10.3303/CET1438060.

FURTHER READING

Bisaria, V.S., Ghose, T.K., 1981. Cranberry processing waste for solid state fungal inoculant production. Enzyme and Microbial Technology 3, 90–104.

Chandel, A.K., Singh, O.V., Rao, L.V., 2010. Biotechnological applications of hemicellulosic derived sugars: state-of-the-art. In: Singh, O.V., Harvey, S.P. (Eds.), Sustainable Biotechnology: Renewable Resources and New Perspectives. Springer, Dordrecht, Netherlands, pp. 63–81.

Gírio, F.M., Fonseca, C., Carvalheiro, F., Duarte, L.C., Marques, S., Bogel-Łukasik, R., 2010. Hemicelluloses for fuel ethanol: a review. Bioresource Technology 101 (13), 4775–4800.

Martínez, A.T., Rencoret, J., Marques, G., Gutiérrez, A., Ibarra, D., Jiménez-Barbero, J., del Río, J.C., 2008. Monolignol acylation and lignin structure in some nonwoody plants: a 2D NMR study. Phytochemistry 69 (16), 2831–2843.

Miller, J.N., 1986. An introduction to pectins: structure and properties. In: Fishman, M.L., Jem, J.J. (Eds.), Chemistry and Functions of Pectins. ACS Symposium Series, vol. 310. American Chemical Society, Washington, DC.

Park, S., Baker, J.O., Himmel, M.E., Parilla, P.A., Johnson, D.K., 2010. Cellulose crystallinity index: measurement techniques and their impact on interpreting cellulase performance. Biotechnology for Biofuels 3, 10.

Santos, M.M., Rosa, A.S., Dalboit, S., Mitchell, D.A., Kriger, N., 2004. Thermal denaturation: is solid-state fermentation really a good technology for the production of enzymes? Bioresource Technology 93 (3), 261–268. ISSN:0960-8524.

Sun, J., Hu, X., Zhao, G., Wu, J., Wang, Z., Chen, F., Liao, X., 2007. Characteristics of thin layer infrared drying of apple pomace with and without hot air pre-drying. Food Science and Technology International 13 (2), 91–97.

UN Food and Agriculture Organization, 2012. Production of Kiwi (Fruit) by Countries. https://en.wikipedia. org/wiki/Kiwifruit.

CHAPTER 3

Bioprocesses for Enzyme Production Using Agro-Industrial Wastes: Technical Challenges and Commercialization Potential

M. Kapoor, D. Panwar, G.S. Kaira
CSIR–Central Food Technological Research Institute, Mysuru, India

INTRODUCTION

Agricultūra is the Latin word for *agriculture*, in which *ager* means "field," and *cultūra* means "cultivation." India produced 106.54, 95.91, 43.05, 19.27, 264.77, 32.86, 350.02 million tonnes of rice, wheat, coarse cereals, pulses, food grains, oil seeds and sugarcane, respectively during 2013–14 (Directorate of Economics and Stastistics, 2013, http://eands.dacnet.nic.in/). Agricultural maneuvers and practices contribute significantly to various industrial sectors, such as food, feed, textiles, etc. (Rodríguez-Couto, 2008) but also result in the generation of huge quantities (1.3 billion tons/year, Food and Agricultural Oraganization, 2013) of agricultural wastes/residues. These wastes are simply burnt or dumped to rot, which poses a threat to human health and the environment. On the contrary, renewable and inexpensive agro-wastes with abundant carbon (lignin, cellulose, hemicellulose, starch, and pectin) and nitrogen reserves can act as an ideal starting material for the cost-effective generation of different industrial products. However, lack of established commercial usage and lacunas in infrastructure facilities to process agro-wastes toward value addition act as major deterrents toward effective utilization. Many recent initiatives by government organizations in close association with farmers and scientists are targeting optimal utilization of agro-wastes by using advanced tools of modern biology and ligno–cellulose biotechnology. Value addition of agro-wastes by converting them to useful products is currently one of the foremost areas of basic and applied research in many private and public funded laboratories (Anwar et al., 2014).

Wheat bran, wheat straw, rice straw, corn cob, sugarcane bagasse, fruit pomace, and oil cake (eg, groundnut, mustard, flaxseed, soy) represent the major agro-residues. They are primarily composed of cellulose (40–50%), hemicellulose (20–30%), lignin (10–25%), and pectin (~35%) (Malherbe and Cloete, 2002; Kumar et al., 2009; Voragen et al., 2009; Ahmed et al., 2011) (Table 3.1). Cellulose, the most abundant biopolymer, is majorly

Agro-Industrial Wastes as Feedstock for Enzyme Production
ISBN 978-0-12-802392-1
http://dx.doi.org/10.1016/B978-0-12-802392-1.00003-4

Table 3.1 Biochemical Composition of Lignocellulosic Residues

Lignocellulosic Residue	Cellulose (%)	Hemicellulose (%)	Lignin (%)	Ash (%)	References
Nut shells	25–30	25–30	30–40	NR	Howard et al. (2004)
Corn cobs	45	35	15	1.36	Howard et al. (2004), Prasad et al. (2007), McKendry (2002)
Rice straw	32.1	24	18	NR	Howard et al. (2004), Prasad et al. (2007), McKendry (2002)
Switch grass	45	31.4	12	NR	Howard et al. (2004)
Sugar cane bagasse	32–44	27–32	19–24	4.5–9	Rowell (1992)
Wheat straw	29–35	26–32	16–21	NR	Rowell (1992), Prasad et al. (2007), McKendry (2002)
Barley straw	31–34	24–29	14–15	5–7	Rowell (1992)
Oat straw	31–37	27–38	16–19	6–8	Rowell (1992), Martin (2012)
Rye straw	33–35	27–30	16–19	2–5	Rowell (1992), Stewart et al. (1997), Reguant and Rinaudo (2000), Hon (2000)
Bamboo	26–43	15–26	21–31	1.7–5	Rowell (1992), Stewart et al. (1997), Reguant and Rinaudo (2000), Hon (2000)
Esparto grass	33–38	27–32	17–19	6–8	Rowell (1992), Stewart et al. (1997), Reguant and Rinaudo (2000), Hon (2000)
Sabai grass	NR	23.9	22.0	6.0	Rowell (1992), Stewart et al. (1997), Reguant and Rinaudo (2000), Hon (2000)
Elephant grass	22	24	23.9	6	Rowell (1992), Stewart et al. (1997), Reguant and Rinaudo (2000), Hon (2000)

Table 3.1 Biochemical Composition of Lignocellulosic Residues—Cont'd

Lignocellulosic Residue	Cellulose (%)	Hemicellulose (%)	Lignin (%)	Ash (%)	References
Seed flax bast fiber	47	25	23	5	Rowell (1992), Stewart et al. (1997), Reguant and Rinaudo (2000), Hon (2000)
Kenaf bast fiber	31–39	22–23	15–19	2–5	Rowell (1992), Stewart et al. (1997), Reguant and Rinaudo (2000), Hon (2000)
Jute bast fiber	45–53	18–21	21–26	0.5–2	Rowell (1992), Stewart et al. (1997), Reguant and Rinaudo (2000), Hon (2000)
Leaf fiber sisal (agave)	43–56	21–24	7–9	0.6–1.1	Rowell (1992), Stewart et al. (1997), Reguant and Rinaudo (2000), Hon (2000)
Banana waste	13.2	14.8	14	11.4	John et al. (2006)
Cassava waste	NR	NR	NR	4.2	John et al. (2006)
Brewers spent grain	16.8	28.4	27.8	4.6	Mussatto et al. (2008)
Wheat bran	10–15	30	4–8	NR	Beaugrand et al. (2004)
Orange peel waste	9.2	10.5	NR	NR	Rivas et al. (2008)
Rice bran	9–12.8	8.7–11.4	NR	NR	Ramezanzadeh et al. (2000)
Apple pomace	5	NR	$23.5 \, g \, kg^{-1}$	1.1	Kosmala et al. (2011), Dhillon et al. (2012a)
Reed grass	74.6	6.7	13.1	NR	Philippoussis and Diamantopoulou (2011)
Bean stalk	80.3	5.7	8.4	NR	Philippoussis and Diamantopoulou (2011)
Barley bran	23	32.7	21.4	NR	Cruz et al. (2000)
Corn leaves	37.6	34.5	12.6	NR	Cruz et al. (2000)

NR: not reported.

composed of β-1,4 linked-D-glucopyranose units with an average molecular weight of ~100,000 Da (Himmel et al., 2007). Cellulose molecules have a strong tendency to form intra- and intermolecular hydrogen bonds and provide rigidity to the plant cell wall. Hemicelluloses, the second-most abundant biopolymer, are noncovalently associated with cellulose micro-fibrils and have an average molecular weight of <30,000 Da. Depending on their backbone sugar composition, they are distinguished as xylans, mannans, glucans, galactans, arabinans, and arabinogalactans. Lignin comprises phenyl-propane, methoxy groups, and non-carbohydrate polyphenolic substances, which are commonly linked by ether bonds. Cell wall architecture studies show that lignin acts as glue by filling the gap between and around the cellulose and hemicellulose fractions (Hamelinck et al., 2005; Calvo-Flores and Dobado, 2010; Jiang et al., 2010; Menon and Rao, 2012). Pectins with a backbone of α-1,4 linked D-galacturonic acid units are diverse biopolymers present in the middle lamellae and primary cell wall of plant cells and are responsible for maintaining the structural integrity (Kohli and Gupta, 2015).

The biological catalyst "enzymes" are exceptional in nature as they are bestowed with the ability to catalyze multifaceted reactions, transformations with exceptional stereo-selectivity, regio-selectivity, and chemo-selectivity. In the past century, substantial improvement has been made through exemplary research on developing a better understanding on working of enzymes and their production technologies. The ever increasing demand for enzymes with better substrate specificity, stability, solubility, and to fit the needs of a process in industrial sectors, such as detergent, paper, food, feed, and pharmaceuticals has been the key driving force for the development of low-cost and efficient microbial enzyme production technologies (Rodwell and Kennelly, 1993; Beg et al., 2001; Kapoor et al., 2001; Kaira et al., 2015; Srivastava and Kapoor, 2015).

MAJOR AGRO-INDUSTRIAL RESIDUES/WASTES
Agricultural Crop Residues and By-Products
Rice Straw
Rice straw is produced in huge quantities worldwide (annual production 731 million tons) [Oceania 1.7 million tons, Europe 3.9 million tons, Africa 20.9 million tons, America 37.2 million tons, Asia 667.6 million tons (Karimi et al., 2006)]. Rice straw cannot be recycled in soil due to limited time (20–25 days) left before sowing of succeeding crops, such as wheat.

Wheat Straw
Wheat straw, a by-product obtained after harvesting of wheat grains, has an annual global production of 529 million tons (Govumoni et al., 2013). Wheat straw is composed of internodes (57 ± 10%), nodes (10 ± 2%), leaves (18 ± 3%), chaffs (9 ± 4%), and rachis (6 ± 2%). Compositional analysis of wheat straw has shown the presence of cellulose (34–40%), hemicellulose (20–25%), and lignin (20%) (Rodriguez-Gomez et al., 2012).

Corn Cobs

Maize, being one of the major commercial agriculture crops, is grown in large quantities across the globe. Corn cob (central core of maize ear) (yield: $1.42–1.53\,dry\,t\,ha^{-1}$) and corn husk (leafy outer covering of maize) are the two major by-products and constitute 20–30% of the maize plant (Samanta et al., 2012). It has been estimated that 18 kg of corn cob are produced from 100 kg of corn ear. The world wide production of corn cob in the year 2008 was 144 million tonnes (da Silva et al., 2015). Corn cobs are a rich source of cellulose (27.71%) and hemicellulose (38.78%) but also contain a significant amount of lignin (9.4%) (Shinners et al., 2007).

Wheat Bran

The annual global production of wheat in 2013 was ~713.2 million tons (FAOSTAT, 2013a). Wheat bran, the outer layer of the wheat kernel, is obtained as a by-product (15–20%, w/w) from wheat milling industries.

Current Usage/Disposal Techniques

Broadly, agricultural crop residues and by-products are disposed by industries or farmers in the following ways. (1) Burning: farmers normally resort to burning of agro-residues, such as rice straw, wheat straw, and corn cobs which leads to (a) soot particles that cause health problems (asthma or other respiratory disorders), (b) greenhouse gases, and (c) loss of important plant nutrients (N, P, K, and S) (Kerstetter and Lyons, 2001; Laura and Porcar, 2009). (2) Dumping: agro-residues, such as rice straw, corn cobs, and wheat straw are left unattended in either agricultural fields or open areas leading to rotting and associated environmental issues. (3) Feed incorporation: agro-residues, such as wheat bran, which is rich in non-starch carbohydrates (~58%), starch (~19%), and crude protein (~18%), are supplemented to feed meant for ruminants and poultry (Sun et al., 2008; Xie et al., 2008; Muazu and Stegemann, 2015).

Food Wastes

Overproduction, physical damage during harvesting, infestation due to microbes, insects, and pests, postharvest handling, improper transportation and storage, food processing, distribution, and consumer usage pattern are the major reasons for food wastage (Gooch et al., 2010). Food wastes are rich in fermentable sugars, such as glucose (21–39%), fructose (0.7–25%), and mixed sugars (17–36%) along with cellulose and hemicellulose (Choi et al., 2015).

Oil Cakes

Oil cakes/meals are produced after removal of oil from oilseeds. Edible oil cakes are derived from sunflower, sesame, soy bean, coconut, mustard, palm kernel, groundnut, cottonseed, canola, olive, and rapeseed, while nonedible oil cakes are derived from jatropha, pongamia, sal, babassu, and karanja (Ramachandran et al., 2007). In India, 32.88 million tons of oilseeds were produced in 2013–14 (Directorate of Economics and Stastistics, 2013, http://eands.dacnet.nic.in/) which contributed ~6.3% toward the world's oil cake production. The composition and nutritional content of oil cakes are quite variable due to differences in the methods of oil extraction, storage conditions, and quality of oilseeds.

Sugarcane Bagasse

Sugarcane is grown in around 200 tropical and subtropical countries with an annual production of ~1.91 Gt in 2012–13 (FAOSTAT, 2013b). Sugarcane processing industries generate fibrous fraction (bagasse) and harvest residue (straw) as the major by-products (Ortiz and Oliveira, 2014). Sugarcane bagasse is composed of cellulose (52%), hemicellulose (20%), and lignin (24%), while straw is composed of cellulose (33–45%), hemicellulose (18–30%), lignin (17–41%), ash (2–12%), and extractives (5–17%) (Costa et al., 2013).

Fruit and Vegetable Waste

The fruit and vegetable processing industries along with agricultural activities generate enormous quantities of wastes leading to significant commercial losses and severe environmental hazards. The major sources of fruit and vegetable wastes are preliminary cleaning operations, eg, peeling, trimming, and cutting, etc. Overproduction of fruits and vegetables coupled with improper transportation and storage also contributes significantly toward generation of such wastes.

Apples are produced in enormous quantities worldwide (76.37 million tons) (FAO, 2012). Most of the apples produced are consumed in fresh form (71%), and around 20% are used by beverage and other food industries for value-added products, such as apple juice concentrate, ready-to-serve apple juice, apple cider, wine and vermouth, apple purees and jams, and dried apple products. The solid waste (~30% of the original fruit content) obtained after juice extraction is called pomace, which is a heterogeneous mixture of peel, core, seed, calyx, stem, and soft tissue (Vendruscolo et al., 2008). Apples that are infested by pests or infected by microorganisms are the other sources for pomace. Compositional analysis of apple pomace has shown the presence of carbohydrates (480–838 g kg^{-1}), reducing sugars (108–150 g kg^{-1}), protein (29–57 g kg^{-1}), and lignin (23.5 g kg^{-1}) (Dhillon et al., 2012a). The composition of pomace varies due to varietal differences, type of processing used for juice extraction, and the number of times fruits are pressed for juice extraction.

The processing of grapes, the most widely cultivated fruit crop in the world, in juice and wine industries generates a waste consisting of peel, seed, stalk, and pomace. The annual production of grape pomace is around 240 million kg in Spain, Italy, and France (Bustamante et al., 2008). Biochemically, grape pomace is composed of carbohydrates (8% in the seeds, 13% in the skin), fats (10% in pips and 4% in the skin), fiber (50% of the total mass), proteins (8% and 14% in pips and skin, respectively), and mineral salts (Botella et al., 2005). Significant quantities of orange (5.0 MT) and kinnow (0.4 MT) are produced in India on an annual basis (Khandelwal et al., 2006; FAOSTAT, 2012). The processing and consumption of these citrus fruits (Oberoi et al., 2010) generates waste in the form of peel, seeds, and pulp (Wilkins et al., 2007). These wastes are rich in carbohydrates, proteins, reducing sugars, hemicellulose, and pectin (Rivas et al., 2008).

Mango is one of the popular fruits worldwide, and Asia accounts for around 77% of the global mango production, followed by the Americas (13%) and Africa (9%) (FAOSTAT, 2007). India produced around 18,002 MT of mango in 2012–13 (Indian Horticulture Database, 2013). Mango peel is rich in pectic substances and can be used for production of value-added products (Srirangarajan and Shrikhande, 1977). Similarly, potato is produced and consumed worldwide. Many fast food industries are primarily based on potato-based products. Potato waste basically is comprised of potatoes wasted in storage or transit and residues originating from food processing units. These wastes are formed of cellulose, hemicellulose, and lignin. Potato peel, in particular, contains 46.0% (w/v) total sugar and 3.0% (w/v) reducing sugar, 24.3% (w/v) starch, and 3.5% (w/v) protein (Shukla and Kar, 2006). Tomato is another well-known vegetable crop and is grown in many countries. Tomato processing industries produce millions of tons of waste called tomato pomace, which consists of tomato peel and seeds as well as some pulp. The pomace is rich in proteins, lipids, and carbohydrates including pectins, amino acids, carotenoids, and minerals (Alvarado et al., 2001; Knoblich et al., 2005; Del Valle et al., 2006). Similarly, carrot pomace is produced in significant quantities after juice extraction in many industries (Chau et al., 2004).

Current Usage/Disposal Techniques

Food wastes easily are rotten by microorganisms due to their high moisture content and favorable composition (significant proportions of assimilable carbohydrates other than cellulose). They are currently disposed in municipal sanitary landfills, which results in release of greenhouse gases (Dhillon et al., 2013), or simply left to rot, causing environmental/health problems due to high chemical oxygen demand (Velmurugan and Ramanujam, 2011). Fruit pomace and oil cakes are used as low-cost ingredients in feed meant for swine, poultry, and fish (Ramachandran et al., 2007). Sugarcane bagasse is utilized by sugar mills and ethanol distilleries for their energy requirements (Rabelo et al., 2011). Some food wastes are used as low-cost feed stock for bioprocesses.

ENZYMES: THE BIOLOGICAL TOOLS FOR INDUSTRIAL APPLICATIONS

Enzymes, which work under mild conditions, have proven themselves as a viable option when compared to their chemical counterparts in varied processes pertaining to dairy, brewing, wine and juice, fats and oils, baking, detergents, textiles, leather, pulp and paper,

Table 3.2 Major Industrial Enzymes and Their Commercial Applications

			Production Statistics		
Enzyme	EC Number	Industrial Applications	Percent of world enzyme market or market share	Approximate cost (US$ kg⁻¹)	Commercial Manufacturers
Cellulases	3.2.1.4, 3.2.1.91, 3.2.1.176, 3.2.1.21	Biofuels, pulp and paper, textiles, food and beverages, detergents, and animal feed	20	15.67–90	Altech[a], Novozymes[b], Dupont–Genencor[c], Enzyme Bio Systems[p], bioWORLD[q], Merck KGaA[r],
Laccases	1.10.3.2, 1.11.1.14 1.11.1.13	Bioremediation, lignocellulose conversion, textiles, biosensors, food, pharmaceuticals, organic synthesis, and paper	NA	1–99	AB enzymes[d], Lignozyme GmbH[e], Novozymes[b], Zytex Pvt Ltd[f], Novo Nordisk[g], Enzyme Bio Systems[p], Sekisui diagnostics[s]
Pectinases	3.2.1.15, 3.1.1.11 4.2.2.10, 4.2.2.2	Fruit and wine, pectic waste water treatment, degumming of plant bast fibers, textile processing, and animal feed	10	1–105	Novozymes[b], Dupont-Danisco[h], Dupont-Genencor[c], DSM[i], BASF[j], Novartis[k], Biocon[l], Enzyme Bio Systems[p], bioWORLD[q], Merck KGaA[r], Noor enzymes[u]
Xylanases	3.2.1.8, 3.2.1.37	Textiles, food, feed, baking, bioconversion of lignocelluloses, and pulp and paper, clarification of must and juices, and liquefaction of fruits and vegetables	200 million dollars	10–80	Altech[a], Dupont-Genencor[c], Novo Nordisk[g], Biocon[l], AB enzymes[e], Thomas Swan[m], Novozymes[b], Enzyme Development Corporation[n], Enzyme Bio Systems[p], Prozomix[v]
Proteases	3.4.11.1, 3.4.11.6 3.4.17.1, 3.4.17.2 3.4.21.1, 3.4.22.2 3.4.23.15, 3.4.21.62 3.4.24.27, 3.4.21.4	Detergents, brewing, meat, photography, dairy, food processing, leather, feed, paper, and pharmaceuticals	60	3–30	Novozymes[b], Dupont-Danisco[h], Dupont-Genencor[c], DSM[i], BASF[j], Enzyme Bio Systems[p]

α-amylases	3.2.1.1	Starch hydrolysis, detergents, baking, biofuels, pharmaceuticals, and fine chemicals	25–33	1500–10,000	Novozymes, Dupont-Danisco[h], Dupont-Genencor[c], DSM[i], BASF[j], PMP Fermentation Products[o], Novozymes[b], Enzyme Bio Systems[p], bioWORLD[q], Prozomix[v]
Mannanases	3.2.1.78, 3.2.1.25, 3.2.1.21, 3.1.1.6, 3.2.1.22	Feed, oil and gas drilling, bio-bleaching of softwood pulps, oil extraction, coffee clarification, textiles, and production of prebiotic mannooligosaccharides	NA	1–100	Novozymes[b], Dupont-Danisco[h], Dupont-Genencor[c], DSM[i], BASF[j], Elanco[s], Enzyme Bio Systems[p], Prozomix[v]

[a]Kentucky, USA.
[b]Bagsvaerd, Denmark.
[c]Rochester, New York, USA.
[d]Darmstadt, Germany.
[e]Germany.
[f]Mumbai, India.
[g]Bagsvaerd, Denmark.
[h]Copenhagen, Denmark.
[i]Heerlen, Netherlands.
[j]Ludwigshafen, Germany.
[k]Basel, Switzerland.
[l]Bengaluru, India.
[m]Consett, UK.
[n]New York, USA.
[o]Illinois, USA.
[p]Chagrin Falls, Ohio, USA.
[q]Dublin, Ohio, USA.
[r]Darmstadt, Germany.
[s]Greenfield, USA.
[u]Kolkata, India.
[v]Haltwhistle, UK.

and personal care industries (Table 3.2). Earlier, methodologies used for producing enzymes were not well defined. The advent of fermentation technology viz SmF and SSF made it possible to produce enzymes in huge quantities and well-characterized form. Newer developments, such as recombinant DNA technology, protein engineering, and directed evolution, have further revolutionized the development of industrial enzymes.

BIOPROCESSES FOR ENZYME PRODUCTION USING AGRO-INDUSTRIAL WASTES

Submerged Fermentation (SmF)

SmF has been defined as the cultivation of microorganisms in a nutrient medium with an excess of free flowing water (Singhania et al., 2010). In SmF, microorganisms are grown on soluble (xylan, pectin, mannan) or insoluble substrates (wheat bran, rice bran, wheat straw) that are dissolved or submerged in liquid media. SmF has gained immense importance over the past several years for the production of enzymes and secondary metabolites on an industrial scale owing to its strict control on fermentation parameters, consistent productivity, and easy downstream processing (Vaidyanathan et al., 1999). Batch, continuous, and fed-batch are the major modes in SmF processes. Batch mode represents fermentation of sterilized nutrient solution in a closed vessel by the desirable microbial culture. The advantages of such a process are low cost and simple infrastructure for process control. However, low productivity and feed-back inhibition are the major disadvantages. Continuous mode involves a tightly regulated, highly productive bioreactor system (chemostat or turbidostat) where the sterile substrate and other medium components are fed at a particular rate, and an equivalent amount of product is harvested in a time-dependent manner. Fed-batch fermentation is characterized by the intermittent feeding and maintenance of optimum concentrations of required growth substrates with continuous harvesting of product (Lim and Shin, 2013).

SOLID-STATE FERMENTATION (SSF)

SSF is the cultivation of microorganisms on solid, moist substrates in the absence of a free aqueous phase (Pandey, 2003). It is a three-phase heterogeneous process, comprising solid, liquid, and gaseous phases, which offers colossal benefits for microbial bioprocesses and product development (Thomas et al., 2013). Microorganisms, such as fungi and yeast with lower water activity requirements (a_w 0.5–0.6) are best suited for SSF. However, bacterial cultures requiring high water activity (a_w 0.8–0.9) tend to be less suitable. Agrowastes are the prime matrices in SSF processes as they provide microorganisms an environment similar to their natural habitat, leading to higher cell mass and product yield within a short period of time (Thomas et al., 2013). Many agro-wastes have been used for producing enzymes [polygalacturonase (Kapoor et al., 2000; Kapoor and Kuhad,

2002; Gupta et al., 2008), pectinases (Martins et al., 2002), cellulases (Oberoi et al., 2010; Dhillon et al., 2011a, 2012a,b; Palaniyandi et al., 2014; Saratale et al., 2014; Vijayaraghavan et al., 2015), and xylanases (Carmona et al., 2005; Kapoor et al., 2007; Dhillon et al., 2011b, 2012c; Panwar et al., 2014), amylases, ligninases (Couto and Sanromán, 2005), inulinases (Mazutti et al., 2010), chitinases (Binod et al., 2005; Dhillon et al., 2011c), and phytases (Chantasartrasamee et al., 2005; Roopesh et al., 2006)] from microorganisms. SSF as compared to SmF appears to be a more advantageous process due to low energy expenditures, simple instrumentation, economic feasibility, and sustainability (Table 3.3). Traditionally, cost-effective tray reactors have been used for producing enzymes (cellulase, xylanase, phytase, and protease) in SSF using agro-residues as they require low maintenance and labor (Hooge et al., 2010; Dhillon et al., 2011a, 2011b). However, control of process parameters, such as temperature in tray reactors is not feasible. Today, many commercial establishments are using agitated solid-state systems based on a rotating drum reactor with improved sterility and better control over temperature for producing enzymes (WO 1994/EP 0683815 B1; US, 2003/6664095 B1).

COMMERCIALIZATION POTENTIAL

Understanding the commercial potential by stringent cost and benefits analysis is pivotal for the success of enzyme production by SmF and SSF. There are very few studies where attempts were made to compare the cost of enzyme production in SmF and SSF using agro-residues. An economic analysis for production of lipase by *Penicillium restrictum* in SmF and SSF indicated that the cost of generating a lipase concentrate of

Table 3.3 Comparison of Process Parameters Among SSF and SmF for Production of Enzymes Using Agro-Wastes (Zhuang et al., 2007; Hansen et al., 2015)

Process Parameter	SSF	SmF
Water requirement for fermentation	Small	Large
Water requirement for downstream processing	Large	Small
Capital investment	Low	High
Machinery	Simple	Complex
Energy requirement	Low	High
Substrate nature	Raw	Defined
Product yield	High	Low
Process automation	Low	High
Temperature maintenance	Poor	Good
Cultivation period	Long	Short
Online monitoring	No	Yes
Process control	No	Yes
Substrate particle size range	Broad (cm to inches)	Narrow (mm to cm)
Agitation (rpm)	3–150	100–500

$100\,m^3$ per year in SmF was 78% higher as compared to the SSF process (Castilho et al., 2000b). A drastic cost difference was also observed between SmF ($40.36 kg^{-1}) and SSF ($15.67 kg^{-1}) for producing cellulase from *Clostridium thermocellum* (Zhuang et al., 2007). Castro et al. (2010) investigated the production of amylases from *Aspergillus awamori* IOC-3914 in SSF using babassu cake and reported a significant decrease in the production cost. However, polygalacturonase production from a mutant of *Aspergillus carbonarius* using SmF required lower capital investment (15–24%) compared to SSF (Nakkeeran et al., 2012).

EFFECT OF FERMENTATION PARAMETERS

Process optimization in SmF and SSF processes for enzyme production is aimed at maintaining optimum and homogenous reaction conditions, consistent product quality, minimizing microbial stress exposure, and enhancing metabolic accuracy. For each individual enzyme, a comprehensive and detailed process optimization and characterization strategy involving process design, type of fermentation, and elucidation of relevant parameters influencing enzyme yield needs to be deciphered.

CARBON AND NITROGEN SOURCE

In fermentation processes meant for enzyme production, selection of appropriate carbon and nitrogen sources to augment microbial growth and metabolism is a prime requirement. Carbon sources, such as corn starch, glucose, sucrose, and molasses and nitrogen sources, such as yeast extract, peptone, tryptone, ammonium sulfate, etc., are costly growth substrates that are commonly used for enzyme production. A number of reports over the past two decades have clearly shown the potential of cost-effective wheat bran, corncobs, rice straw, wheat straw, sugarcane bagasse, corn steep liquor, and oil cakes as growth substrates for producing enzymes from bacterial and fungal sources. Wheat bran has been a universal choice for production of various enzymes, such as α-amylase (Babu and Satyanarayana, 1995; Bhanja et al., 2007; Baysal et al., 2008; Dhillon et al., 2011b; Maity et al., 2015), lipase (Mahadik et al., 2002), xylanase (Beg et al., 2000; Kapoor et al., 2007, 2008; Dhillon et al., 2011b; Panwar et al., 2014; Scholl et al., 2015), cellobiase (Tsao et al., 2000; Rajoka et al., 2006), protease (Prakasham et al., 2006; Meena et al., 2013), and chitinase (Nampoothiri et al., 2004; Patidar et al., 2005; Kumar et al., 2012; Patil and Jadhav, 2014) as it contains an optimal intermix of hemicellulose and digestible nitrogen (Khurana et al., 2007) (Table 3.4). In SSF, agro-residues have been used for enzyme production after supplementing them with mineral salt solution (Babu and Satyanarayana, 1995; Madeira et al., 2012; Ruiz et al., 2012; Selwal and Selwal, 2012; Gonzalez et al., 2013; Panwar et al., 2014; Scholl et al., 2015) (Table 3.4).

Table 3.4 A Comparative Account of Process Parameters Used for Producing Enzymes From Microorganisms Using Agro-Wastes as Substrate

Enzyme	Microbial Source	Agro-Waste	Pretreatment	Fermentation Process	Cultivation Conditions	Fermentation Vessel	Amount of Substrate (g)	Enzyme Production IU g⁻¹ (SSF) or IU mL⁻¹ (SmF)	References
α-amylase	Bacillus coagulans	WB	NR	SSF	pH 7, 50°C	250 mL EF	10	26,350	Babu and Satyanarayana (1995)
	Bacillus megaterium	WB	NR	SSF	pH 4, 30°C	Tray bioreactor	50–400	15,480	Bhanja et al. (2007)
	Bacillus sp.	WB and Lentil husk	NR	SSF	pH 10, 37°C	250 mL EF	30	216,000	Baysal et al. (2008)
	Aspergillus awamori	BC	Grinding and sieving	SSF	Moisture 70%, 30°C	Tray bioreactor	2.5	40.5	Castro et al. (2010)
	Bacillus subtilis	Starchy substrates	Sun drying	SSF	Moisture 50%, 30°C	250 mL EF	10	2135	Pavithra et al. (2014)
	B. subtilis ATCC 6633	WB	NR	SSF	Moisture 50%, 40°C	100 mL EF	4	1730	Maity et al. (2015)
Lipase	Penicillium restrictum	BC	Grinding and sieving	SSF	Moisture 70%, pH 5–6, 30°C	PP beakers	10	30.3	Gombert et al. (1999)
	Aspergillus niger	WB	NR	SSF	Moisture 75%, 30°C	500 mL EF	10	630	Mahadik et al. (2002)
	Rhizopus homothallicus	SB	Sieving	SSF	Moisture 75%, pH 6.5, 40°C	Glass columns	20	826	Rodriguez et al. (2006)
	Rhizopus oryzae	SB	Sieving	SSF	pH 8, 45°C, humidity of 80%	Tray bioreactor	Optimum filling	215.16	Vaseghi et al. (2013)
	A. niger	Two-phase olive mill waste	NR	SSF	Moisture 75%, pH 5.6, 25°C	500 mL EF	30	18.67	Salgado et al. (2014)
Pectinase	A. niger	Coffee pulp	NR	SSF	Moisture 60%, pH 5.5, 25°C	Glass cylinders	20	228	Antier et al. (1993)
	A. niger	Soy and WB	NR	SSF	Moisture 40%, 30°C	PP beakers	10	26	Castilho et al. (2000a)
	Thermoascus aurantiacus	Orange, SB and WB	Commercial processing	SSF	Moisture 25–67%, pH 10–11, 55°C	250 mL EF	5	43	Martins et al. (2002)

Continued

Table 3.4 A Comparative Account of Process Parameters Used for Producing Enzymes From Microorganisms Using Agro-Wastes as Substrate—cont'd

Enzyme	Microbial Source	Agro-Waste	Pretreatment	Fermentation Process	Cultivation Conditions	Fermentation Vessel	Amount of Substrate (g)	Enzyme Production IU g^{-1} (SSF) or IU mL^{-1} (SmF)	References
	A. niger	LPP	Milling and sieving	SSF	Moisture 70%, pH 5, 30°C	Column-tray bioreactor	3	2183.1 U L^{-1}	Ruiz et al. (2012)
	Bacillus pumilus	Sugar beet, pulp, Apple and Grape pomace and WB	NR	SmF	pH 8, 30°C, 150 rpm	250 mL EF	5	31.8	Tepe and Dursun (2014)
Xylanase	Streptomyces sp.	WB and Eucalyptus kraft pulp	NR	SmF	pH 8, 37°C, 200 rpm	250 mL EF	50 mL	82	Beg et al. (2000)
	Paecilomyces thermophila	WS	Chopping, grinding, and milling	SSF	pH 6.9, 50°C	250 mL EF	5	18,580	Yang et al. (2006)
	B. pumilus	WB	NR	SSF	Moisture 99%, 37°C	Tray	700	44,000	Kapoor et al. (2007)
	B. pumilus	WB and WS	NR	SmF	pH 9, 37°C	5 L fermenter	1% (w/v)	4000	Kapoor et al. (2008)
	A. niger and Trichoderma reseei	Rice straw and WB	Sieving	SSF	95% relative humidity; 30°C ± 1	Tray	500	3106.34 IU gds^{-1}	Dhillon et al. (2011b)
	A. niger NRRL 567	Apple pomace	Drying and milling	SSF	70% relative humidity; 30°C ± 1	Tray	500	4868 IU gds^{-1}	Dhillon et al. (2012a)
	A. niger NRRL-567	Apple pomace	NR	SSF	85% relative humidity; 30°C ± 1	Tray	500	3578.8 IU gds^{-1}	Dhillon et al. (2012c)
	Bacillus sp.	WB	NR	SSF	37°C, pH 8	250 mL EF	10	980,000	Panwar et al. (2014)
	Penicillium echinulatum S1M29	WB and Elephant grass	Steam explosion	SSF	95% relative humidity; 28–30°C	100 mL EF	1	571	Scholl et al. (2015)

Enzyme	Microorganism	Substrate	Pretreatment	Fermentation type	Conditions	Reactor	Scale	Activity	References
Laccase	Cyathus olla	Canola roots	Mill grinding	SSF	pH 4.5, 25°C	1 L LEF	10	7110 mU mg^{-1} protein	Carnelley et al. (2002)
	Panus tigrinus	Olive mill wastewater	NR	SmF	500 rpm, pH 5.3, 28°C	3 L Stirred tank reactors	2 L	4600 U L^{-1}	Fenice et al. (2003)
	Pycnoporus cinnabarinus	SB	NR	SmF	25°C	Vapor phase bioreactor	Optimum filling	90	Meza et al. (2006)
	Trametes hirsute	Orange peels	Soaking and acid treatment	SSF	Moisture 90%, pH 4–4.5, 30°C	Tray bioreactor	150	11000 U L^{-1}	Rosales et al. (2007)
	Coriolus versicolor	Tomato pomace	NR	SmF	25°C and 150 rpm	1 L LEF	200 mL	362	Freixo do Rosário et al. (2008)
	Trametes versicolor ATCC 20869	Brewer's spent grain	NR	SSF	90% relative humidity, 30°C ± 1	Tray	500	13506.2 IU gds^{-1}	Dhillon et al. (2012d)
	T. versicolor ATCC 20869	Brewer's spent grain	NR	SSF	80% relative humidity, 28–30°C	Tray	400	13,506.2	Dhillon et al. (2012e)
	Trametes pubescens	Coffee and Soybean pod husk	Hypochlorite treatment	Semisolid	pH 4.5, 30°C	1 L LEF	15	204	Gonzalez et al. (2013)
Cellulase	T. reesei	Willow	Steam pretreated	SmF	350–700 rpm, pH 5.5–6, 30°C	22 L Fermenter	10 g L^{-1}	108 FPU g^{-1}	Reczey et al. (1996)
	B. subtilis	Banana fruit stalk	Acid and alkali hydrolysis	SSF	Moisture 65%, pH 7, 35°C	250 mL EF	10	2.80	Krishna (1999)
	Trichoderma harzianum	Oil palm empty fruit bunches	Oven drying, chopping, and grinding	SSF	Moisture 60%, pH 5.1, 32°C	50 L Rotary drum bioreactor	4 kg	10.1 FPA gds^{-1}	Alam et al. (2009)
	T. reesei Rut C-30	Kinnow pulp and WB	Oven drying, grinding, and sieving	SSF	30°C	250 mL EF	10	13.4 IU gds^{-1}	Oberoi et al. (2010)

Continued

Table 3.4 A Comparative Account of Process Parameters Used for Producing Enzymes From Microorganisms Using Agro-Wastes as Substrate—cont'd

Enzyme	Microbial Source	Agro-Waste	Pretreatment	Fermentation Process	Cultivation Conditions	Fermentation Vessel	Amount of Substrate (g)	Enzyme Production IU g⁻¹ (SSF) or IU mL⁻¹ (SmF)	References
	Aspergillus fumigatus	Rice, Wheat, and Cotton straw, Corn stover, and cob	Chopping, milling, and sieving	SSF	Moisture 65%, pH 4, 50°C	1 L EF	20	146 FPU g⁻¹	Liu et al. (2011)
	A. niger	Apple pomace	NR	SSF	55% relative humidity, 30°C ±1	Tray	500	64.18	Dhillon et al. (2011a)
	A. niger NRRL-567	Apple pomace	NR	SSF	55% relative humidity, 30°C ±1	Tray	500	146.5 IU gds⁻¹	Dhillon et al. (2011c)
	A. niger NRRL 2001	Apple pomace	Drying and milling	SSF	70% relative humidity, 30°C ±1	Tray	500	544.7 IU gds⁻¹	Dhillon et al. (2012a)
	A. niger NRRL-567	Apple pomace	NR	SSF	75% relative humidity, 30°C ±1	500 mL EF	40	172.31 IU gds⁻¹	Dhillon et al. (2012b)
	Macrophomina phaseolina	Apple pomace	NR	SSF	70% relative humidity, 30°C ±1	500 mL EF	50	278.11 IU gds⁻¹	Kaur et al. (2012)
	Hypocrea koningii	Rice straw	Chopping and milling	SSF	pH 5, 35°C,	500 mL EF	15	44.15 FPU g⁻¹	Palaniyandi et al. (2014)
	Phanerochaete chrysosporium	Grass powder	Milled, sieved, and mild acid	SSF	pH 5, 30°C	250 mL EF	5	244.60	Saratale et al. (2014)
	Bacillus halodurans IND18	Cow dung	Dried and powdered	SSF	pH 8, 60°C ±1	100 mL in EF	3	4140	Vijayaraghavan et al. (2015)

Enzyme	Organism	Substrate	Pretreatment		Conditions				Reference
		and WB			and 100 rpm				(1999)
	A. niger	WB and CB	NR	SSF	Moisture 70%, pH 5.5, 35°C	500 mL EF	100	215	Tsao et al. (2000)
	A. niger	CB	Acid and heat	SSF	pH 4.8, 30°C	1000 mL EF	5 cm thick bed	438	Shen and Xia (2004)
	A. niger	CB, WB, and Grass	NR	SSF	pH 6.5, 28°C	250 mL EF	3	562	Rajoka et al. (2006)
Protease	Bacillus sp.	WB and Lentil husk	NR	SSF	Moisture 140%, pH 10, 37°C, 200 rpm	250 mL EF	60%	429	Uyar and Baysal (2004)
	Bacillus sp.	WB and Green gram husk	Sieving	SSF	pH 9, 30°C	250 mL EF	10	35,000	Prakasham et al. (2006)
	B. subtilis	Grass and potato peel	Hot water	SSF	Moisture 100%, pH 8, 50°C	Tray reactor	100	NR	Mukherjee et al. (2008)
	Aspergillus oryzae	Curcas seed cake	Drying and grinding	SSF	Moisture 45%, 30°C	250 mL EF	25	14,273	Thanapimmetha et al. (2012)
	Pseudomonas aeruginosa	WB	NR	SSF	pH 9, 45°C	250 mL EF	20	528	Meena et al. (2013)
Endo-mannanase	B. subtilis	WB	NR	SmF	37°C, pH 7	250 mL EF	35 g L^{-1}	37.1	Helow et al. (1997)
	Pichia pastoris	Corn steep liquor	NR	SmF	NR	50 L fermenter	1350	5132	Zheng et al. (2012)
	Bacillus sp. CFR1601	Defatted coconut	NR	SSF	pH 6, 45°C	250 mL EF	5	198	Srivastava and Kapoor (2013)
	Bacillus sp. CFR1601	Green gram husk and Sunflower oil cake	NR	SmF	pH 6, 45°C, 180 rpm	250 mL EF	0.5% and 4% (w/v)	25.6	Srivastava and Kapoor (2014)
	Hi-Control Escherichia coli BL21 (DE3)	Defatted flax seed meal	Autoclaving and centrifugation	SmF	pH 7.0, 37°C, 250 rpm	250 mL EF	NR	5926	Kaira et al. (2016)
Tannase	Lactobacillus sp. ASR-S1	Coffee husk	NR	SSF	Moisture 50%, pH 5, 37°C	250 mL EF	5	0.85 U gds^{-1}	Sabu et al. (2006)
	A. fumigatus MA	Jamun leaves	Drying and grinding	SSF	Moisture 1:1, pH 5, 25°C	250 mL EF	10	174.32 U g^{-1}	Manjit et al. (2008)

Continued

Table 3.4 A Comparative Account of Process Parameters Used for Producing Enzymes From Microorganisms Using Agro-Wastes as Substrate—cont'd

Enzyme	Microbial Source	Agro-Waste	Pretreatment	Fermentation Process	Cultivation Conditions	Fermentation Vessel	Amount of Substrate (g)	Enzyme Production IU g⁻¹ (SSF) or IU mL⁻¹ (SmF)	References
	Paecilomyces variotii	Orange pomace	Grounded and sieved	SSF	Moisture 59%, 30°C	250 mL EF	5	$5000\,U\,gds^{-1}$	Madeira et al. (2012)
	Penicillium atramentosum	Keekar leaves	Drying and grinding	SmF	pH 7.5, 30°C, 150 rpm	100 mL EF	3%	34.7	Selwal and Selwal (2012)
	A. niger JMU-TS528	Tea stalks	Drying and grinding	SSF	pH 6.0, 28°C	250 mL EF	5	$62\,U\,g^{-1}$	Wang et al. (2013)
Chitinase	*Trichoderma harzianum*	WB	NR	SSF	Moisture 65%, pH 4.5, 30°C	250 mL EF	5	$3.18\,U\,gds^{-1}$	Nampoothiri et al. (2004)
	Penicillium chrysogenum PPCS 1	WB	Sun-dried	SSF	Moisture 80%, pH 5, 40°C	150 mL EF	5	3809 units	Patidar et al. (2005)
	Trichoderma asperellum UTP-16	WB	NR	SSF	pH 5.1, 35°C	250 mL EF	5	$4.01\,U\,gds^{-1}$	Kumar et al. (2012)
	Penicillium ochrochloron MTCC 517	WB	NR	SSF	pH 7, 30°C	250 mL EF	5	$2443.23\,U\,g^{-1}$	Patil and Jadhav (2014)
α-galactosidase	*Bacillus stearothermophilus* (NCIM 5146)	Soybean meal	NR	SmF	pH 7, 60°C, 200 rpm	250 mL EF	2%	2.0	Gote et al. (2004)
	Streptomyces griseoloalbus	Soya bean flour	NR	SSF	Moisture 40%, pH 7, 30°C	250 mL EF	10	$117\,U\,g^{-1}$	Anisha et al. (2008)
	Lactobacillus agilis LPB 56	Soybean vinasse	Alcoholic fermentation of soybean molasses	SmF	pH 6.5, 30°C, 150 rpm	14 L bioreactor	30%	$11.07\,U\,mL^{-1}$	Sanada et al. (2009)
β-galactosidase	*K. marxianus*	Whey and cauliflower waste	Drying, grinding, and sieving	SmF	pH 5, 28°C, 100 rpm	500 mL EF	40%	$1.61\,IU\,mg^{-1}$ dry weight	Oberoi et al. (2008)
	Kluyveromyces marxianus MTCC 1389	Whey	NR	SmF	pH 5, 28°C, 100 rpm	500 mL EF	40%	$1.6\,IU\,mg^{-1}$ dry weight	Bansal et al. (2008)

AERATION

Aeration shows diverse effects on microbial fermentation processes by influencing the metabolic pathways (Çalík et al., 1998). In SmF, aeration in growth vessels or flasks is commonly provided by shaking of fermentation medium at variable speeds. In bioreactors, shaking by impellors and bubbling of air through spargers caters to the oxygen requirement. Different aeration conditions have been examined for enzyme production through SmF while using agro-wastes (Naby et al., 1999; Gote et al., 2004; Sanada et al., 2009; Selwal and Selwal, 2012; Srivastava and Kapoor, 2014) (Table 3.4). In SSF, it is quite difficult to maintain sufficient aeration for microbial cultures due to its static nature. However, few workers have shown the role of forced flow of moistened air on the solid substrate bed (Meien et al., 2004). Recently, Gassara et al. (2013) studied the role of different types of agitation (1) continuous agitation; (2) continuous, discontinuous, continuous agitation; and (3) discontinuous, continuous, discontinuous agitation and aeration (0.87, 1.25, 1.66 vvm) on ligninolytic enzyme production by solid-state cultures of *Phanerochaete chrysosporium* BKM-F-1767 growing on apple pomace.

pH

pH of the fermentation medium affects the transport of various metabolic products, for example, enzymes across the cell membrane besides affecting microbial physiology. It has been shown that growth rate of bacteria increases or decreases monotonically with pH (Tan et al., 1998). Typically, pH of the fermentation medium supplemented with agro-residues is set in a broad range to understand its role on enzyme production from microorganisms, for example, pH 4–11, xylanase (Kapoor et al., 2008); pH 5–8, and pH 3–8, tannase (Selwal and Selwal, 2012; Wang et al., 2013); pH 7–12, pectinase (Martins et al., 2002); pH 4–10, pH 4–8, and pH 2–10, cellulase (Krishna, 1999; Palaniyandi et al., 2014; Saratale et al., 2014), and pH 4–7 and pH 4–12, chitinase (Patidar et al., 2005; Kumar et al., 2012) (Table 3.4). The control of pH during the SmF process is done by the addition of an appropriate acid/base. However, for SSF processes, there are no electrodes that are capable of recording the pH due to lack of free water (Jiang et al., 2012). Substrate formulations, especially nitrogen sources, buffering capacity of different components employed, or use of a buffer that has no deleterious influence on biological activity can help in overcoming the problem of pH variability to some extent during SSF processes (Gowthaman et al., 2001).

TEMPERATURE

The temperature has a direct consequence on various biological processes by affecting enzyme activity, protein denaturation, and growth of microorganisms (Betts et al., 1988). Therefore, cultivation temperature in SmF and SSF for enzyme production from microbial sources using agro-residues has been optimized by many authors, eg, α-amylase

[Maity et al., 2015 (30–60°C)], tannase [Sabu et al., 2006 (25–40°C); Selwal and Selwal, 2012 (20–45°C); Wang et al., 2013 (20–33°C)], cellulase [Krishna, 1999 (20–55°C)], chitinase [Kumar et al., 2012 (27–40°C)] and lipase [Vaseghi et al., 2013 (25–50°C)] (Table 3.4). Various mathematical models have also evolved for explaining the role of temperature on microbial growth kinetics (Zhu and Chen, 2015). In SmF, adequate technological advances have been made for temperature maintenance. However, SSF processes still lack precise control over the maintenance of homogenous temperature. The conventional convection or conductive cooling devices are inadequate for dissipating metabolic heat due to the poor thermal conductivity of solid substrates and static nature of fermentation in SSF. Forced bed aeration has shown some positive results with respect to temperature control in SSF (Sato et al., 1982; Yadav, 1988; Raimbault, 1998; Gowthaman et al., 2001).

INOCULUM AGE AND SIZE

The size of the inoculum substantiates the ability of a microbial population to initiate growth in a particular nutrient environment. A lower level of inoculum may lead to longer lag phase and delay enzyme synthesis (Sen and Swaminathan, 2004). On the other hand, a higher number of cells results in rapid proliferation and exhaustion of substrate (Sabu et al., 2006). Thus, optimal inoculum size is crucial for microbial fermentation processes, and many researchers have explored the role of inoculum size on enzyme production using agro-residues, eg, xylanase [Kapoor et al., 2008 (0.1–1.75%, v/v); Panwar et al., 2014 (5–20%, v/v)], tannase [Sabu et al., 2006 (0.5–5.0 mL); Wang et al., 2013 (0.5–3.0 mL)], cellulase [Krishna, 1999 (5–40%, v/w)], protease [Uyar and Baysal, 2004 (10–35%); Mukherjee et al., 2008 (10–100%, v/w); Thanapimmetha et al., 2012 (1–10%)], and α-galactosidase [Anisha et al., 2008 $((0.9 - 3) \times 10^6 \, \text{CFU} \, \text{g}^{-1})$] (Table 3.4).

SUBSTRATE PRETREATMENT

Agro-industrial residues are resistant toward microbial colonization due to the presence of crystal and nonporous lignocellulosics as the major components. Pretreatment methodologies are mainly aimed at breaking down lignin structure, disrupting crystalline structure of cellulose, and increasing the surface area of biomass for its accessibility toward microbes (Yang et al., 2006; Castro et al., 2010; Ruiz et al., 2012; Scholl et al., 2015). Pretreatment of agro-residues using physical (mechanical comminution), chemical [ozonolysis, acid hydrolysis, alkaline hydrolysis, oxidative delignification, organosolv process (Ghedalia and Miron, 1981; Thring et al., 1990; Sivers and Zacchi, 1995)], physicochemical [steam explosion, ammonia fiber explosion, CO_2 explosion (Dale and Moreira, 1982)] and pyrolysis [decomposition at temperatures greater than 300°C (Kilzer and Broido, 1965; Shafizadeh and Bradbury, 1979)], and biological means has shown improved utilization of agro-residues for microbial enzyme production (Ucar and Fengel, 1988; Xiang et al., 2003; Mosier et al., 2005).

TECHNICAL CHALLENGES AND NEW TRENDS IN ENZYME PRODUCTION USING AGRO-RESIDUES

SSF Processes

The major challenges observed in SSF for enzyme production are heat dispersal, scale-up, and biomass determination.

HEAT DISPERSAL

On the commencement of SSF, uniformity of temperature and concentration of oxygen throughout the substrate are reasonable. But as the fermentation progresses, solid materials/matrices become compact, leading to low thermal conductivities, poor heat removal, and enormous heat accumulation. As a consequence, metabolic activities of the cultivated microorganism (growth, spore formation, and germination) are affected along with the possibility of product denaturation (Durand, 2003; Pandey, 2003). Various approaches have been formulated to tackle heat generation in SSF systems, which include modification in the bioreactor design and employing packed-bed or column-type bioreactors (Duran et al., 2011; Dey and Banerjee, 2012; Dilipkumar et al., 2013). These bioreactors introduce air through a sieve and offer better heat and mass transfer, accurate measurement of the gaseous environment (CO_2 and O_2), and higher substrate loading per bioreactor volume (Durand, 2003; Thomas et al., 2013).

Other modifications in packed-bed or column-type bioreactors, such as the multiplex gas sampler (Sandoval et al., 2012), laterally aerated moving bed bioreactor (Wong et al., 2011), and counter-current bioreactor for continuous fermentation process (Varzakas et al., 2008) and process optimization (Alani et al., 2009; Duran et al., 2011; Aguirre et al., 2014) are potentiating the utilization of SSF processes in the production of commercial metabolites including enzymes.

SCALE-UP

In SSF, most workers have reported enzyme production from microorganisms using agro-residues at the flask or tray level by using few grams of solid substrates. Tray- or flask-based SSF is difficult to scale-up due to its limited capacity and poor aeration and temperature control. To commercialize SSF processes for enzyme production, it is pertinent to develop bioreactors that can take up high quantities of solid substrates and produce the maximum titer of enzymes. Many researchers have investigated bioreactor configurations, which include, eg, fixed-bed bioreactors (Duran et al., 2011; Silveira et al., 2014), aerated trays (Montero et al., 2011), rotary drums (Thomas et al., 2013), different operation modes [eg, batch (Zúñiga et al., 2013), fed-batch (Astolfi et al., 2011), and continuous (Lagemaat and Pyle, 2004)], and mixing strategies [eg, intermittent mixing (Flodman and Noureddini, 2013)] for SSF using agro-residues for enzyme production.

The packed-bed reactor appears to be a promising configuration; it helps with the following: (1) analyzing the global evolution of process on an empirical basis; (2) determining environmental parameters for controlling the temperature and moisture of solid medium, and (3) studying mass and heat transfer phenomena and oxygen diffusion (Durand, 2003). A packed-bed bioreactor configuration has been successfully tested for inulinase production from *Kluyveromyces marxianus* NRRL Y-7571. The best conditions to use for inulinase production ($463\,U\,gds^{-1}$) were inlet air temperature and volumetric flow rate of 30°C and $3\,m^3\,h^{-1}$, respectively (Mazutti et al., 2010). Pectinase production by *Aspergillus niger* Aa-20 using lemon peel pomace as a support and carbon source has been carried out in a column-tray bioreactor at an air flow rate of $194\,mL\,min^{-1}$ (Ruiz et al., 2012). Pitol et al. (2016) reported maximum pectinase productivity of 1840 $U\,kg^{-1}\,h^{-1}$ after 10 h using a 40-cm high bed containing 27 kg of wheat bran and 3 kg of sugarcane bagasse. Fixed-bed bioreactors with forced aeration have also been proposed recently as a promising alternative. A cylindrical fixed-bed bioreactor with forced aeration using babassu cake as the raw material was used for the production of a pool of industrially relevant enzymes by *Aspergillus awamori* IOC-3914 (Castro et al., 2015).

BIOMASS DETERMINATION

The estimation of biomass is essential for kinetic studies and understanding the physiological behavior of culture. However, separation of biomass has been a bottleneck in SSF processes. Indirect methods, such as estimation of biomass components (chitin, glucosamine, ergosterol, protein, nucleic acid) (Sharma et al., 1977; Koliander et al., 1984; Matcham et al., 1985), measurement of biological activity (ATP, respiration rate, immunological activity) (Brezonik et al., 1975; Frankland et al., 1981; Bengtsson and Rundgren, 1983; Barak and Chet, 1986), DNA assay, light scattering (Kennedy et al., 1992) carbon dioxide evolution rate (Desgranges et al., 1991), and light reflectance (Murthy et al., 1993) have been applied to determine microbial growth. Recently, the coupling of dynamic imaging and computational modeling has shown potential in the online determination of fungal biomass in SSF. This technique has advantages of low cost, accuracy, and good adjustability (Yingyi et al., 2012).

SmF Processes

SmF processes, due to their established scale-up facilities, face fewer challenges than SSF processes for enzyme production. However, usage of agricultural residues as substrates in the production medium may pose difficulties with respect to reproducibility due to the heterogenic nature of substrate apart from difficulties faced with in situ sterilization processes (Hansen et al., 2015).

CONCLUSIONS

It is imperative that a sound balance between the production of food and ecological awareness will promote the sustainability of production systems with continuous economic growth. As waste output and by-products are inseparable parts of all industrial sectors and their improper disposal have pancontinental environmental consequences, methodologies for recovery and reusage of such by-products are increasingly being disseminated (Laufenberg et al., 2003). Globally, many industries generate huge quantities of agro-industrial wastes, which accumulate over a period of time and lead to environmental safety issues. There is an urgent need to change global perception toward agro-wastes as they can be effectively utilized in an eco-friendly manner for sustainable productivity and generation of value-added products. Many industrial and academic laboratories are focusing on utilization of agro-wastes for enzyme production as the number of industrial processes requiring enzymes and new fields for enzyme applications are growing exponentially. SmF and SSF represent fundamental techniques for producing enzymes on an industrial scale using agro-wastes. Most of the commercial success with respect to producing enzymes has been attained with the aid of SSF technology. However, there is a need to focus on improving the scale-up facilities in SSF by the development of robust designs and configurations of bioreactors, process automation, and online monitoring of parameters. It is envisaged that in the near future the colossal potential of agro-wastes as cost-effective substrates for enzyme production and allied technologies will be realized, and this will help in creating integrated green technologies with zero environmental impact.

REFERENCES

Aguirre, C.A.P., Bastos, R.G., Carvalho, A.L., Alegre, R.M., 2014. The influence of process parameters in production of lipopeptide iturin A using aerated packed bed bioreactors in solid-state fermentation. Bioprocess and Biosystems Engineering 37 (8), 1569–1576.

Ahmed, I., Zia, M.A., Iftikhar, T., Iqbal, H.M., 2011. Characterization and detergent compatibility of purified protease produced from *Aspergillus niger* by utilizing agro wastes. BioResources 6 (4), 4505–4522.

Alam, M.Z., Mamun, A.A., Qudsieh, I.Y., Muyibi, S.A., Salleh, H.M., Omar, N.M., 2009. Solid state bioconversion of oil palm empty fruit bunches for cellulase enzyme production using a rotary drum bioreactor. Biochemical Engineering Journal 46 (1), 61–64.

Alani, F., Grove, J.A., Anderson, W.A., Moo-Young, M., 2009. Mycophenolic acid production in solid-state fermentation using a packed-bed bioreactor. Biochemical Engineering Journal 44 (2), 106–110.

Alvarado, A., Pacheco-Delahaye, E., Hevia, P., 2001. Value of a tomato byproduct as a source of dietary fiber in rats. Plant Foods for Human Nutrition 56 (4), 335–348.

Anisha, G.S., Sukumaran, R.K., Prema, P., 2008. Evaluation of α-galactosidase biosynthesis by *Streptomyces griseoloalbus* in solid-state fermentation using response surface methodology. Letters in Applied Microbiology 46 (3), 338–343.

Antier, P., Minjares, A., Roussos, S., Raimbault, M., Viniegra-Gonzalez, G., 1993. Pectinase-hyper producing mutants of *Aspergillus niger* C28B25 for solid-state fermentation of coffee pulp. Enzyme and Microbial Technology 15 (3), 254–260.

Anwar, Z., Gulfraz, M., Irshad, M., 2014. Agro-industrial lignocellulosic biomass a key to unlock the future bio-energy: a brief review. Journal of Radiation Research and Applied Sciences 7 (2), 163–173.

Astolfi, V., Joris, J., Verlindo, R., Oliveira, J.V., Maugeri, F., Mazutti, M.A., et al., 2011. Operation of a fixed-bed bioreactor in batch and fed-batch modes for production of inulinase by solid-state fermentation. Biochemical Engineering Journal 58, 39–49.

Babu, K.R., Satyanarayana, T., 1995. α-Amylase production by thermophilic *Bacillus coagulans* in solid state fermentation. Process Biochemistry 30 (4), 305–309.

Bansal, S., Oberoi, H.S., Dhillon, G.S., Patil, R.T., 2008. Production of β-galactosidase by *Kluyveromyces marxianus* MTCC 1388 using whey and effect of four different methods of enzyme extraction on β-galactosidase activity. Indian Journal of Microbiology 48 (3), 337–341.

Barak, R., Chet, I., 1986. Determination, by fluorescein diacetate staining, of fungal viability during myco-parasitism. Soil Biology and Biochemistry 18 (3), 315–319.

Baysal, Z., Uyar, F., Doğru, M., Alkan, H., 2008. Production of extracellular alkaline α-amylase by solid state fermentation with a newly isolated *Bacillus* sp. Preparative Biochemistry and Biotechnology 38 (2), 184–190.

Beaugrand, J., Reis, D., Guillon, F., Debeire, P., Chabbert, B., 2004. Xylanase-mediated hydrolysis of wheat bran: evidence for subcellular heterogeneity of cell walls. International Journal of Plant Sciences 165 (4), 553–563.

Beg, Q.K., Bhushan, B., Kapoor, M., Hoondal, G.S., 2000. Production and characterization of thermostable xylanase and pectinase from *Streptomyces* sp. QG-11-3. Journal of Industrial Microbiology and Biotechnology 24 (6), 396–402.

Beg, Q., Kapoor, M., Mahajan, L., Hoondal, G.S., 2001. Microbial xylanases and their industrial applications: a review. Applied Microbiology and Biotechnology 56 (3–4), 326–338.

Bengtsson, G., Rundgren, S., 1983. Respiration and growth of a fungus, *Mortierella isabellina*, in response to grazing by *Onychiurus armatus* (Collembola). Soil Biology and Biochemistry 15 (4), 469–473.

Betts, W.B., Dart, R.K., Ball, M.C., 1988. Degradation of larch wood by *Aspergillus flaws*. Transactions of the British Mycological Society 91 (2), 227–232.

Bhanja, T., Rout, S., Banerjee, R., Bhattacharyya, B.C., 2007. Comparative profiles of α-amylase production in conventional tray reactor and GROWTEK bioreactor. Bioprocess and Biosystems Engineering 30 (5), 369–376.

Binod, P., Pusztahelyi, T., Nagy, V., Sandhya, C., Szakács, G., Pócsi, I., Pandey, A., 2005. Production and purification of extracellular chitinases from *Penicillium aculeatum* NRRL 2129 under solid-state fermentation. Enzyme and Microbial Technology 36 (7), 880–887.

Botella, C., de Ory, I., Webb, C., Cantero, D., Blandino, A., 2005. Hydrolytic enzymes production by *Aspergillus awamori* on grape pomace. Biochemical Engineering Journal 26, 100e106.

Brezonik, P.L., Browne, F.X., Fox, J.L., 1975. Application of ATP to plankton biomass and bioassay studies. Water Research 9 (2), 155–162.

Bustamante, M.A., Moral, R., Paredes, C., Pérez-Espinosa, A., Moreno-Caselles, J., Pérez-Murcia, M.D., 2008. Agrochemical characterisation of the solid byproducts and residues from the winery and distillery industry. Waste Management 28 (2), 372–380.

Çalık, P., Çalık, G., Özdamar, T.H., 1998. Oxygen transfer effects in serine alkaline protease fermentation by *Bacillus licheniformis*: use of citric acid as the carbon source. Enzyme and Microbial Technology 23 (7), 451–461.

Calvo-Flores, F.G., Dobado, J.A., 2010. Lignin as renewable raw material. ChemSusChem 3 (11), 1227–1235.

Carmona, E.C., Fialho, M.B., Buchgnani, É.B., Coelho, G.D., Brocheto-Braga, M.R., Jorge, J.A., 2005. Production, purification and characterization of a minor form of xylanase from *Aspergillus versicolor*. Process Biochemistry 40 (1), 359–364.

Carnelley, T.C.S., Szpacenko, A., Tewari, J.P., Palcic, M.M., 2002. Enzymatic activity of *Cyathus olla* during solid state fermentation of canola roots. Phytoprotection 83 (1), 31–40.

Castilho, L.R., Medronho, R.A., Alves, T.L., 2000a. Production and extraction of pectinases obtained by solid state fermentation of agroindustrial residues with *Aspergillus niger*. Bioresource Technology 71 (1), 45–50.

Castilho, L.R., Polato, C.M., Baruque, E.A., Sant'Anna, G.L., Freire, D.M., 2000b. Economic analysis of lipase production by *Penicillium restrictum* in solid-state and submerged fermentations. Biochemical Engineering Journal 4 (3), 239–247.

Castro, A.M.D., Carvalho, D.F., Freire, D.M.G., Castilho, L.D.R., 2010. Economic analysis of the production of amylases and other hydrolases by *Aspergillus awamori* in solid-state fermentation of babassu cake. Enzyme Research.

Castro, A.M., Castilho, L.R., Freire, D.M., 2015. Performance of a fixed-bed solid-state fermentation bioreactor with forced aeration for the production of hydrolases by *Aspergillus awamori*. Biochemical Engineering Journal 93, 303–308.

Chantasartrasamee, K., Ayuthaya, D.I.N., Intarareugsorn, S., Dharmsthiti, S., 2005. Phytase activity from *Aspergillus oryzae* AK9 cultivated on solid state soybean meal medium. Process Biochemistry 40 (7), 2285–2289.

Chau, C.F., Chen, C.H., Lee, M.H., 2004. Comparison of the characteristics, functional properties, and in vitro hypoglycemic effects of various carrot insoluble fiber-rich fractions. Lebensmittel-Wissenschaft und Technologie 37 (2), 155–169.

Choi, I.S., Lee, Y.G., Khanal, S.K., Park, B.J., Bae, H.J., 2015. A low-energy, cost-effective approach to fruit and citrus peel waste processing for bioethanol production. Applied Energy 140, 65–74.

Costa, S.M., Mazzola, P.G., Silva, J.C.A.R., Pahl, R., Pessoa Jr., A., Costa, S.A., 2013. Use of sugarcane straw as a source of cellulose for textile fiber production. Industrial Crops and Products 42, 189–194.

Couto, S.R., Sanromán, M.A., 2005. Application of solid-state fermentation to ligninolytic enzyme production. Biochemical Engineering Journal 22 (3), 211–219.

Cruz, J.M., Domínguez, J.M., Domínguez, H., Parajó, J.C., 2000. Preparation of fermentation media from agricultural wastes and their bioconversion into xylitol. Food Biotechnology 14 (1–2), 79–97.

da Silva, J.C., de Oliveira, R.C., da Silva Neto, A., Pimentel, V.C., dos Santos, A.D.A., 2015. Extraction, addition and characterization of hemicelluloses from corn cobs to development of paper properties. Procedia Materials Science 8, 793–801.

Dale, B.E., Moreira, M.J., 1982. Freeze-explosion technique for increasing cellulose hydrolysis. In: Biotechnology and Bioengineering Symposium; (United States), vol. 12. Colorado State University, Fort Collins. No. CONF-820580-.

Del Valle, M., Cámara, M., Torija, M.E., 2006. Chemical characterization of tomato pomace. Journal of the Science of Food and Agriculture 86 (8), 1232–1236.

Desgranges, C., Vergoignan, C., Georges, M., Durand, A., 1991. Biomass estimation in solid state fermentation I. Manual biochemical methods. Applied Microbiology and Biotechnology 35 (2), 200–205.

Dey, T.B., Banerjee, R., 2012. Hyperactive α-amylase production by *Aspergillus oryzae* IFO 30103 in a new bioreactor. Letters in Applied Microbiology 54 (2), 102–107.

Dhillon, G.S., Brar, S.K., Kaur, S., Metahni, S., M'hamdi, N., 2012a. Lactoserum as a moistening medium and crude inducer for fungal cellulase and hemicellulase induction through solid-state fermentation of apple pomace. Biomass and Bioenergy 41, 165–174.

Dhillon, G.S., Brar, S.K., Valero, J.R., Verma, M., 2011a. Bioproduction of hydrolytic enzymes using apple pomace waste by *A. niger*: applications in biocontrol formulations and hydrolysis of chitin/chitosan. Bioprocess and Biosystems Engineering 34 (8), 1017–1026.

Dhillon, G.S., Kaur, S., Brar, S.K., 2012e. In-vitro decolorization of recalcitrant dyes through an ecofriendly approach using laccase from *Trametes versicolor* grown on brewer's spent grain. International Biodeterioration & Biodegradation 72, 67–75.

Dhillon, G.S., Kaur, S., Brar, S.K., 2013. Perspective of apple processing wastes as low-cost substrates for bioproduction of high value products: a review. Renewable and Sustainable Energy Reviews 27, 789–805.

Dhillon, G.S., Kaur, S., Verma, M., 2011c. Chitinolytic and chitosanolytic activities from crude cellulase extract produced by *A. niger* grown on apple pomace through koji fermentation. Journal of Microbiology and Biotechnology 21 (12), 1312–1321.

Dhillon, G.S., Kaur, S., Brar, S.K., Verma, M., 2012b. Potential of apple pomace as a solid substrate for fungal cellulase and hemicellulase bioproduction through solid-state fermentation. Industrial Crops and Products 38, 6–13.

Dhillon, G.S., Kaur, S., Brar, S.K., Verma, M., 2012d. Flocculation and haze removal from crude beer using in-house produced laccase from *Trametes versicolor* cultured on brewer's spent grain. Journal of Agricultural and Food Chemistry 60 (32), 7895–7904.

Dhillon, G.S., Kaur, S., Brar, S.K., Gassara, F., Verma, M., 2012c. Improved xylanase production using apple pomace waste by *Aspergillus niger* in Koji fermentation. Engineering in Life Sciences 12 (2), 198–208.

Dhillon, G.S., Oberoi, H.S., Kaur, S., Bansal, S., Brar, S.K., 2011b. Value-addition of agricultural wastes for augmented cellulase and xylanase production through solid-state tray fermentation employing mixed-culture of fungi. Industrial Crops and Products 34 (1), 1160–1167.

Dilipkumar, M., Rajamohan, N., Rajasimman, M., 2013. Inulinase production in a packed bed reactor by solid state fermentation. Carbohydrate Polymers 96 (1), 196–199.

Directorate of Economics and Stastistics, 2013. Department of Agriculture and Corporation of Farmers Welfare, Ministry of Agriculture and Farmers Welfare. Government of India. Retrieved from: http://eands.dacnet.nic.in/PDF/Agricultural-Statistics-At-Glance2014.pdf.

Duran, R., Luis, V., Rodriguez, R., Aguilar, C.N., 2011. Optimization of tannase production by *Aspergillus niger* in solid-state packed-bed bioreactor. Journal of Microbiology and Biotechnology 21 (9), 960–967.

Durand, A., 2003. Bioreactor designs for solid state fermentation. Biochemical Engineering Journal 13 (2), 113–125.

FAOSTAT, 2007. Food and Agriculture Organization of the United Nations; Production, Crops, Mango. Retrieved from: http://faostat3.fao.org.

FAOSTAT, 2012. Food and Agriculture Organization of the United Nations; Production, Crops, Orange. Retrieved from: http://faostat3.fao.org.

FAOSTAT, 2013a. Food and Agriculture Organization of the United Nations; Production, Crops, Wheat. Retrieved from: http://faostat3.fao.org.

FAOSTAT, 2013b. Food and Agriculture Organization of the United Nations; Production, Crops, Sugar Cane. Retrieved from: http://faostat3.fao.org.

FAO-utilization of Fruit and Vegetable Wastes as Livestock Feed and as Substrates for Generation of Other Value-added Products, 2012. Retrieved from: http://www.fao.org/docrep/018/i3273e/i3273e.pdf.

Fenice, M., Sermanni, G.G., Federici, F., D'Annibale, A., 2003. Submerged and solid-state production of laccase and Mn-peroxidase by *Panus tigrinus* on olive mill wastewater-based media. Journal of Biotechnology 100 (1), 77–85.

Flodman, H.R., Noureddini, H., 2013. Effects of intermittent mechanical mixing on solid-state fermentation of wet corn distillers grain with *Trichoderma reesei*. Biochemical Engineering Journal 81, 24–28.

Food and Agricultural Organization, 2013. Food Waste Harms Climate, Water, Land and Biodiversity – New FAO Report. Retrieved from: http://www.fao.org/news/story/en/item/196220/icode/.

Frankland, J.C., Bailey, A.D., Gray, T.R.G., Holland, A.A., 1981. Development of an immunological technique for estimating mycelial biomass of *Mycena galopus* in leaf litter. Soil Biology and Biochemistry 13 (2), 87–92.

Freixo do Rosário, M., Karmali, A., Frazão, C., Arteiro, J.M., 2008. Production of laccase and xylanase from *Coriolus versicolor* grown on tomato pomace and their chromatographic behaviour on immobilized metal chelates. Process Biochemistry 43 (11), 1265–1274.

Gassara, F., Ajila, C.M., Brar, S.K., Tyagi, R.D., Verma, M., Valero, J., 2013. Influence of aeration and agitation modes on solid-state fermentation of apple pomace waste by *Phanerochaete chrysosporium* to produce ligninolytic enzymes and co-extract polyphenols. International Journal of Food Science and Technology 48 (10), 2119–2126.

Ghedalia, D.B., Miron, J., 1981. The effect of combined chemical and enzyme treatments on the saccharification and in vitro digestion rate of wheat straw. Biotechnology and Bioengineering 23 (4), 823–831.

Gombert, A.K., Pinto, A.L., Castilho, L.R., Freire, D.M., 1999. Lipase production by *Penicillium restrictum* in solid-state fermentation using babassu oil cake as substrate. Process Biochemistry 35 (1), 85–90.

Gonzalez, J.C., Medina, S.C., Rodriguez, A., Osma, J.F., Alméciga-Díaz, C.J., Sánchez, O.F., 2013. Production of *Trametes pubescens* laccase under submerged and semi-solid culture conditions on agro-industrial wastes. PLoS One 8 (9), e73721.

Gooch, M., Felfel, A., Marenick, N., 2010. Food Waste in Canada, Value Chain Management Chain. Retrieved from: www.vcmtools.ca/.

Gote, M., Umalkar, H., Khan, I., Khire, J., 2004. Thermostable α-galactosidase from *Bacillus stearothermophilus* (NCIM 5146) and its application in the removal of flatulence causing factors from soymilk. Process Biochemistry 39 (11), 1723–1729.

Govumoni, S.P., Koti, S., Kothagouni, S.Y., Venkateshwar, S., Linga, V.R., 2013. Evaluation of pretreatment methods for enzymatic saccharification of wheat straw for bioethanol production. Carbohydrate Polymers 91 (2), 646–650.

Gowthaman, M.K., Krishna, C., Moo-Young, M., 2001. Fungal solid state fermentation-an overview. Applied Mycology and Biotechnology 1, 305–352.

Gupta, S., Kapoor, M., Sharma, K.K., Nair, L.M., Kuhad, R.C., 2008. Production and recovery of an alkaline exo-polygalacturonase from *Bacillus subtilis* RCK under solid-state fermentation using statistical approach. Bioresource Technology 99 (5), 937–945.

Hamelinck, C.N., Van Hooijdonk, G., Faaij, A.P., 2005. Ethanol from lignocellulosic biomass: techno-economic performance in short-, middle-and long-term. Biomass and Bioenergy 28 (4), 384–410.

Hansen, G.H., Lübeck, M., Frisvad, J.C., Lübeck, P.S., Andersen, B., 2015. Production of cellulolytic enzymes from ascomycetes: comparison of solid state and submerged fermentation. Process Biochemistry 50 (9), 1327–1341.

Helow, E.R., Sabry, S.A., Khattab, A.A., 1997. Production of α-mannanase by *B. subtilis* from agro-industrial by-products: screening and optimization. Antonie van Leeuwenhoek 71 (3), 189–193.

Himmel, M.E., Ding, S.Y., Johnson, D.K., Adney, W.S., Nimlos, M.R., Brady, J.W., Foust, T.D., 2007. Biomass recalcitrance: engineering plants and enzymes for biofuels production. Science 315 (5813), 804–807.

Hon, D.N.S., 2000. Pragmatic approaches to utilization of natural polymers: challenges and opportunities. Natural Polymers and Agrofibers Based Composites 1–14.

Hooge, D.M., Pierce, J.L., McBride, K.W., Rigolin, P.J., 2010. Meta-analysis of broiler chicken trials using diets with or without Allzyme® SSF enzyme complex. International Journal of Poultry Science 9, 819–823.

Howard, R.L., Abotsi, E., Van Rensburg, E.J., Howard, S., 2004. Lignocellulose biotechnology: issues of bioconversion and enzyme production. African Journal of Biotechnology 2 (12), 602–619.

Indian Horticulture Database, 2013. Retrieved from: http://nhb.gov.in/area-pro/Indian%20Horticulture%202013.pdf.

Jiang, G., Nowakowski, D.J., Bridgwater, A.V., 2010. A systematic study of the kinetics of lignin pyrolysis. Thermochimica Acta 498 (1), 61–66.

Jiang, H., Liu, G., Mei, C., Yu, S., Xiao, X., Ding, Y., 2012. Rapid determination of pH in solid-state fermentation of wheat straw by FT-NIR spectroscopy and efficient wavelengths selection. Analytical and Bioanalytical Chemistry 404 (2), 603–611.

John, F., Monsalve, G., Medina, P.I.V., Ruiz, C.A.A., 2006. Ethanol production of banana shell and cassava starch. Dyna Universidad Nacional de Colombia 73 (150), 21–27.

Kaira, G.S., Dhakar, K., Pandey, A., 2015. A psychrotolerant strain of *Serratia marcescens* (MTCC 4822) produces laccase at wide temperature and pH range. AMB Express 5 (1), 1.

Kaira, G.S., Panwar, D., Kapoor, M., 2016. Recombinant endo-mannanase (ManB-1601) production using agro-industrial residues: development of economical medium and application in oil extraction from copra. Bioresource technology 209, 220–227.

Kapoor, M., Kuhad, R.C., 2002. Improved polygalacturonase production from *Bacillus* sp. MG-cp-2 under submerged (SmF) and solid state (SSF) fermentation. Letters in Applied Microbiology 34 (5), 317–322.

Kapoor, M., Beg, Q.K., Bhushan, B., Dadhich, K.S., Hoondal, G.S., 2000. Production and partial purification and characterization of a thermo-alkali stable polygalacturonase from *Bacillus* sp. MG-cp-2. Process Biochemistry 36 (5), 467–473.

Kapoor, M., Beg, Q.K., Bhushan, B., Singh, K., Dadhich, K.S., Hoondal, G.S., 2001. Application of an alkaline and thermostable polygalacturonase from *Bacillus* sp. MG-cp-2 in degumming of ramie (*Boehmeria nivea*) and sunn hemp (*Crotalaria juncea*) bast fibres. Process Biochemistry 36 (8), 803–807.

Kapoor, M., Kapoor, R.K., Kuhad, R.C., 2007. Differential and synergistic effects of xylanase and laccase mediator system (LMS) in bleaching of soda and waste pulps. Journal of Applied Microbiology 103 (2), 305–317.

Kapoor, M., Nair, L.M., Kuhad, R.C., 2008. Cost-effective xylanase production from free and immobilized *Bacillus pumilus* strain MK001 and its application in saccharification of *Prosopis juliflora*. Biochemical Engineering Journal 38 (1), 88–97.

Karimi, K., Emtiazi, G., Taherzadeh, M.J., 2006. Ethanol production from dilute-acid pretreated rice straw by simultaneous saccharification and fermentation with *Mucor indicus, Rhizopus oryzae*, and *Saccharomyces cerevisiae*. Enzyme and Microbial Technology 40, 138–144.

Kaur, S., Dhillon, G.S., Brar, S.K., Chauhan, V.B., 2012. Carbohydrate degrading enzyme production by plant pathogenic mycelia and microsclerotia isolates of *Macrophomina phaseolina* through koji fermentation. Industrial Crops and Products 36 (1), 140–148.

Kennedy, M.J., Thakur, M.S., Wang, D.I.C., Stephanopoulos, G.N., 1992. Estimating cell concentration in the presence of suspended solids: a light scatter technique. Biotechnology and Bioengineering 40 (8), 875–888.

Kerstetter, J.D., Lyons, J.K., 2001. Wheat Straw for Ethanol Production in Washington: A Resource, Technical, and Economic Assessment. Washington State University Cooperative Extension. Retrieved from: http://www.energy.wsu.edu/documents/renewables/WheatstrawForEthanol.pdf.

Khandelwal, P., Vijay, K., Das, N., Tyagi, S.M., 2006. Development of process for preparation of pure and blended kinnow wine without debittering kinnow mandarin juice. Internet Journal of Food Safety 8, 24–29.

Khurana, S., Kapoor, M., Gupta, S., Kuhad, R.C., 2007. Statistical optimization of alkaline xylanase production from *Streptomyces violaceoruber* under submerged fermentation using response surface methodology. Indian Journal of Microbiology 47 (2), 144–152.

Kilzer, F.J., Broido, A., 1965. Speculations on nature of cellulose pyrolysis. Pyrodynamic 2 (2–3), 151.

Knoblich, M., Anderson, B., Latshaw, D., 2005. Analyses of tomato peel and seed byproducts and their use as a source of carotenoids. Journal of the Science of Food and Agriculture 85 (7), 1166–1170.

Kohli, P., Gupta, R., 2015. Alkaline pectinases: a review. Biocatalysis and Agricultural Biotechnology 4 (3), 279–285.

Koliander, B., Hampel, W., Roehr, M., 1984. Indirect estimation of biomass by rapid ribonucleic acid determination. Applied Microbiology and Biotechnology 19 (4), 272–276.

Kosmala, M., Kołodziejczyk, K., Zduńczyk, Z., Juśkiewicz, J., Boros, D., 2011. Chemical composition of natural and polyphenol-free apple pomace and the effect of this dietary ingredient on intestinal fermentation and serum lipid parameters in rats. Journal of Agricultural and Food Chemistry 59 (17), 9177–9185.

Krishna, C., 1999. Production of bacterial cellulases by solid state bioprocessing of banana wastes. Bioresource Technology 69 (3), 231–239.

Kumar, D.P., Singh, R.K., Anupama, P.D., Solanki, M.K., Kumar, S., Srivastava, A.K., et al., 2012. Studies on exo-chitinase production from *Trichoderma asperellum* UTP-16 and its characterization. Indian Journal of Microbiology 52 (3), 388–395.

Kumar, P., Barrett, D.M., Delwiche, M.J., Stroeve, P., 2009. Methods for pretreatment of lignocellulosic biomass for efficient hydrolysis and biofuel production. Industrial and Engineering Chemistry Research 48 (8), 3713–3729.

Lagemaat, J.V.D., Pyle, D.L., 2004. Solid-state fermentation: a continuous process for fungal tannase production. Biotechnology and Bioengineering 87 (7), 924–929.

Laura, D.E., Porcar, M., 2009. Rice straw management: the big waste. Biofuel Bioproducts Biorefining 4 (2), 154–159.

Laufenberg, G., Kunz, B., Nystroem, M., 2003. Transformation of vegetable waste into value added products: (A) the upgrading concept; (B) practical implementations. Bioresource Technology 87 (2), 167–198.

Lim, H.C., Shin, H.S., 2013. Fed-batch Cultures: Principles and Applications of Semi-batch Bioreactors. Cambridge University Press.

Liu, D., Zhang, R., Yang, X., Wu, H., Xu, D., Tang, Z., Shen, Q., 2011. Thermostable cellulase production of *Aspergillus fumigatus* Z5 under solid-state fermentation and its application in degradation of agricultural wastes. International Biodeterioration and Biodegradation 65 (5), 717–725.

Madeira Jr., J.V., Macedo, J.A., Macedo, G.A., 2012. A new process for simultaneous production of tannase and phytase by *Paecilomyces variotii* in solid-state fermentation of orange pomace. Bioprocess and Biosystems Engineering 35 (3), 477–482.

Mahadik, N.D., Puntambekar, U.S., Bastawde, K.B., Khire, J.M., Gokhale, D.V., 2002. Production of acidic lipase by *Aspergillus niger* in solid state fermentation. Process Biochemistry 38 (5), 715–721.

Maity, S., Mallik, S., Basuthakur, R., Gupta, S., 2015. Optimization of solid state fermentation conditions and characterization of thermostable alpha amylase from *Bacillus subtilis* (ATCC 6633). Journal of Bioprocessing and Biotechniques 5, 4.

Malherbe, S., Cloete, T.E., 2002. Lignocellulose biodegradation: fundamentals and applications. Reviews in Environmental Science and Biotechnology 1 (2), 105–114.

Manjit, K.S., Yadav, A., Aggarwal, N.K., Kumar, K., Kumar, A., 2008. Tannase production by *Aspergillus fumigatus* MA under solid-state fermentation. World Journal of Microbiology and Biotechnology 24 (12), 3023–3030.

Martin, A.M. (Ed.), 2012. Bioconversion of Waste Materials to Industrial Products. Springer Science & Business Media.

Martins, E.S., Silva, D., Da Silva, R., Gomes, E., 2002. Solid state production of thermostable pectinases from thermophilic *Thermoascus aurantiacus*. Process Biochemistry 37 (9), 949–954.

Matcham, S.E., Jordan, B.R., Wood, D.A., 1985. Estimation of fungal biomass in a solid substrate by three independent methods. Applied Microbiology and Biotechnology 21 (1–2), 108–112.

Mazutti, M.A., Zabot, G., Boni, G., Skovronski, A., De Oliveira, D., Di Luccio, M., et al., 2010. Kinetics of inulinase production by solid-state fermentation in a packed-bed bioreactor. Food Chemistry 120 (1), 163–173.

McKendry, P., 2002. Energy production from biomass (part 1): overview of biomass. Bioresource Technology 83 (1), 37–46.

Meena, P., Tripathi, A.D., Srivastava, S.K., Jha, A., 2013. Utilization of agro-industrial waste (wheat bran) for alkaline protease production by *Pseudomonas aeruginosa* in SSF using Taguchi (DOE) methodology. Biocatalysis and Agricultural Biotechnology 2 (3), 210–216.

Meien, O.F.V., Luz Jr., L.F., Mitchell, D.A., Pérez-Correa, J.R., Agosin, E., Fernández-Fernández, M., Arcas, J.A., 2004. Control strategies for intermittently mixed, forcefully aerated solid-state fermentation bioreactors based on the analysis of a distributed parameter model. Chemical Engineering Science 59 (21), 4493–4504.

Menon, V., Rao, M., 2012. Trends in bioconversion of lignocellulose: biofuels, platform chemicals & biorefinery concept. Progress in Energy and Combustion Science 38 (4), 522–550.

Meza, J.C., Sigoillot, J.C., Lomascolo, A., Navarro, D., Auria, R., 2006. New process for fungal delignification of sugar-cane bagasse and simultaneous production of laccase in a vapor phase bioreactor. Journal of Agricultural and Food Chemistry 54 (11), 3852–3858.

Montero, A.F., Esparza-Isunza, T., Saucedo-Castañeda, G., Huerta-Ochoa, S., Gutiérrez-Rojas, M., Favela-Torres, E., 2011. Improvement of heat removal in solid-state fermentation tray bioreactors by forced air convection. Journal of Chemical Technology and Biotechnology 86 (10), 1321–1331.

Mosier, N., Wyman, C., Dale, B., Elander, R., Lee, Y.Y., Holtzapple, M., Ladisch, M., 2005. Features of promising technologies for pretreatment of lignocellulosic biomass. Bioresource Technology 96 (6), 673–686.

Muazu, R.I., Stegemann, J.A., 2015. Effects of operating variables on durability of fuel briquettes from rice husks and corn cobs. Fuel Processing Technology 133, 137–145.

Mukherjee, A.K., Adhikari, H., Rai, S.K., 2008. Production of alkaline protease by a thermophilic *Bacillus subtilis* under solid-state fermentation (SSF) condition using *Imperata cylindrica* grass and potato peel as low-cost medium: characterization and application of enzyme in detergent formulation. Biochemical Engineering Journal 39 (2), 353–361.

Murthy, M.R., Thakur, M.S., Karanth, N.G., 1993. Monitoring of biomass in solid state fermentation using light reflectance. Biosensors and Bioelectronics 8 (1), 59–63.

Mussatto, S.I., Fernandes, M., Milagres, A.M., Roberto, I.C., 2008. Effect of hemicellulose and lignin on enzymatic hydrolysis of cellulose from brewer's spent grain. Enzyme and Microbial Technology 43 (2), 124–129.

Naby, M.A., Osman, M.Y., Abdel-Fattah, A.F., 1999. Purification and properties of three cellobiases from *Aspergillus niger* A20. Applied Biochemistry and Biotechnology 76 (1), 33–44.

Nakkeeran, E., Gowthaman, M.K., Umesh-Kumar, S., Subramanian, R., 2012. Techno-economic analysis of processes for *Aspergillus carbonarius* polygalacturonase production. Journal of Bioscience and Bioengineering 113 (5), 634–640.

Nampoothiri, K.M., Baiju, T.V., Sandhya, C., Sabu, A., Szakacs, G., Pandey, A., 2004. Process optimization for antifungal chitinase production by *Trichoderma harzianum*. Process Biochemistry 39 (11), 1583–1590.

Oberoi, H.S., Bansal, S., Dhillon, G.S., 2008. Enhanced β-galactosidase production by supplementing whey with cauliflower waste. International Journal of Food Science & Technology 43 (8), 1499–1504.

Oberoi, H.S., Chavan, Y., Bansal, S., Dhillon, G.S., 2010. Production of cellulases through solid state fermentation using kinnow pulp as a major substrate. Food and Bioprocess Technology 3 (4), 528–536.

Ortiz, P.S., Oliveira, de, S., 2014. Energy analysis of pretreatment processes of bioethanol production based on sugarcane bagasse. Energy 76, 130–138.

Palaniyandi, S.A., Yang, S.H., Suh, J.W., 2014. Cellulase production and saccharification of rice straw by the mutant strain *Hypocrea koningii* RSC1. Journal of Basic Microbiology 54 (1), 56–65.

Pandey, A., 2003. Solid-state fermentation. Biochemical Engineering Journal 13 (2), 81–84.

Panwar, D., Srivastava, P.K., Kapoor, M., 2014. Production, extraction and characterization of alkaline xylanase from *Bacillus* sp. PKD-9 with potential for poultry feed. Biocatalysis and Agricultural Biotechnology 3 (2), 118–125.

Patidar, P., Agrawal, D., Banerjee, T., Patil, S., 2005. Optimisation of process parameters for chitinase production by soil isolates of *Penicillium chrysogenum* under solid substrate fermentation. Process Biochemistry 40 (9), 2962–2967.

Patil, N.S., Jadhav, J.P., 2014. Enzymatic production of N-acetyl-ᴅ-glucosamine by solid state fermentation of chitinase by *Penicillium ochrochloron* MTCC 517 using agricultural residues. International Biodeterioration and Biodegradation 91, 9–17.

Pavithra, S., Ramesh, R., Aarthy, M., Ayyadurai, N., Gowthaman, M.K., Kamini, N.R., 2014. Starchy substrates for production and characterization of *Bacillus subtilis* amylase and its efficacy in detergent and breadmaking formulations. Starch-Stärke 66 (11–12), 976–984.

Philippoussis, A., Diamantopoulou, P., 2011. Agro-food industry wastes and agricultural residues conversion into high value products by mushroom cultivation. In: Proceedings of the 7th International Conference on Mushroom Biology and Mushroom Products (ICMBMP7), France, pp. 4–7.

Pitol, L.O., Biz, A., Mallmann, E., Krieger, N., Mitchell, D.A., 2016. Production of pectinases by solid-state fermentation in a pilot-scale packed-bed bioreactor. Chemical Engineering Journal 283, 1009–1018.

Prakasham, R.S., Rao, C.S., Sarma, P.N., 2006. Green gram husk—an inexpensive substrate for alkaline protease production by *Bacillus* sp. in solid-state fermentation. Bioresource Technology 97 (13), 1449–1454.

Prasad, S., Singh, A., Joshi, H.C., 2007. Ethanol as an alternative fuel from agricultural, industrial and urban residues. Resources, Conservation and Recycling 50 (1), 1–39.

Rabelo, S.C., Carrere, H., Filho, R.M., Costa, A.C., 2011. Production of bioethanol, methane and heat from sugarcane bagasse in a biorefinery concept. Bioresource Technology 102 (17), 7887–7895.

Raimbault, M., 1998. General and microbiological aspects of solid substrate fermentation. Electronic Journal of Biotechnology 1 (3), 26–27.

Rajoka, M.I., Akhtar, M.W., Hanif, A., Khalid, A.M., 2006. Production and characterization of a highly active cellobiase from *Aspergillus niger* grown in solid state fermentation. World Journal of Microbiology and Biotechnology 22 (9), 991–998.

Ramachandran, S., Singh, S.K., Larroche, C., Soccol, C.R., Pandey, A., 2007. Oil cakes and their biotechnological applications - a review. Bioresource Technology 98 (10), 2000–2009.

Ramezanzadeh, F.M., Rao, R.M., Prinyawiwatkul, W., Marshall, W.E., Windhauser, M., 2000. Effects of microwave heat, packaging, and storage temperature on fatty acid and proximate compositions in rice bran. Journal of Agricultural and Food Chemistry 48 (2), 464–467.

Reactor for sterile solid-state fermentation methods. EP. Patent. WO1994/EP 0683815 B1.

Reczey, K., Szengyel, Z.S., Eklund, R., Zacchi, G., 1996. Cellulase production by *T. reesei*. Bioresource Technology 57 (1), 25–30.

Reguant, J., Rinaudo, M., 2000. Fibres lignocellulosiques. En Iniciation á la chimie et á la physico-chimie macromoleculares. Les Polymères Naturels: Structure, Modifications, Applications 13.

Rivas, B., Torrado, A., Torre, P., Converti, A., Domínguez, J.M., 2008. Submerged citric acid fermentation on orange peel autohydrolysate. Journal of Agricultural and Food Chemistry 56 (7), 2380–2387.

Rodriguez, J.A., Mateos, J.C., Nungaray, J., González, V., Bhagnagar, T., Roussos, S., Baratti, J., 2006. Improving lipase production by nutrient source modification using *Rhizopus homothallicus* cultured in solid state fermentation. Process Biochemistry 41 (11), 2264–2269.

Rodríguez-Couto, S., 2008. Exploitation of biological wastes for the production of value-added products under solid-state fermentation conditions. Biotechnology Journal 3 (7), 859–870.

Rodriguez-Gomez, D., Lehmann, L., Schultz-Jensen, N., Bjerre, A.B., Hobley, T.J., 2012. Examining the potential of plasma-assisted pretreated wheat straw for enzyme production by *Trichoderma reesei*. Applied Biochemistry and Biotechnology 166 (8), 2051–2063.

Rodwell, V.W., Kennelly, P.J., 1993. Enzymes: general properties. Harper's Biochemistry 68.

Roopesh, K., Ramachandran, S., Nampoothiri, K.M., Szakacs, G., Pandey, A., 2006. Comparison of phytase production on wheat bran and oilcakes in solid-state fermentation by *Mucor racemosus*. Bioresource Technology 97 (3), 506–511.

Rosales, E., Couto, S.R., Sanromán, M.A., 2007. Increased laccase production by *Trametes hirsuta* grown on ground orange peelings. Enzyme and Microbial Technology 40 (5), 1286–1290.

Rowell, R.M., 1992. Opportunities for Lignocellulosic Materials and Composites. American Chemical Society, pp. 12–27.

Ruiz, H.A., Rodríguez-Jasso, R.M., Rodríguez, R., Contreras-Esquivel, J.C., Aguilar, C.N., 2012. Pectinase production from lemon peel pomace as support and carbon source in solid-state fermentation column-tray bioreactor. Biochemical Engineering Journal 65, 90–95.

Sabu, A., Augur, C., Swati, C., Pandey, A., 2006. Tannase production by *Lactobacillus* sp. ASR-S1 under solid-state fermentation. Process Biochemistry 41 (3), 575–580.

Salgado, J.M., Abrunhosa, L., Venâncio, A., Domínguez, J.M., Belo, I., 2014. Integrated use of residues from olive mill and winery for lipase production by solid state fermentation with *Aspergillus* sp. Applied Biochemistry and Biotechnology 172 (4), 1832–1845.

Samanta, A.K., Senani, S., Kolte, A.P., Sridhar, M., Sampath, K.T., Jayapal, N., Devi, A., 2012. Production and in vitro evaluation of xylooligosaccharides generated from corn cobs. Food and Bioproducts Processing 90 (3), 466–474.

Sanada, C.T., Karp, S.G., Spier, M.R., Portella, A.C., Gouvêa, P.M., Yamaguishi, C.T., et al., 2009. Utilization of soybean vinasse for α–galactosidase production. Food Research International 42 (4), 476–483.

Sandoval, J.P., Amaya-Delgado, L., Mateos-Diaz, J.C., Rodriguez, J., Cordova, J., Alba, A., et al., 2012. Multiplex gas sampler for monitoring respirometry in column-type bioreactors used in solid-state fermentation. Biotechnology and Biotechnological Equipment 26 (3), 3031–3038.

Saratale, G.D., Kshirsagar, S.D., Sampange, V.T., Saratale, R.G., Oh, S.E., Govindwar, S.P., Oh, M.K., 2014. Cellulolytic enzymes production by utilizing agricultural wastes under solid state fermentation and its application for biohydrogen production. Applied Biochemistry and Biotechnology 174 (8), 2801–2817.

Sato, K., Nagatani, M., Sato, S., 1982. A method of supplying moisture to the medium in a solid-state culture with forced aeration. Journal of Fermentation Technology 60 (6), 607–610.

Scholl, A.L., Menegol, D., Pitarelo, A.P., Fontana, R.C., Zandoná Filho, A., Ramos, L.P., et al., 2015. Elephant grass pretreated by steam explosion for inducing secretion of cellulases and xylanases by *Penicillium echinulatum* S1M29 solid-state cultivation. Industrial Crops and Products 77, 97–107.

Selwal, M.K., Selwal, K.K., 2012. High-level tannase production by *Penicillium atramentosum* KM using agro residues under submerged fermentation. Annals of Microbiology 62 (1), 139–148.

Sen, R., Swaminathan, T., 2004. Response surface modeling and optimization to elucidate and analyze the effects of inoculum age and size on surfactin production. Biochemical Engineering Journal 21 (2), 141–148.

Shafizadeh, F., Bradbury, A.G.W., 1979. Thermal degradation of cellulose in air and nitrogen at low temperatures. Journal of Applied Polymer Science 23 (5), 1431–1442.

Sharma, P.D., Fisher, P.J., Webster, J., 1977. Critique of the chitin assay technique for estimation of fungal biomass. Transactions of the British Mycological Society 69 (3), 479–483.

Shen, X., Xia, L., 2004. Production and immobilization of cellobiase from *Aspergillus niger* ZU-07. Process Biochemistry 39 (11), 1363–1367.

Shinners, K.J., Adsit, G.S., Binversie, B.N., Digman, M.F., Muck, R.E., Weimer, P.J., 2007. Single-pass, split-stream harvest of corn grain and stover. Transactions of the American Society of Agricultural and Biological Engineers 50 (2), 355–363.

Shukla, J., Kar, R., 2006. Potato peel as a solid state substrate for thermostable α-amylase production by thermophilic Bacillus isolates. World Journal of Microbiology and Biotechnology 22 (5), 417–422.

Silveira, C.L., Mazutti, M.A., Salau, N.P., 2014. Modeling the microbial growth and temperature profile in a fixed-bed bioreactor. Bioprocess and Biosystems Engineering 37 (10), 1945–1954.

Singhania, R.R., Sukumaran, R.K., Patel, A.K., Larroche, C., Pandey, A., 2010. Advancement and comparative profiles in the production technologies using solid-state and submerged fermentation for microbial cellulases. Enzyme and Microbial Technology 46 (7), 541–549.

Sivers, M.V., Zacchi, G., 1995. A techno-economical comparison of three processes for the production of ethanol from pine. Bioresource Technology 51 (1), 43–52.

Solid state fermentation. USA, Patent. US 2003/6664095 B1.

Srirangarajan, A.N., Shrikhande, A.J., 1977. Characterization of mango peel pectin. Journal of Food Science 42 (1), 279–280.

Srivastava, P.K., Kapoor, M., 2013. Extracellular endo-mannanase from Bacillus sp. CFR 1601: economical production using response surface methodology and downstream processing using aqueous two phase system. Food and Bioproducts Processing 91 (4), 672–681.

Srivastava, P.K., Kapoor, M., 2014. Cost-effective endo-mannanase from Bacillus sp. CFR 1601 and its application in generation of oligosaccharides from guar gum and as detergent additive. Preparative Biochemistry and Biotechnology 44 (4), 392–417.

Srivastava, P.K., Kapoor, M., 2015. Recombinant GH-26 endo-mannanase from Bacillus sp. CFR 1601: biochemical characterization and application in preparation of partially hydrolysed guar gum. LWT-Food Science and Technology 64 (2), 809–816.

Stewart, D., Azzini, A., Hall, A.T., Morrison, I.M., 1997. Sisal fibres and their constituent non-cellulosic polymers. Industrial Crops and Products 6 (1), 17–26.

Sun, X., Liu, Z., Qu, Y., Li, X., 2008. The effects of wheat bran composition on the production of biomass-hydrolyzing enzymes by Penicillium decumbens. Applied Biochemistry & Biotechnology 146 (1), 119–128.

Tan, Y., Wang, Z.X., Marshall, K.C., 1998. Modeling pH effects on microbial growth: a statistical thermodynamic approach. Biotechnology and Bioengineering 59 (6), 724–731.

Tepe, O., Dursun, A.Y., 2014. Exo-pectinase production by Bacillus pumilus using different agricultural wastes and optimizing of medium components using response surface methodology. Environmental Science and Pollution Research 21 (16), 9911–9920.

Thanapimmetha, A., Luadsongkram, A., Titapiwatanakun, B., Srinophakun, P., 2012. Value added waste of Jatropha curcas residue: optimization of protease production in solid state fermentation by Taguchi DOE methodology. Industrial Crops and Products 37 (1), 1–5.

Thomas, L., Larroche, C., Pandey, A., 2013. Current developments in solid-state fermentation. Biochemical Engineering Journal 81, 146–161.

Thring, R.W., Chornet, E., Overend, R.P., 1990. Recovery of a solvolytic lignin: effects of spent liquor/acid volume ratio, acid concentration and temperature. Biomass 23 (4), 289–305.

Tsao, G.T., Xia, L., Cao, N., Gong, C.S., 2000. Solid-state fermentation with Aspergillus niger for cellobiase production. In: Twenty-first Symposium on Biotechnology for Fuels and Chemicals. Humana Press, pp. 743–749.

Ucar, G., Fengel, D., 1988. Characterization of the acid pretreatement for the enzymatic hydrolysis of wood. Holzforschung-International Journal of the Biology, Chemistry, Physics and Technology of Wood 42 (3), 141–148.

Uyar, F., Baysal, Z., 2004. Production and optimization of process parameters for alkaline protease production by a newly isolated Bacillus sp. under solid state fermentation. Process Biochemistry 39 (12), 1893–1898.

Vaidyanathan, S., Macaloney, G., Vaughan, J., McNeil, B., Harvey, L.M., 1999. Monitoring of submerged bioprocesses. Critical Reviews in Biotechnology 19 (4), 277–316.

Varzakas, T.H., Roussos, S., Arvanitoyannis, I.S., 2008. Glucoamylases production of Aspergillus niger in solid state fermentation using a continuous counter-current reactor. International Journal of Food Science & Technology 43 (7), 1159–1168.

Vaseghi, Z., Najafpour, G.D., Mohseni, S., Mahjoub, S., 2013. Production of active lipase by Rhizopus oryzae from sugarcane bagasse: solid state fermentation in a tray bioreactor. International Journal of Food Science & Technology 48 (2), 283–289.

Velmurugan, B., Ramanujam, R.A., 2011. Anaerobic digestion of vegetable wastes for biogas production in a fed-batch reactor. International Journal of Emerging Sciences 1 (3), 478–486.

Vendruscolo, F., Albuquerque, P.M., Streit, F., Esposito, E., Ninow, J.L., 2008. Apple pomace: a versatile substrate for biotechnological applications. Critical Reviews in Biotechnology 28 (1), 1–12.

Vijayaraghavan, P., Vincent, S.P., Dhillon, G.S., 2015. Solid-substrate bioprocessing of cow dung for the production of carboxymethyl cellulase by *Bacillus halodurans* IND18. Waste Management.

Voragen, A.G., Coenen, G.J., Verhoef, R.P., Schols, H.A., 2009. Pectin, a versatile polysaccharide present in plant cell walls. Structural Chemistry 20 (2), 263–275.

Wang, F., Ni, H., Cai, H.N., Xiao, A.F., 2013. Tea stalks–a novel agro-residue for the production of tannase under solid state fermentation by *Aspergillus niger* JMU-TS528. Annals of Microbiology 63 (3), 897–904.

Wilkins, M.R., Suryawati, L., Maness, N.O., Chrz, D., 2007. Ethanol production by *Saccharomyces cerevisiae* and *Kluyveromyces marxianus* in the presence of orange-peel oil. World Journal of Microbiology and Biotechnology 23 (8), 1161–1168.

Wong, Y.P., Saw, H.Y., Janaun, J., Krishnaiah, K., Prabhakar, A., 2011. Solid-state fermentation of palm kernel cake with *Aspergillus flavus* in laterally aerated moving bed bioreactor. Applied Biochemistry and Biotechnology 164 (2), 170–182.

Xiang, Q., Lee, Y.Y., Pettersson, P.O., Torget, R.W., 2003. Heterogeneous aspects of acid hydrolysis of α-cellulose. Applied Biochemistry and Biotechnology 105–108.

Xie, X.S., Cui, S.W., Li, W., Tsao, R., 2008. Isolation and characterization of wheat bran starch. Food Research International 41 (9), 882–887.

Yadav, J.S., 1988. SSF of wheat straw with alcaliphilic *Coprinus*. Biotechnology and Bioengineering 31 (5), 414–417.

Yang, S.Q., Yan, Q.J., Jiang, Z.Q., Li, L.T., Tian, H.M., Wang, Y.Z., 2006. High-level of xylanase production by the thermophilic *Paecilomyces themophila* J18 on wheat straw in solid-state fermentation. Bioresource Technology 97 (15), 1794–1800.

Yingyi, D., Wang, L., Chen, H., 2012. Digital image analysis and fractal-based kinetic modelling for fungal biomass determination in solid-state fermentation. Biochemical Engineering Journal 67, 60–67.

Zheng, J., Zhao, W., Guo, N., Lin, F., Tian, J., Wu, L., Zhou, H., 2012. Development of an industrial medium and a novel fed-batch strategy for high-level expression of recombinant β-mannanase by *Pichia pastoris*. Bioresource Technology 118, 257–264.

Zhu, S., Chen, G., 2015. Numerical solution of a microbial growth model applied to dynamic environments. Journal of Microbiological Methods 112, 76–82.

Zhuang, J., Marchant, M.A., Nokes, S.E., Strobel, H.J., 2007. Economic analysis of cellulase production methods for bio-ethanol. Applied Engineering in Agriculture 23 (5), 679–687.

Zúñiga, U.F.R., Couri, S., Neto, V.B., Crestana, S., Farinas, C.S., 2013. Integrated strategies to enhance cellulolytic enzyme production using an instrumented bioreactor for solid-state fermentation of sugarcane bagasse. BioEnergy Research 6 (1), 142–152.

FURTHER READING

Livestock industry, oil extractors hail decision to lift customs duty on oil cake, oilmeal. Budget2014. Retrieved from. http://articles.economictimes.indiatimes.com/2014-07-10/news/51301183_1_edible-oil-oil-year-by-mehta.

CHAPTER 4

Industrial Enzymes: Recovery and Purification Challenges

P. Vijayaraghavan[1], S.R.F. Raj[2], S.G.P. Vincent[1]

[1]Manonmaniam Sundaranar University, Kanyakumari, India; [2]Nesamony Memorial Christian College, Kanyakumari, India

MAJOR INDUSTRIAL ENZYMES

Enzymes such as proteases, amylases, cellulases, lipases, glucanases, β-galactosidase, pectin lyase, and ureases are widely used in industry. These enzymes catalyze biological reactions by forming transition state complexes with their substrate, which reduces the activation energy of the reaction. Enzymes are useful in various sectors, such as manufacturing of food and feedstuffs, cosmetics, and medicinal products and are also used as a tool for research and development. Proteases and cellulases are applied widely in the detergent industry, pulp and paper applications, the textile industry, the leather industry, and for biofuel production. Therapeutic enzymes are used in wound healing, lysis of vein thrombosis, and acute therapy of myocardial infarction. Among the 3000 different known enzymes, approximately 150–170 enzymes are used commercially. The global market for industrial enzymes is expected to reach $4.4 billion by 2015; the highest sale of industrial enzymes occurred in the leather processing industry, followed by bioethanol production (Binod et al., 2013).

ENZYME PURIFICATION: A TRIAL AND ERROR STRATEGY

The purification of a specific enzyme is generally difficult but an achievable task. Enzyme purification involves some basic rules, practices, and protocols. The goals of enzyme purification can be achieved by trial-and-error method. Bacterial cell culture extract or supernatant is generally used as the sample source for extracellular enzymes. Many cell disruption techniques have been developed for the liberation of enzymes from the bacterial cells. For extracting intracellular enzymes, the cell wall of the microbe is initially disrupted by chemical methods (detergent, alkali), enzymatic methods (lysozyme), structural damage (osmotic shock, freeze and thaw), and mechanical methods (shear, grinding, or sonication). The clear supernatant of centrifuged cell wall–disrupted samples is used as a crude enzyme. At first, the enzyme liberated from its source should be in an active form. The enzyme to be purified usually forms only a small fraction of the total protein

Agro-Industrial Wastes as Feedstock for Enzyme Production
ISBN 978-0-12-802392-1
http://dx.doi.org/10.1016/B978-0-12-802392-1.00004-6

95

in the crude extract. There are two main important steps involved in purifying enzymes. The foremost step is to eliminate the nonprotein contaminants, followed by the isolation of enzyme from other sources. It is easier to remove the contaminants than to isolate the enzyme. The enzyme protein content is generally 0.1% of the total protein of the crude extract. To purify this small quantity of enzyme to homogeneity, 99.9% of the other protein should be removed from the protein pool with little enzyme activity loss. This can be a difficult task because enzymes are easily denatured by heating, foaming, inactivated by organic solvents (particularly at high temperature), by drying at room temperature, and by concentrated acids or bases. The aqueous solutions containing enzymes are generally prone to microbial growth. Hence, aseptic conditions of equipment and avoidance of contamination are vital for successful purification of enzymes. A single purification step is not adequate to purify an enzyme completely. There are several different procedures that are used widely for enzyme purification. The key to success for the enzyme purification process is the selection of the most appropriate purification step. Generally, either salt precipitation (ammonium sulfate precipitation), organic solvent precipitation (acetone or ethanol), or isoelectric precipitation (polyethylene glycol) are used in an early stage of enzyme purification. Chromatographic procedures, such as ion exchange, gel filtration, adsorption chromatography, or affinity chromatography are employed after the enzyme has been partially purified by one of the herein mentioned precipitation techniques. Although, many protocols have been proposed for the purification of enzymes, the establishment of an enzyme purification procedure for the particular source must be done largely by trial and error.

PURIFICATION OF ENZYMES LARGELY DEPENDS ON ITS APPLICATION

The purity of an enzyme required depends on the purpose for which the enzyme is needed. For an enzyme that is to be used in the detergent industry (eg, alkaline protease), relatively partially purified enzyme or crude enzyme is sufficient. However, if the enzyme is aimed for therapeutic use (streptokinase or urokinase) it must be extremely pure (>99.9% purity) and purification must be done in several subsequent steps, including fractionation and chromatography. Urease is generally used to determine the urea content of the biological samples. The commercial production of urease is from the Jack bean meal and extracted using acetone (32%, v/v) (Sumner, 1926). The raw material was mixed with acetone (32%, v/v), and this procedure was carried out under ice cold conditions. The mixture was then subjected to vacuum filtration and stored at 2°C overnight for crystallization. Further, the urease crystal formed was centrifuged at $10,000 \times g$ for 20 min under ice cold conditions and recovered. This enzyme was assayed using Nessler's method (Sumner, 1955) and lyophilized. For diagnostic purposes, this single-step crystallization is generally followed by addition of stabilizers to the enzyme to maintain the activity in liquid state. Therapeutic enzymes generally require >99% purity for administration by injection (Sofer and Hagel, 1997). Streptokinase is widely used to treat cardiovascular diseases. This therapeutic enzyme requires ultrapurification. Ammonium sulfate was initially used to fractionate streptokinase, followed by gel permeation column chromatography, diethylaminoethyl (DEAE)-cellulose column chromatography (Banerjee et al., 2004). Affinity chromatography has also been used for the purification of streptokinase (Liu et al., 1999). The bacterial fibrinolytic enzymes dissolve the fibrin net of blood clot and may have wide application in the treatment of cardiovascular diseases. Recently, bacterial fibrinolytic enzyme was purified by the combination of ammonium sulfate precipitation, ion exchange, gel permeation, and affinity chromatography (Vijayaraghavan and Vincent, 2014). The important task of enzyme purification process is not only to remove the contaminants but also to concentrate the desired enzyme from the protein pool. Procedures and conditions used in the purification process of one enzyme may differ from another.

CHALLENGES IN ENZYME PURIFICATION

There are no set protocols for the purification of different enzymes. The purification protocol or the sequence will vary according to the source and the properties of the enzyme. Generally, more than one step (salt or solvent precipitation and chromatography procedures) of purification is followed for the purification of enzymes. In each step of enzyme purification, the level or fold of purification is generally increased. It is very important to minimize enzyme activity losses throughout the purification process; therefore the use of minimal steps during the purification is important as losses can occur in every step of enzyme purification. Enzyme purification strategy can be made more efficient if the properties, such as optimal pH, temperature, isoelectric point, and molecular weight of the enzyme are known. In order to fully characterize the biochemical properties of an enzyme, it should be free from contaminants, such as nucleic acid, cell membrane components, or other proteins. In enzyme purification process the parameters, such as isoelectric point, molecular mass of the protein (Andrews et al., 1994), and aqueous two-phase system (Gu and Glatz, 2007) for protein hydrophobicity characterization are critically important.

The primary objective of enzyme purification is to remove the contaminants from the crude enzyme extract. These contaminants may interfere with subsequent enzyme purification steps or with enzyme assays. The purity of enzyme can be monitored in each step of enzyme purification by the presence of a single band on polyacrylamide gel electrophoresis (PAGE) or a single peak in a high performance liquid chromatography (HPLC) analysis. Fig. 4.1 shows the native PAGE separated protease from *Bacillus* sp. This purification was achieved after the crude enzyme was fractionated with solid ammonium sulfate (30–80% saturation), dialysis, followed by Sephadex G-75 gel-filtration chromatography. The fractions obtained from gel filtration column were subjected to protease assay, and the active fractions were analyzed for homogeneity on native PAGE.

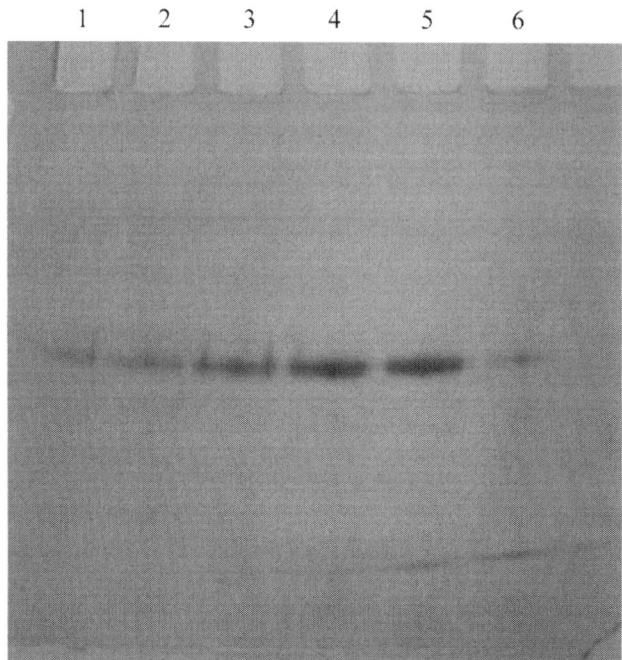

Figure 4.1 Sodium dodecyl sulfate polyacrylamide gel (12%) shows the homogeneity of the enzyme. The gel was stained with Coomassie and then destained. All lanes are the enzyme fractions obtained from Sephadex G-75 gel-filtration chromatography.

This gel was stained with Coomassie Brilliant Blue and showed homogeneity of enzyme. Analysis of protein on PAGE is simple and more cost-effective than HPLC analysis.

Ideally, one never achieves 100% purity of enzyme or protein. After chromatographic methods of enzyme purification, the purity of the enzyme may vary between 90% and 99.99%. If there is a single band on an SDS PAGE or polyacrylamide gel, we cannot actually claim it as a pure enzyme (100% purity). A single protein band detected on the PAGE simply means that there are no other proteins of equivalent abundance present in that fraction. The enzyme purification process is technically challenging because the enzymes are often present at very low concentrations, with many impurities, and similarity of the properties of product and contaminants, poorly characterized properties with respect to physiochemical characters and thermodynamic properties, flow properties, enzyme denaturation, thermoliability, and pH sensitivity etc. Protein estimation is prerequested in all stages of enzyme purification.

MONITORING THE PROGRESS OF ENZYME PURIFICATION

Batch manufacturing record (BMR) will help to monitor the progress of the entire enzyme purification. BMR also examines the quality of the chemicals and solvents used in enzyme purification. The chemicals, solvents, and quality of distilled water, and sophisticated instruments are also equally important in enzyme purification process to achieve the goal. The enzyme assay and total protein estimation should be carried out in each step of enzyme purification. The total protein content of the sample is to be expressed in terms of mg protein/ml sample. Enzyme activity of the sample is expressed as units of enzyme activity/ml. Specific activity (units of activity/mg protein) measures the purity of an enzyme and generally increases in each step of enzyme purification.

$$\text{Specific activity} = \frac{\text{Total activity of enzyme}}{\text{Total protein}}$$

Decreased specific activity may be due to the poorly designed protocol. In this case, it is necessary to redesign the enzyme purification protocol (precipitation with other salts/solvents, changing chromatography types, etc.). The percentage yield measures the activity retained after each step of enzyme purification. Enzyme purification fold is also usually increased. But it should be noted that it is not mandatory to increase the specific activity of enzyme and fold purification. This decreased specific activity is one of the reasons for enzyme instability. Enzyme instability may be due to the loss of an activating cofactor or inactivation of enzyme (proteolysis being most likely). Some enzymes (eg, cellulases) needed Mg^{2+}, Ca^{2+}, or other divalent metal ions for optimal activity. These divalent ions are leached out during the purification steps, depending on the buffer used for enzyme purification. The following diagram summarizes the overview of enzyme purification (Fig. 4.2). The purification fold is considered as one in the first step of enzyme purification, and the yield is generally assumed as 100%.

IMPORTANT STEPS IN ENZYME PURIFICATION

Several hundred proteins have been purified in their active form on the basis of various characteristics, such as size, solubility, charge, and affinity. Usually, protein mixtures are subjected to a series of separation steps to yield a pure protein. At each step in the enzyme purification, the enzyme activity is assayed and the protein concentration is determined. Substantial quantities of purified enzymes, generally in milligrams, are needed to purify a single protein. A variety of enzyme purification techniques are available. Most proteins are less soluble at high salt concentrations, an effect called "salting out." The salt concentration at which a protein precipitates differs for one protein to another. Hence, salting out can be used to fractionate proteins. This is the initial step of enzyme purification and is followed by dialysis. Proteins can be separated from small molecules by dialysis through a semipermeable membrane, such as a cellulose membrane with pores. This technique is useful for removing a salt or other small molecule, but it will not distinguish between proteins effectively. The dialyzed sample is applied on ion exchange or gel-filtration chromatography or followed by any other chromatography to achieve higher degree of purification. Proteins can be separated on the basis of their net

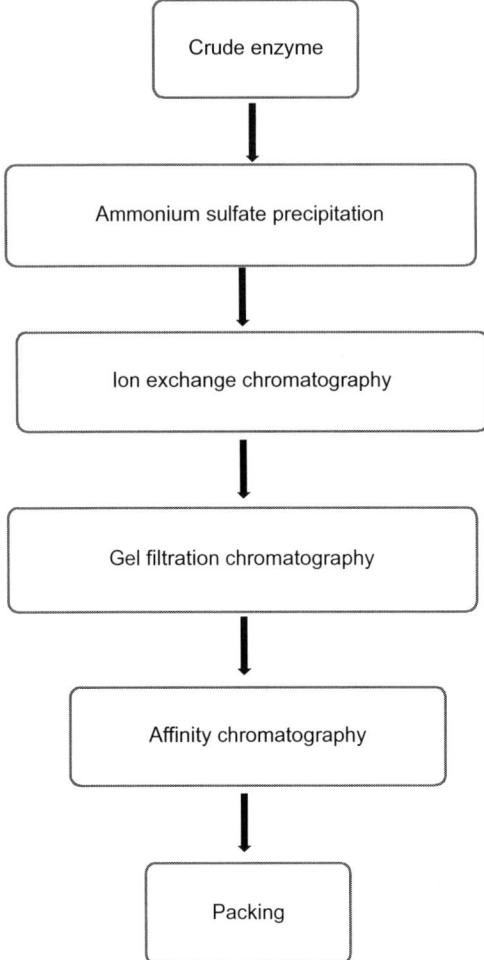

Figure 4.2 Overview of enzyme purification.

charge by ion-exchange chromatography (IEC). Hydrophobic interaction chromatography (HIC) is an excellent complement to size exclusion and IEC during difficult separations, particularly those in which the impurities are of similar isoelectric point or molecular weight. Further, proteins and other molecules with hydrophobic surface properties are attracted to the relatively mild hydrophobic surface of an HIC resin under high-salt aqueous conditions but are released with decreasing salt concentration. Affinity chromatography is another powerful and generally applicable means of purifying proteins. For example, casein–agarose affinity matrix is widely used for the purification of proteases. Table 4.1 summarizes the various methods of enzyme purification based on their characteristic features.

Table 4.1 Important Methods of Downstream Processes and Their Working Principle for Enzyme Purification

Methods	Working Principles
Precipitation	Solubility
Ion-exchange chromatography	Ionic charge
Hydrophobic-interaction chromatography	Hydrophobicity
Gel-filtration chromatography	Molecular sieve
Affinity chromatography	Affinity constant

DETERMINATION OF ENZYME ACTIVITY BY SIMPLE METHODS

Simple enzyme assays are preferred for the continuous monitoring of enzyme during purification. Enzyme assay is usually carried out through the direct or indirect determination of the rate of digestion of the specific substrate, at different time intervals, after a given period of time. Generally, different methods are followed for the enzyme assays. The dinitro salicylate (DNS) method detects the reducing sugars liberated by the action of hydrolase enzymes on carbohydrates, under specific pH and temperatures (Bailey, 1988). Based on the source of enzyme, the pH and temperature of enzyme assay parameter vary. A second method for the determination of reducing sugar is based on the use of colored substrates, such as Congo Red (Wood, 1981). A third method refers to the use of radial diffusion techniques on agar gels. Chemically colored polysaccharides are added to the agar gel, and the enzyme to be assayed is dropped on little wells made on the agar. Enzyme activity can be followed visually as when the enzyme acts, concentric discoloration rings are formed around the wells, and these can be measured directly correlating to the concentration and activity of the enzyme. Casein is generally used as the substrate for the determination of protease activity (Chopra and Mathur, 1985). Gelatin, collagen, and casein derivatives covalently linked with Remazol Brilliant Blue R (RBB) are also used for protease assay (Wolf and Wirth, 1996). Recently, bromocresol green (BCG)-based radial diffusion technique on agar gel was proposed for protease assay (Vijayaraghavan and Vincent, 2013). A carboxymethyl-substituted soluble chitin and a reprecipitated colloidal chitin derivative were covalently linked with Remazol Brilliant Violet 5R and was used for chitinase assay (Wirth and Wolf, 1990). A novel high-throughput colorimetric urease activity assay was compared to the Nessler method (Okyay and Rodrigues, 2013). Some of the enzyme assay procedure is time consuming. For example, the typical protease assay procedure takes more than 2 h. In this case, casein–agar plates may be the alternative to assay the large number of fractions. Usually 1% (w/v) casein incorporated in agar medium is used to make these plates. Different fractions were placed on each well cut on the casein agar plates and incubated at room temperature for 2 h. This kind of plate-screening technique helps to handle a large number of fractions or samples within a short period (Fig. 4.3). This plate assay facilitates the identification of active column fractions.

FRACTIONATION OF ENZYMES WITH AMMONIUM SULFATE AND OTHER SOLVENTS

Ammonium sulfate is an ideal salt for the fractionation of enzymes owing to its high solubility, low toxicity to enzymes, and low cost and serves as a preservative. As the saturation of the salt increases, different enzyme proteins precipitate from solution and can be recovered. In order to ensure an efficient precipitation, a minimal concentration of 1.0 mg/ml of protein is required. Hence, measuring protein content of the supernatant or crude extract is important before proceeding. Large concentrations of salts (10%, 20%, 30%, 40%, 50%, 60%, 70%, 80%, 90%, and 100% saturation levels) are generally used to precipitate enzymes from crude extract. Ammonium sulfate is the salt of choice because it can be used in even lower concentration than other salts and it preserves the biological activity of enzymes. Ammonium sulfate is not a preferred precipitating agent for some alkaline proteases. Acetone was widely used to precipitate alkaline proteases at various concentrations (Horikoshi, 1971; Yamagata et al., 1995; Kim et al., 1996). Enzyme precipitation can also be achieved by the use of water-soluble neutral polymers, such as polyethylene glycol (Larcher et al., 1996).

ENZYME INACTIVATION AND CONTROL

An enzyme purification scheme is generally considered successful if the resulting enzyme is pure and active. Therefore, one of the key considerations in working with an enzyme is to prevent it from inactivation. Proteolytic degradation could be avoided by working at low temperature (<4°C) and working quickly. The typical enzyme purification procedure should be completed within one week. Avoiding vigorous

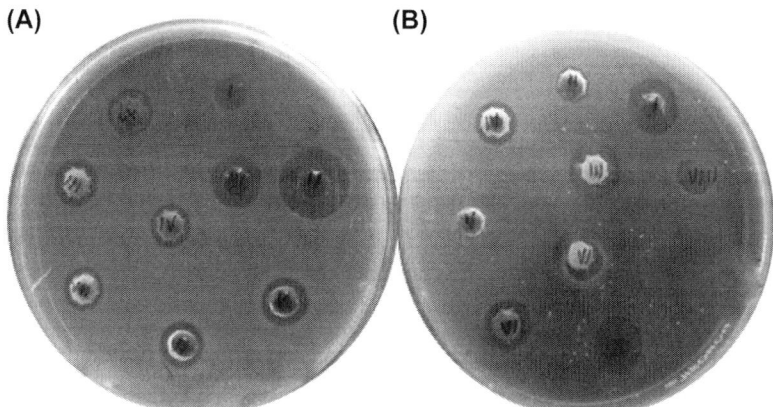

(A)   **(B)**

Figure 4.3 Simple protease assay using casein-agar plate method. Fractions were loaded into the casein-agar plate wells and incubated at room temperature for >2 h. A halo zone around the well indicated protease activity.

stirring or undue exposure to oxygen and addition of reducing agent could prevent oxidation. Generally, physiological buffers used to maintain pH and a chelating agent like ethylenediaminetetraacetic acid help to protect activity loss against heavy-metal ions. Addition of glycerol (5%, v/v) stabilizes most proteins and reduces adsorption to the walls of the container. Repeat freezing and thawing should be avoided. Buffer exchange should be done slowly. Dialysis and gel-filtration chromatography are the preferred choice for buffer exchange. The enzyme activity is continuously monitored by enzyme assay and zymographic analysis. Zymography helps to compare the enzyme activity of fractions obtained from all stages of enzyme purification. Generally, SDS-PAGE gel copolymerized with substrate was used to determine the enzyme activity. For protease activity determination, casein (1%, w/v) was copolymerized with SDS-PAGE (12%). After electrophoresis, the purified protease showed more protease activity on SDS-PAGE (Fig. 4.4). In the next well, the denatured enzyme was loaded (sample kept at 37°C for one week). This resulted in loss of enzyme activity, which can be prevented by storing the enzyme at lower temperatures (<4°C).

Figure 4.4 Zymographic analysis of protease from *Bacillus* sp.: 0.1% casein was incorporated with SDS-PAGE and co-polymerized; 10 µg protein was loaded and run at 4°C for 4 h. The gel was further rinsed with buffer (sodium phosphate buffer, pH 7.4, 0.1 M) for 30 min and incubated with Triton X-100 (2.5%) for 1 h. The renatured enzyme was further incubated with buffer for 4 h and stained with Coomassie Brilliant Blue R-250 and destained. The proteolytic activity appeared as a colorless band. Lane 1: enzyme activity of the purified enzyme; Lane 2: enzyme activity of the denatured enzyme.

CHROMATOGRAPHIC STEPS

Chromatographic method of enzyme purification is often employed after the fractionation of enzyme by solvent precipitation or salting out procedure. A single chromatographic step is not sufficient to get the required level of purity. Hence, combinations of chromatographic steps are followed for effective enzyme purification. IEC is the most common chromatographic method for enzyme purification. It is often effective at earlier stages in the fractionation of enzymes. The enzyme solution is added to an insoluble polymer, eg, cellulose, containing ionic groups whose charge will determine the types of mobile ion (cation or anion) it attracts. Proteins whose net charge is opposite to that of the ion exchange material will bind and be displaced from solution. A subsequent change in pH or an introduction of salt can alter the electrostatic forces, allowing the protein to be released again into solution. The most frequently employed ion exchangers are the DEAE group in anion exchange and the carboxymethyl (CM) in cation exchange. Gel filtration is the second most frequently used purification method for some of the industrial enzymes. Affinity chromatography has been used to purify the therapeutic enzymes or analytic enzymes. HIC has also been used in some cases with the most popular hydrophobic adsorbents being octyl or phenyl functional groups. Heparin and concanavalin A legends (Farooqui et al., 1994) are employed for the purification of fungal, mammalian lipases (Tombs and Blake, 1982; Aires–Barros and Cabral, 1991). In adsorption chromatography, the adsorbent hydroxyapatite is generally used for enzyme purification (Taipa et al., 1992). Although gel filtration has the lower capacity for loaded protein, it can be used at an early stage in the purification or as one of the last steps for fine polishing of the protocol. Affinity methods can be applied at an early stage, but as the materials are expensive, the less costly ion exchange and gel filtration are usually preferred after the precipitation step.

Lipases have been extensively purified to homogeneity and crystallized. Purification of lipases generally depended on nonspecific techniques, such as ammonium sulfate precipitation, gel filtration, hydrophobic-interaction chromatography, and IEC. Affinity chromatography was used in some cases to reduce the number of individual purification steps needed (Woolley and Peterson, 1994). Ion exchange and gel filtration is used to purify a lipase enzyme from *Pseudomonas putida* 3SK (Lee and Rhee, 1993). Most commercial applications of lipases do not require highly pure enzyme. Excessive purification is expensive and reduces overall recovery of the enzyme (Chisti, 1998). Alkaline proteases are not generally bound to anion exchangers because these enzymes are generally positively charged (Fujiwara et al., 1993). However, cation exchangers are widely used for the purification of alkaline protease, and the bounded proteins were eluted from the column by increasing NaCl or KCl concentration or pH gradient (Tsuchiya et al., 1992). Size-exclusion chromatography is preferable for the purification of alkaline proteases. Casein–agarose affinity chromatography also has been used for the purification of protease from *Bacillus* sp. (Kobayashi et al., 1996). Protease was purified from *Bacillus* sp. by the combination of salting out and chromatographic procedure. All purification steps were carried out at 4°C. The crude enzyme was fractionated with ammonium sulfate (30–80% saturation) and kept at 4°C for 4 h. The precipitate was collected by centrifugation and dissolved in minimal volume of double-distilled water and dialyzed. The highly active fraction was subjected to DEAE cellulose column (1.5 × 10 cm) that was previously equilibrated with 50 mM Tris–HCl buffer (pH 7.8). The adsorbed proteins were eluted with a linear gradient of sodium chloride solution in a range of 0–0.75 M. Fractions with high-protease activity were pooled and concentrated by ultrafiltration using a 10 kDa MW cutoff membrane. The concentrated sample was further applied on preequilibrated Sephadex G-75 gel-filtration column (0.6 × 50 cm) and eluted with 50 mM Tris–HCl buffer (pH 7.8). The active fractions were pooled and loaded on casein–agarose affinity column (1 × 5 cm). Bounded proteins were eluted with a linear gradient of sodium chloride (0–1.0 M) in Tris buffer (pH 7.8). Active fractions of protease were subjected to determine the molecular weight using SDS-PAGE. The molecular weight of the purified protease was found to be 43 kDa (Fig. 4.5).

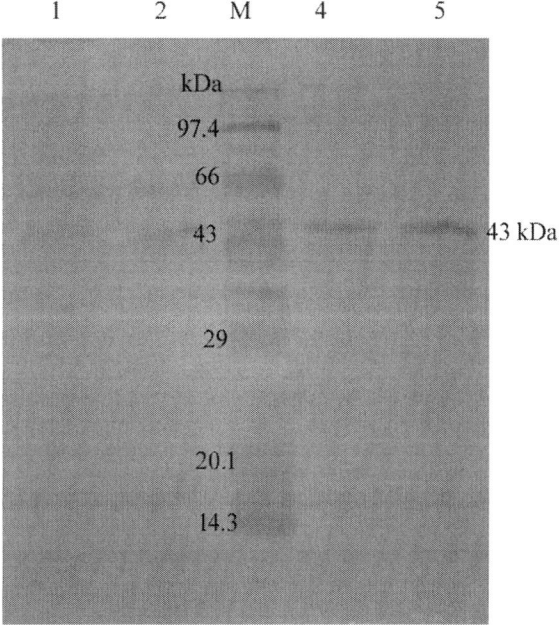

Figure 4.5 SDS-PAGE separated protease from *Bacillus* sp. Lane 1: crude protease; Lane 2: ion-exchanged fraction; Lane M: protein marker; Lane 4: Sephadex G-75 gel-filtrated fraction; and Lane 5: affinity fraction.

ULTRAFILTRATION

Ultrafiltration (UF) is one of the membrane separation processes and is widely used for the recovery of enzymes (Bohdziewicz, 1996). UF separation process is less inexpensive than other molecules and it also prevents the loss of enzyme activity. UF is widely used to purify and concentrate enzymes, such as proteases. The main disadvantage of UF is the membrane clogging due to the precipitates formed by the final product. This clogging can be eliminated by treatment with proteases, acids, alkalines, or with detergent (Kumar and Takagi, 1999). Cross-flow membrane filtration has been used in the downstream processing of lipases, namely, for microbial cell removal and concentration of the supernatant-containing lipases. The possibility of *Pseudomonas fluorescens* lipase purification by using ultrafiltration capillary membranes was studied by Sztajer and Bryjak (1989). Membrane filtration is useful to separate species in aqueous solution by shape, size, and charge. Repeated or reconcentrations are used to exchange solvent or to remove salt (Schratter, 2004). An extracellular protease was purified from *Penicillium chrysogenum* by various chromatographic steps and concentrated by UF. After purification of enzyme by UF, the active enzyme fraction was observed in the cutoff range between 8 and 50 kDa. The molecular weight of the enzyme was detected using 12.5% SDS-PAGE (Benito et al., 2002). Recently, *Bacillus licheniformis* alkaline protease was also concentrated by UF. Protease activity was concentrated to 4.5 times the initial culture supernatant. This recovered alkaline thermostable protease by UF can be directly used in the detergent industry (Bezawada et al., 2011).

AUTOMATION OF PROTEIN PURIFICATION

Several automated protocols have been developed recently for the purification of proteins. The automated protocols are mainly based on affinity purification approach (Steen et al., 2006) or purification with the use of dual column protocol including affinity chromatography (Sigrell et al., 2003; Bhikhabhai et al., 2005). Recently, Camper and Viola (2009) purified protein using fully automated protein purification system. It carried out all operations automatically including sample loading, column washing, sample elution, and peak collection steps for ion exchange, metal affinity, hydrophilic interaction, and gel-filtration chromatography. A combined approach was carried out to improve large-scale production of tobacco Etch virus protease by a fully automated two-step purification protocol (Blommel and Fox, 2007). The algorithmic methods of enzyme purification involve the application of mathematical models and computer programming to predict, optimize, and scale up enzyme purification process. These kinds of robustness studies will ensure consistency and accuracy. In algorithmic or model-based methods, the number of experiments is kept to the minimum. Nowadays, the laboratory-scale experiments have been replaced by high-throughput experimentation (HTE), mainly high-throughput screening. HTE miniaturizes the experiments and reduces experimental costs and time (Bensch et al., 2005; Wiendah et al., 2008). The algorithmic or model-based approaches are not well suited for the purification of commercial enzymes. Recently, a model-based hybrid approach was proposed for protein purification process by Nfor et al. (2008). HTE is highly useful for the fractionation and characterization of crude enzyme mixture and to study the thermodynamic properties and physiochemical properties of the target protein and other protein contaminants.

To achieve a high degree of purity in the purification of enzymes for therapeutic or analytical application, several chromatographic steps are required. There are a range of techniques available including anion and cation exchange, gel filtration, HIC, and affinity chromatography. Two algorithms were developed for effective purification of enzymes and proteins. The first algorithm IS used to select the most effective purification method to separate a protein from a mixture based on its physiochemical properties. The second algorithm is used to predict the concentration and number of contaminants in each step of enzyme or protein purification (Asenjo and Andrews, 2009). Vasquez-Alvarez et al. (2001) developed a mathematical model–based optimization approach for the purification of protein. This mathematical model is based on mixed integer linear programming for the optimal selection and sequencing of purification steps. The main objective of this mathematical model is to maximize product purity.

Recently, stimulated moving bed (SMB) chromatography was proposed for the purification of pharmaceutical products in laboratory and industrial scale. SMB chromatography offers high productivity, purity and it reduces solvent consumption. "Triangle theory" is widely used to control the process in SMB technology (Rajendran et al., 2009). In SMB process, the stationary phase material is filled over a number of columns and the columns are connected to each other. This SMB system avoids product loss (Sá Gomes and Rodrigues, 2012). Recently, SMB column was developed by KNAUER, Germany, for commercial applications. Recombinant proteins were purified by combining continuous matrix-assisted refolding and purification by tandem SMB size-exclusion chromatography (Wellhoefer et al., 2014). This SMB chromatography is a limelight in protein and enzyme purification process in an industrial scale.

CONCLUSIONS AND FUTURE PERSPECTIVES

Current prospects of commercial enzymes and future demand are increasing dramatically. Simple, conventional method of enzyme purification is the key for the production of enzymes in commercial scale. Purification of commercial enzymes with increasing specific activity is a challenging and important task. This can be achieved by optimizing the enzyme purification process. Introduction of robotics technology and synthesis of new chromatographic media for multiple use could improve the scaling up of enzyme production and minimize the unit price of the commercial enzymes. Recent development on the purification of recombinant proteins using SMB size-exclusion chromatography increased the productivity and also decreased the production cost considerably. In the near future, SMB size-exclusion chromatography and IEC could lead the production of enzymes in an industrial scale.

ABBREVIATIONS

BCG Bromocresol green
BMR Batch manufacturing record
CM Carboxy methyl
DEAE Diethylaminoethyl
DNS Dinitro salicylate
EDTA Ethylenediaminetetraacetic acid
HIC Hydrophobic-interaction chromatography
HPLC High-performance liquid chromatography
HTE High-throughput experimentation
IEC Ion-exchange chromatography
MW Molecular weight
PAGE Polyacrylamide gel electrophoresis
RBB Remazol Brilliant Blue
SDS Sodium dodecyl sulfate
SMB Stimulated moving bed chromatography
UF Ultra filtration

ACKNOWLEDGMENTS

The author P. Vijayaraghavan gratefully acknowledges the Council for Scientific and Industrial Research, India, for providing a Senior Research Fellowship (Ref: 09/652(0024)/2012 EMR-1).

REFERENCES

Aires-Barros, M.R., Cabral, J.M.S., 1991. Selective separation and purification of two lipases from *Chromobacterium viscosum* using AOT reversed micelles. Biotechnology and Bioengineering 38, 1302–1307.
Andrews, A.T., Noble, I., Keeratipibul, S., Asenjo, J.A., 1994. Physicochemical properties of the matrix proteins of three main culture vehicles. Biotechnology and Bioengineering 44, 29–37.
Asenjo, J.A., Andrews, B.A., 2009. Protein purification using chromatography: selection of type, modelling and optimization of operating conditions. Journal of Molecular Recognition 22 (2), 65–76.
Bailey, M.J., 1988. A note on the use to dinitrosalicyclic acid for determining the products of enzymatic reactions. Applied Microbiology and Biotechnology 29, 494–496.
Banerjee, A., Chisti, Y., Banerjee, U.C., 2004. Streptokinase- a clinically useful thrombolytic agent. Biotechnology Advances 22, 287–307.
Benito, M.J., Rodriguez, M., Nunez, F., Asensio, M.A., 2002. Purification and characterization of an extracellular protease from *Penicillium chrysogenum* Pg222 active against meat proteins. Applied and Environmental Microbiology 68 (7), 3532–3536.
Bensch, M., Wierling, P.S., Lieres, E.V., Hubbuch, J., 2005. High throughput screening of chromatographic phases for rapid process development. Chemical Engineering Technology 28, 1274–1284.
Bezawada, J., Yan, S., John, R.P., Tyagi, R.D., 2011. Recovery of *Bacillus licheniformis* alkaline protease from supernatant of fermented wastewater sludge using ultrafiltration and its characterization. Biotechnology Research International 2011, 11 Article ID 238549.

Bhikhabhai, R., Sjoberg, A., Hedkvist, L., Galin, M., 2005. Production of milligram quantities of affinity-tagged proteins using automated multistep chromatographic purification. Journal of Chromatography 1080, 83–92.

Binod, P., Palkhiwala, P., Gaikaiwari, R., Nampoothiri, M., 2013. Industrial enzymes-present status and future perspectives for India. Journal of Scientific and Industrial Research 72, 271–286.

Blommel, P.G., Fox, B.G., 2007. A combined approach to improving large-scale production of tobacco Etch virus protease. Protein Expression and Purification 55 (1), 53–68.

Bohdziewicz, J., 1996. Ultrafiltration of technical amylolytic enyzmes. Process Biochemistry 31, 185–191.

Camper, D.V., Viola, R.E., 2009. Fully automated protein purification. Analytical Biochemistry 393, 176–181.

Chisti, Y., 1998. Strategies in downstream processing. In: Subramanian, G. (Ed.). Subramanian, G. (Ed.), Bioseparation and Bioprocessing: A Handbook, vol. 2. Wiley-VCH, New York, pp. 3–30.

Chopra, A.K., Mathur, D.K., 1985. Purification and characterization of heat-stable protease from *Bacillus stearothermophilus* RM-67. Journal of Dairy Science 68, 3202–3211.

Farooqui, A.A., Yang, H.C., Horrock, L.A., 1994. Purification of lipases, phospholipases and kinases by heparin – sepharose chromatography. Journal of Chromatography 673 (2), 149–158.

Fujiwara, N., Masui, A., Imanaka, T., 1993. Purification and properties of the highly thermostable alkaline protease from an alkaliphilic and thermophilic *Bacillus* sp. Journal of Biotechnology 30, 245–256.

Gu, Z., Glatz, C.E., 2007. A method for three-dimensional protein characterization and its application to a complex plant (corn) extract. Biotechnology and Bioengineering 97, 1158–1169.

Horikoshi, K., 1971. Production of alkaline enzymes by alkalophilic microorganism. Part I. alkaline protease produced by *Bacillus* No. 221. Agricultural and Biological Chemistry 35, 1407–1414.

Kim, W., Choi, K., Kim, Y., Park, H., 1996. Purification and Characterization of a fibrinolytic enzyme produced from *Bacillus* sp. strain CK 11-4 screened from Chungkook-Jang. Applied Environtol Microbiology 62, 2482–2488.

Kobayashi, T., Hakamada, Y., Hitomi, J., Koike, K., Ito, S., 1996. Purification of alkaline proteases from *Bacillus* strain and their possible interrelationship. Applied Microbiology and Biotechnology 45, 63–71.

Kumar, C.G., Takagi, H., 1999. Microbial alkaline proteases: from a bioindustrial viewpoint. Biotechnology Advances 17, 561–594.

Larcher, G., Cimon, B., Symoens, F., Tronchin, G., 1996. A 33 kDa serine proteinase from *Scedosporium apiospermum*. Biochemical Journal 315, 119–126.

Lee, S.Y., Rhee, J.S., 1993. Production and partial purification of a lipase from *Pseudomonas putida* 3SK. Enzyme and Microbial Technology 15, 617–623.

Liu, L., Houng, A., Tsai, J., Chowdhry, S., 1999. The fibronectin motif in the NH2-terminus of streptokinase plays a critical role in fibrin-independent plasminogen activation. Circulation 100 (Suppl. S), 1.

Nfor, B.K., Ahamed, T., Dedem, G.V., Wielen, L.A.M., 2008. Design strategies for integrated protein purification processes: challenges, progress and outlook. Journal of Chemical Technology and Biotechnology 83, 124–132.

Okyay, T.O., Rodrigues, D.F., 2013. High throughput colorimetric assay for rapid urease activity quantification. Journal of Microbiological Methods 95, 324–326.

Rajendran, A., Paredes, G., Mazzotti, M., 2009. Simulated moving bed chromatography for the separation of enantiomers. Journal of Chromatography A 1216 (4), 709–738.

Sa Gomes, P., Rodrigues, A.E., 2012. Simulated moving bed chromatography: from concept to proof-of-concept. Chemical Engineering and Technology 35 (1), 17–34.

Schratter, P., 2004. Purification and concentration by ultrafiltration. Methods in Molecular Biology 244, 101–116.

Sigrell, J.A., Eklund, P., Galin, M., Hedkvist, L., 2003. Automated multi-dimensional purification of tagged proteins. Journal of Structure and Functional Genomics 4, 109–114.

Sofer, G., Hagel, L., 1997. Purification design, optimization, and scale-up. In: Sofer, G., Hagel, L. (Eds.), Handbook of Process Chromatography: A Guide to Optimization, Scale-up, and Validation. Academic Press Ltd, London, pp. 27–113.

Steen, J., Uhlen, M., Hober, S., Ottosson, J., 2006. High-throughput protein purification using an automated set-up for high-yield affinity chromatography. Protein Expression and Purification 46, 173–178.

Sumner, J.B., 1926. The isolation and crystallization of the enzyme urease: preliminary paper. Journal of Biological Chemistry 69, 435–441.

Sumner, J.B., 1955. In: Colwick, S.P., Kaplan, N.O. (Eds.). Colwick, S.P., Kaplan, N.O. (Eds.), Methods in Enzymology, vol. 2. Academic Press, New York, p. 378.

Sztajer, H., Bryjak, M., 1989. Capillar membranes for purification of *Pseudomonas fluorescens* lipase. Bioprocess Engineering 4, 257–259.

Taipa, M.A., Aires-Barros, M.R., Cabral, J.M.S., 1992. Purification of lipases. Journal of Biotechnology 26, 111–142.

Tombs, M.P., Blake, G.G., 1982. Stability and inhibition of *Aspergillus* and *Rhizopus* lipases. Biochemistry Biophysics Acta 700, 81–89.

Tsuchiya, E., Uno, M., Kiguchi, A., Masuoka, K., 1992. The *Saccharomyces cerevisiae* NPS1 gene, a novel CDC gene which encodes a 160 kDa nuclear protein involved in G2 phase control. EMBO Journal 11 (11), 4017–4026.

Vasquez-Alvarez, E., Lienqueo, M.E., Pinto, J.M., 2001. Optimal synthesis of protein purification processes. Biotechnology Progress 17, 685.

Vijayaraghavan, P., Vincent, S.G.P., 2013. A simple method for the detection of protease activity on agar plates using bromocresolgreen dye. Journal of Biochemical Technology 4 (3), 628–630.

Vijayaraghavan, P., Vincent, S.G.P., 2014. Statistical optimization of fibrinolytic enzyme production using agroresidues by *Bacillus cereus* IND1 and its thrombolytic activity in vitro'. BioMed Research International 2014, 11 Article ID 725064.

Wellhoefera, M., Sprinzla, W., Hahna, R., Jungbauera, A., 2014. Continuous processing of recombinant proteins: integration of refolding and purification using simulated moving bed size-exclusion chromatography with buffer recycling. Journal of Chromatography A 1337, 48–56.

Wiendahl, M., Wierling, P.S., Nielsen, J., Christensen, D.F., 2008. High throughput screening for the design and optimization of chromatographic processes – miniaturization, automation and parallelization of breakthrough and elution studies. Chemical Engineering Technology 31, 893–903.

Wirth, S.J., Wolf, G.A., 1990. Dye-labelled substrates for the assay and detection of chitinase and lysozyme activity. Journal of Microbiological Methods 12, 197–205.

Wolf, G.A., Wirth, S.F., 1996. Soluble, dye-labelled substrates for a micro-plate assay of proteinase activity. Journal of Microbiological Methods 25, 337–342.

Wood, P.J., 1981. The use of dye-polysaccharide interactions in β-D-glucanase assay. Carbohydrate Research 94, C19.

Woolley, P., Peterson, S.B., 1994. Lipases—their Structure, Biochemistry and Applications. Cambridge University Press, Cambridge, pp. 103–110.

Yamagata, Y., Ishiki, K., Ichishima, E., 1995. Subtilisin Sendai from alkalophilic Bacillus sp.: molecular and enzymatic properties of the enzyme and molecular cloning and characterization of the gene, apr S. Enzyme and Microbial Technology 17, 653–663.

CHAPTER 5

Low-Cost Enzymes and Their Applications in Bioenergy Sector*,‡

V.L. Queiroz, A.T. Awan, L. Tasic
State University of Campinas, Campinas, Brazil

LIGNOCELLULOSE BIOMASS AND 2G ETHANOL

Lignocellulose Biomass

Agricultural and agro-industrial waste salvage is growing rapidly as a consequence of efforts for diminishing the environmental impacts caused by industrial and urban activities (Tsukamoto et al., 2013b). These wastes are composed of lignocellulosic biomass and serve as an attractive feedstock for 2G-ethanol production because they are cost effective, renewable, and abundant. Moreover, crops such as corn and sugarcane are primarily used as food and feedstock, therefore being unable to meet the global demand of bioethanol production (Sarkar et al., 2012). Conventionally, sugarcane bagasse is utilized as feedstock for bioethanol production (Soccol et al., 2010). Other agro-industrial residues are important to consider, for example, orange waste (Awan et al., 2013), rice straw, wheat straw, and corn straw (Sarkar et al., 2012) as these are major agricultural wastes in terms of their quantity as well as availability (Awan et al., 2013).

The structural material in the cell wall of woody and nonwoody plants (Hon and Shiraishi, 2000; Menon and Rao, 2012) is known as lignocellulose, which provides support, strength, and shape to the plant. This biomass is being explored for 2G-ethanol production, and its chemical composition directly influences the type and the amount of enzymes employed for hydrolysis. The perfect enzyme combination in an enzyme cocktail is not only needed for successful biomass digestion in an industrial process but also required and constantly produced and secreted in nature by the native microorganisms that habit this biomass. Therefore, it is fundamental to know chemical composition and exact proportions of lignocellulosic biomass components in order to understand how to prepare an adequate enzyme cocktail. Lignocellulosic biomass mainly contains pectin, cellulose, hemicelluloses, and lignin (Wyman, 1996) as is shown in Fig. 5.1. Different parts of the plant have different proportions of these components. The outer wall is mainly composed of carbohydrates and is called primary cell wall. It surrounds growing

---

* Authors contributions: Conceived and designed the structure of the chapter: LT. Wrote the chapter: ALL. Participated in revising the draft: ALL. Contributed materials/analysis tools: ALL.

‡ Conflict of interest: The authors declare no conflict of interest.

Agro-Industrial Wastes as Feedstock for Enzyme Production
ISBN 978-0-12-802392-1
http://dx.doi.org/10.1016/B978-0-12-802392-1.00005-8

111

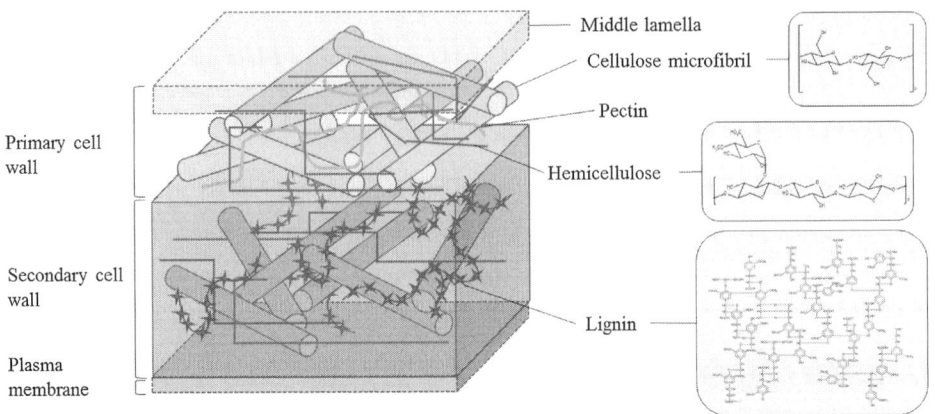

Figure 5.1 Representation of cell wall containing cellulose microfibrils, hemicellulose, pectin, and lignin and their molecular structures.

and dividing plant cells. Next is the much thicker and stronger secondary cell wall, which accounts for most of the lignin part in biomass (Willats et al., 2001; Sticklen, 2008). The secondary wall usually consists of three sublayers: S1 (outer), S2 (middle), and S3 (inner). The formation of these three distinctive layers in secondary cell walls is a result of changes in the orientation of cellulose microfibrils during their deposition. The S1 and S3 layers are typically thin and have cellulose microfibrils oriented in a flat helix relative to the elongation axis of the cell (Zhong and Ye, 2009). The S2 layer is thick and has cellulose microfibrils that account for 75–85% of the total thickness of the cell wall (Plomion et al., 2001). It is the S2 layer that largely determines the mechanical strength of fibers in wood. Hemicelluloses and lignin are also present in each of these layers (Plomion et al., 2001; Zhao et al., 2012).

Pectin is an important component of cell wall and is a polysaccharide rich in galacturonic acid and galacturonic acid methyl ester units. It is a soluble fiber that combines with proteins and other polysaccharides and forms skeletal tissue that is responsible for the chemical stability and physical strength of plants (Willats et al., 2006). *Cellulose* is a high-molecular-weight linear insoluble polymer of D-glucose units connected via β-(1,4)-glycosidic bonds. Cellulose, hemicellulose, and lignin are often found in close association with each other. Usually, cellulose does not occur alone in a free threadlike chain, instead it has a supramolecular structure consisting of both crystalline and amorphous regions made of fibrillary units, and in the majority of cases, the crystalline region is made up of several sheets of long-chain fibers united by both intra- and intermolecular hydrogen bonds (Mansfield et al., 1999).

The structure of cellulose can be cleaved by hydrolysis of β-(1,4)-glycosidic bond catalyzed with acids or enzymes such as cellulases. However, the hydrogen bonds incorporated within long chains of cellulose in a crystalline structure inhibit biomass hydrolysis and demand harsh pretreatment conditions (Gardner and Blackwell, 1974). It has been investigated and widely accepted that the higher the crystalline content of cellulose, the more difficult is the enzymatic attack to hydrolyze this polymer. The amorphous regions, due to their high accessibility to enzymes, can be hydrolyzed easily, while in the crystalline areas enzyme contact efficiency is decreased (Chang and Holtzapple, 2000; Carvalho, 2009).

Unlike cellulose, *hemicellulose polymers* are chemically heterogeneous and mostly amorphous with lower degrees of polymerization. That is why hemicellulose chains are more easily hydrolyzed into monomeric units as compared to cellulose chains (Wyman, 1999). Hemicellulose is a heteropolysaccharide consisting of arabinose, galactose, glucose, mannose, and xylose, with minor amounts of a few other compounds. There are many suitable pretreatments that first help in removing hemicelluloses and later facilitate the cellulose digestion (Saha et al., 2005).

Lignin, however, is not a polysaccharide and does not have a well-defined chemical structure. Lignin is a cross-linked macromolecular material based on a phenyl-propanoid monomer structure (Doherty et al., 2011). Lignin adds strength and rigidity to cell walls and is more resistant to enzyme attack compared to cellulose and other structural polysaccharides (Kirk, 1971; Baurhoo et al., 2008). Lignin is able to form covalent bonds with hemicelluloses. Covalent bonds incorporated in lignin and carbohydrates mostly consist of benzyl esters, benzyl ethers, and phenyl glycosides (Smook, 2002). In this way, lignin provides integrity, structural rigidity, and prevention of swelling of lignocelluloses. It is commonly accepted as one of the major factors responsible for biomass recalcitrance to enzymatic hydrolysis. It implies steric hindrance and prevents fibers from swelling (Mansfield et al., 1999; Carvalho, 2009).

In order to use the lignocellulosic biomass in the production of biofuels, it is important to convert biomass polymers in monomeric sugars, and to accomplish that, pretreatment is a key step to guarantee efficient hydrolysis and high yield of D-glucose. Pretreatment of lignocellulosic materials is s an important step since it reduces the time of hydrolysis (Sun and Cheng, 2002). The main objectives of pretreatment are: reducing the crystallinity of cellulose, lignin removal, increasing the porosity of the lignocellulosic material, and increasing the susceptibility to enzymatic substrate action compared with the original material. The conversion steps of lignocellulosic biomass into 2G ethanol consist of a minimum of four steps:

1. Pretreatment for increasing the degradation capacity of the biomass.
2. Hydrolysis of lignocelluloses to simple monomeric sugars.
3. Fermentation of sugars to 2G ethanol.
4. Recovery of 2G ethanol by distillation/evaporation.

ENZYMES IN BIOMASS HYDROLYSIS

Low-cost Enzymes: Present and Perspectives

Over the past years, scientists from all over the world have invested a great amount of research effort in the area of biomass pretreatment and enzyme hydrolysis, and these efforts have acknowledged lignocellulosic biomass as a potential sustainable source of sugars for biofuel production by fermentation. Several technologies have been developed to allow a cost-competitive process to occur (Himmel et al., 2007). The main barrier to a cost-competitive production of biofuels is the high cost of biomass feedstock and the processes for converting biomass to monomer sugars, including the cost of pretreatment and enzyme hydrolysis, hence augmenting the conversion yield is crucial for counterbalancing feedstock cost (Himmel et al., 2007). The hydrolysis reaction is essential for the production of 2G ethanol. It causes long polymeric chains of cellulose, hemicellulose, and other polymer sugars to break down into simple free monomers such as hexoses and pentoses (Oberoi et al., 2010; Stewart et al., 2006, 2008) that can be readily fermented into bioethanol (Limayem and Ricke, 2012). Enzyme application to biomass has shown distinct advantages over acid-based hydrolysis methods, for example, it occurs at very mild process conditions, gives potentially higher yields, and has no corrosion problem. Therefore, enzyme hydrolysis is considered the most suitable one for future 2G-ethanol production from biomass (Duff and Murray, 1996; Hsu, 1996). Biomass can be hydrolyzed to monosaccharides by applying a combination of enzymes, such as pectinase, cellulase, and β-glucosidase.

Enzymatic hydrolysis of biomass depends on several factors, one of them being the nature of the biomass and the enzymes used. Considering the fact that composition of waste varies significantly, there is no universal recipe for the hydrolysis of any lignocellulosic biomass. One way to reduce the cost of enzyme production is by exploring microorganisms that are naturally involved in degradation of biomass, such as bacteria and fungi. The idea of using *Xanthomonas axonopodis* pv. *citri* (*Xac*) enzymes in a form of an enzyme cocktail were explored recently (Awan et al., 2013). The purpose was to make the enzyme hydrolysis of orange waste less costly to compete with acid hydrolysis, since these bacteria degrade the cell wall of the orange to cause infection; this subject will be further explored in this chapter.

Enzymes in Native Bioflora: Microorganisms Working for Us

As mentioned before, both bacteria and fungi can produce cellulases for the hydrolysis of lignocellulosic material. Compared with fungi, cellulolytic bacteria produce low amounts of cellulolytically active enzymes (Sternberg, 1976; Duff and Murray, 1996). Of all the fungal genera, *Trichoderma* has been most extensively studied for cellulase production (Sternberg, 1976). It produces a complex mixture of cellulase enzymes with high specificity toward β-(1,4)-glucosidic bonds. *Cellulases* are usually a mixture of various cellobiohydrolases and endoglucanases supplemented with beta-glucosidase (Sun and Cheng, 2002). The crystalline cellulose, for example, is hydrolyzed by the synergistic action of endoglucanases and exoglucanases, then the product of cellulase action (cellobiose) is hydrolyzed by β-D-glucosidases (cellobiases), and therefore relieve the system from end product inhibition (Himmel et al., 2007). Following there is a better explanation of the action of endoglucanase, exoglucanase, and β-glucosidase and how each one of them is involved in a different part of the lignocellulosic biomass degradation process as shown in Fig. 5.2.

- Endoglucanase (EG, endo-1,4-D-glucanohydrolase, or EC 3.2.1.4.) attacks regions of low crystallinity in the cellulose fiber, creating free chain ends. Endoglucanases hydrolyze internal bonds in the cellulose chain and break intermolecular bonds between adjacent cellulose chains. These enzymes act mainly on the amorphous parts and cleave glucosidic bonds randomly, generating soluble carbohydrate chains with low degrees of polymerization (Sun and Cheng, 2002; Kumar et al., 2008).
- Exoglucanase or cellobiohydrolase (CBH, 1,4-β-D-glucan cellobiohydrolase, or EC 3.2.1.91) degrades the molecule further by removing cellobiose units from the free chain ends (Sun and Cheng, 2002). These enzymes release mainly cellobiose, but also glucose and small cellodextrins (mainly cellotrioses). One exoglucanase can either act on the reducing or the nonreducing ends or the chains, but microorganisms often produce more than one type of exoglucanase, degrading cellulose chains from both directions (Zhang and Lynd, 2004).
- β-glucosidase (EC 3.2.1.21) hydrolyzes cellobiose that is an end product inhibitor of many cellulases and produces glucose monomers (Sun and Cheng, 2002). Endo- and exoglucanases break down larger oligosaccharides into cellobiose by primary hydrolysis, but the most important step is the action of β-glucosidase, which completes hydrolysis, producing glucose as a sugar monomer that can be used further in the process of fermentation.

In addition to these three major groups of cellulase enzymes, there are other enzymes known to be involved in the cell wall deconstruction like hemicellulases and accessory enzymes, such as hemicellulose debranching, phenolic acid esterase, and perchance lignin degrading and modifying enzymes (Himmel et al., 2007).

Bearing in mind the enzymes that attack hemicellulose, there are a number of known enzymes with this function, such as glucuronidase, acetylesterase, xylanase, β-xylosidase, galactomannanase and glucomannanase (Duff and Murray, 1996). Hemicellulases are produced by many species of bacteria and fungi, as well as by several plants. Today, most commercial hemicellulase preparations are produced by genetically modified *Trichoderma* or *Aspergillus* strains (Mussatto and Teixeira, 2010).

Apart from cellulose and hemicelluloses, pectins are abundant in the soft tissues of citrus fruits, sugar beet pulp, and apple (Numan and Bhosle, 2006). Pectinases are widely

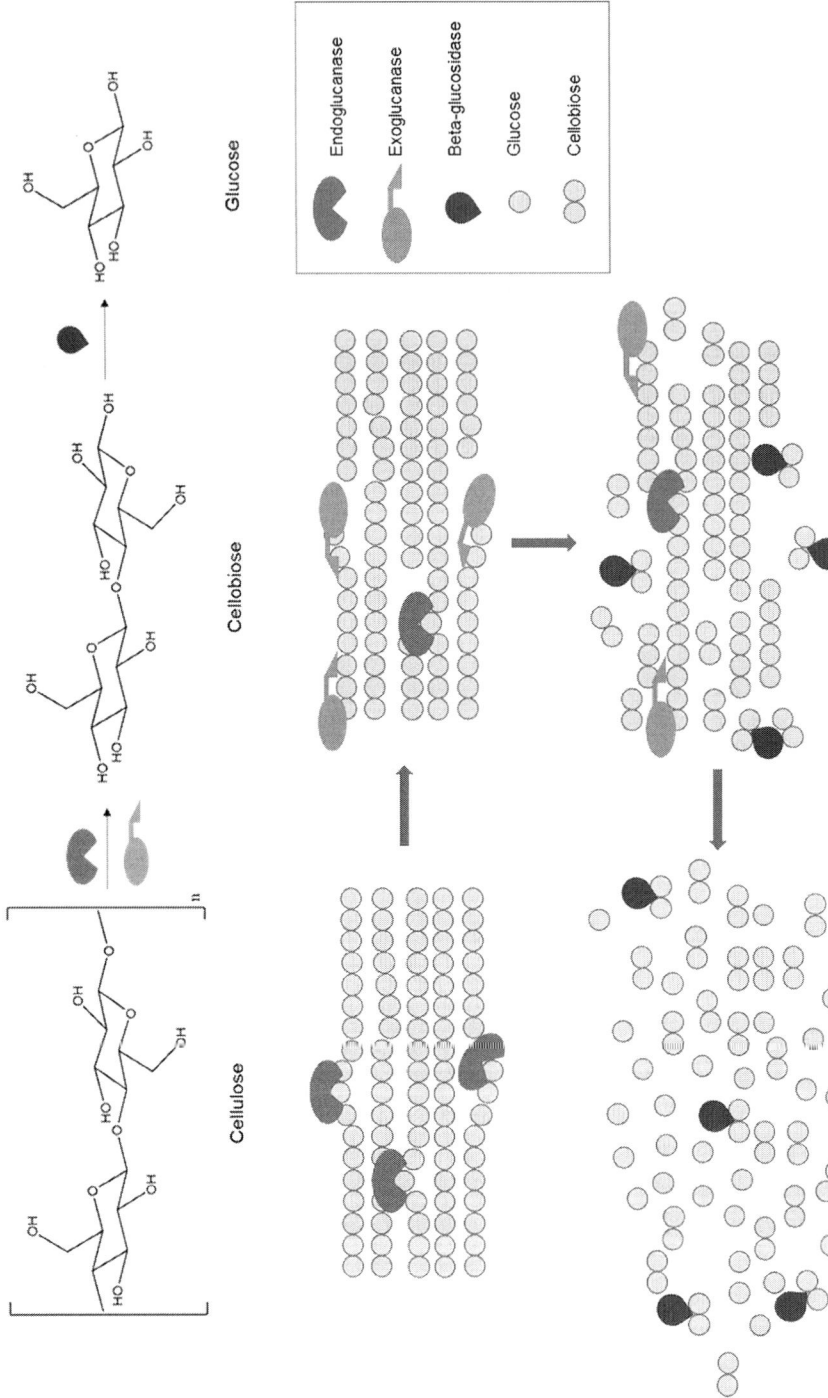

Figure 5.2 Simplified schematic representation of cellulolytic enzyme action. Endoglucanase is acting on amorphous regions of the cellulose microfibrils. Exoglucanase is acting in the free cellulose chain ends producing cellobiose molecules that are further cleaved by beta-glucosidases to yield glucose monomers.

distributed enzymes, such as protopectinases, polygalacturonases, lyases, and pectin ester-ases (Jayani et al., 2005). They are commonly found in bacteria, fungi, and plants.

ORANGE BAGASSE AS AN EXAMPLE OF AGRO-INDUSTRIAL WASTE WITH HIGH REUSABLE VALUE

Composition

There are different types of agricultural wastes that can be used as feedstock to bioethanol production. Here we have chosen the orange waste due to its interesting composition that allows not only the production of bioethanol but also production of very interesting natural by-products in relatively high yields, such as D-limonene, nanocellulose, and hesperidin (Awan et al., 2013; Awan, 2013; Tsukamoto et al., 2013a,b). After juice extraction, around half of the fruit is left as citrus processing waste from oranges, a very interesting low-cost material already being used for producing the first-generation (1G) bioethanol (Grohmann and Baldwin, 1992; Wilkins et al., 2005; Rivas et al., 2008; Edwards and Doran-Peterson, 2012; Tsukamoto et al., 2013a). The 1G ethanol is produced via fermentation from the monosaccharides (mainly fructose and glucose) present in a liquid residue obtained after squeezing the semisolid orange waste that is generated in orange juice factoring process (Awan, 2013; Tsukamoto et al., 2013a).

Orange waste biomass is rich in soluble and insoluble carbohydrates with a very small proportion of lignin when compared with other lignocellulosic biomass (Edwards and Doran-Peterson, 2012; Choi et al., 2013). Due to the great amount of orange waste available at low cost, there is great potential to employ this residue in a more attractive, cost-effective, environmentally friendly, and cleaner manner such as 2G-ethanol production. This can be achieved through fermentation of the carbohydrate monomers generated from this biomass. Enzymatic hydrolysis of biomass for the breakdown of polymeric sugars has been known in the literature but, depending on the nature of the substrate, as the composition of orange waste biomass changes, enzyme behavior and thus hydrolysate composition may also vary significantly (Awan et al., 2013).

In order to harness the maximum value from the orange bagasse, it is essential to understand its chemical composition (Rivas et al., 2008). Briefly, orange bagasse has 75–82% moisture content, soluble sugars, fibers (including cellulose, hemicellulose, lignin, and pectin) along with ash, fats, and proteins (Awan, 2013). The quantitative composition determined by Rivas et al. (2008) is shown in the graphic in Fig. 5.3.

The small quantities of about 4.35% of the peel include organic acids, such as citric acid, malic acid, malonic acid, oxalic acid (which collectively represent about 1%) and vitamin C (ascorbic acid). Bampidis and Robinson (2006) also investigated the composition of orange peel and reported that the dry matter content of orange peel is mainly organic, containing proteins and other short-chain (no more than four carbons) organic acids (Fig. 5.4).

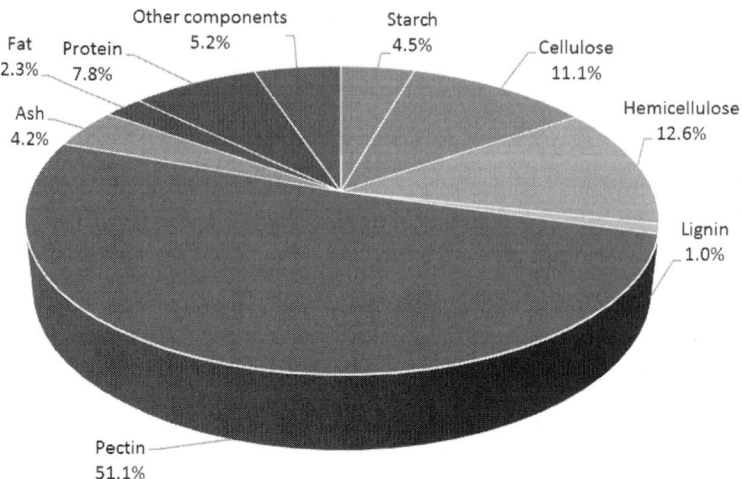

Figure 5.3 Chemical composition of orange peel in percentage on dry matter basis (Rivas et al., 2008).

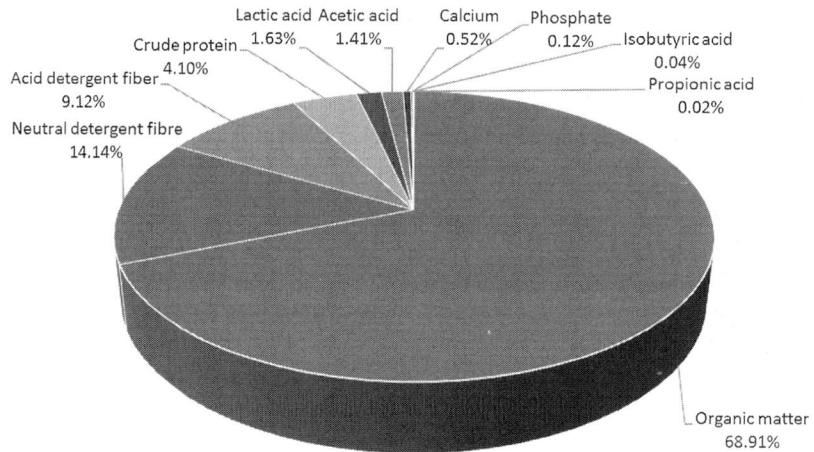

Figure 5.4 Chemical composition of orange peel by dry matter (Bampidis and Robinson, 2006).

It is important to note that just like all other plant products, the chemical composition of oranges varies. It is affected by a number of factors, including climate, technical growing conditions, fruit maturity, harvesting season, rootstock, and fruit variety (Kale and Adsule, 1995). The pH of citrus peel is also variable; it can be as low as 3.6. It should be tested before any application, as neutralization might be required.

Bermejo et al. (2011) carried out analyses of bioactive agents in citrus varieties. They observed that rind contents of different varieties showed similar tendencies for most of the compounds like the flavanone glycosides hesperidin and narirutin, carotenoid, and

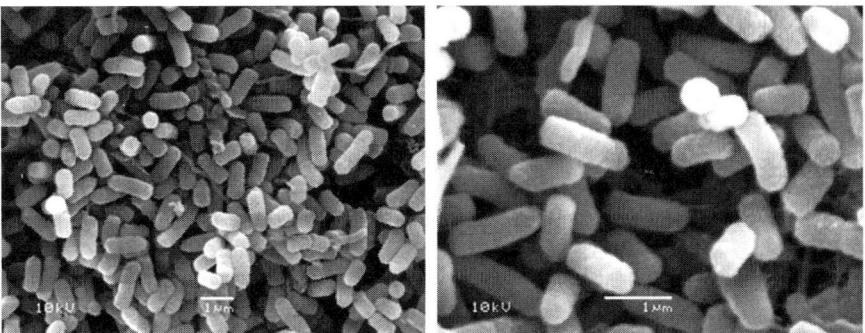

Figure 5.5 Scanning electron micrographs of *Xanthomonas axonopodis* pv. *citri* cells (strain 306).

β-cryptoxanthin. Among all studied varieties, the most abundant terpene found in peel essential oil was limonene, followed by myrcene; calcium and potassium were the dominant macronutrients.

Orange bagasse composition (as shown in Figs. 5.3 and 5.4) shows this biomass potential to be employed in different applications. Extraction of highly valuable natural products, limonene and hesperidin, for example, could transform what is typically considered a problematic substrate into a high-value commodity. Among these, pectin is also very attractive, as well as fermentable sugars that can be used to produce 2G ethanol.

Enzymatic Hydrolysis Using Native Pathogen Proteins: *Xanthomonas axonopodis* pv. *citri*

Various studies reported the use of several bacteria and fungi to produce cocktails of hydrolytic enzymes, and generally fungi are considered a better option in terms of enzyme variety and yields of fermentable sugars after hydrolysis of lignocellulosic biomass (Bakri et al., 2003; Liu et al., 2004; Awan et al., 2013). Nevertheless, *X. axonopodis* pv. *citri* (*Xac*) also produces a cocktail of enzymes that can aid in the degradation of citrus biomass due to different hydrolytic enzyme activities of *Xac* proteins (Alegria et al., 2005; Awan et al., 2013). *Xanthomonas* is an important genus of plant pathogenic gram-negative bacteria (Khater et al., 2007; Tasic et al., 2007; Fattori et al., 2011; Liu et al., 2012) (Fig. 5.5). A great number of citrus diseases are caused by distinct pathovars (pv.) of *Xanthomonas* species (Graham et al., 2004; Schaad et al., 2006). Citrus canker is caused by several pathogenic variants of *Xac*. The *Xac* strain 306 with a suspected origin in southeastern Asia causes the Asiatic type of a canker and is the most widespread and virulent bacteria (Jalan et al., 2011).

Like many other plant diseases caused by bacteria, this pathogen also enters into host plant tissues through their stomatal openings (Cubero et al., 2001; Brunings and Gabriel, 2003). The bacterial attack can be recognized by earliest symptoms that appear on leaves as tiny, slightly raised blister-like lesions. This bacteria then produces hyperplasic and hypertrophic (corky) lesions surrounded by oily or water-soaked margins and a yellow halo on leaves, stems, and fruits (Jalan et al., 2011). These bacterial attacks on the host tissues involve an impressive arsenal of proteins, including pectinases and cellulases and, consequently, their hydrolytic activity. *Xac* (strain 306) genome was completely sequenced in 2002 by da Silva et al. who showed that it has one circular chromosome comprising 5,175,554 base pairs (bp), and two plasmids: pXAC33 (33,699 bp) and pXAC64 (64,920 bp). The transfer of macromolecules into the host body by *Xac* causes infection. Administered by two completely distinct and highly complex multiprotein systems, the type III and the type IV secretion systems (Alegria et al., 2005), *Xac* infection transfer starts. Enzymes and toxins are also secreted by two clusters of type II secretion system (T2SS) that cause degradation of biomass (da Silva et al., 2002; Moreira et al., 2004).

The *Xac* genome transcription can allow the production of target proteins. For example, using well-known molecular biology techniques, the recombinant DNA can be used for production of enzymes, like hydrolases, which is of particular interest due to possible application of these enzymes in the degradation of orange bagasse. First, the gene coding for the desired protein needs to be amplified from genomic DNA of *Xac* by polymerase chain reaction using specific oligonucleotides (Khater et al., 2007; Tasic et al., 2007; Fattori et al., 2011). After that, the product fragment is cloned into the expression vector—usually the vector introduces a tag with six histidine residues in the N- or C-terminal portion of the protein to allow later purification. The protein is then transformed in an adequate host for further expression, which is commonly induced by isopropyl β-D-1-thiogalactopyranoside. Finally, purification of the desired protein can be performed by chromatography and later used for its applications such as biomass degradation.

In 2005 Watt et al. had analyzed the extracellular proteins from *Xanthomonas campestris* pv. *campestris* (*Xcc*). Because *Xcc* and *Xac* share more than 80% of their genes (da Silva et al., 2002), one can extend the reported analysis on *Xcc* to *Xac* proteins as well. In the study, Watt et al. (2005) observed that 53% of the extracellular proteins presented putative secretion peptide in the N-terminal; the analysis scan was done using Signal P software (Nielsen et al., 1997). This is an important factor to be considered when producing recombinant enzymes, since if the protein contains a signal peptide sequence it can be secreted by the host expression, thus, subsequently found in the supernatant.

The genome of *Xac* 306 allows the biosynthesis of enzymes with cellulolytic, hemicellulolytic, and pectinolytic activities (da Silva et al., 2002). The complete genomic sequence shows the presence of gene-related enzymes: three pectate lyases with genes (pel, degenerated pel, and pelB) and two polygalacturonases with genes (peh-1 and pglA) (da Silva et al., 2002). Lin et al. (2010) showed that *Xac* proteins play a vital role in expressing pectinolytic activity. The presence of 12 copies of genes for cellulolytic and hemicellulolytic enzymes has also been reported (Moreira et al., 2004).

Table 5.1 Enzyme Activities of the *Xanthomonas axonopodis* pv. *citri* (*Xac*) Proteins and Commercial Enzymes (Awan et al., 2013)

Enzymes	Commercial	*Xac* Proteins
Pectinase activity	141 U g^{-1}	58 U g^{-1}
Polygalacturonase activity	1.6 U mg^{-1}	78 U mg^{-1}
Cellulase activity	68 FPU mL^{-1}	8 FPU mL^{-1}

This impressive arsenal of hydrolytic proteins triggered the possibility to test *Xac* hydrolytic activity in biomass degradation (Awan et al., 2013). The activities of the commercial and *Xac* enzymes were determined by Awan et al. (2013) and are expressed in international units (IU), and the values are shown in Table 5.1.

The polygalacturonase and pectinase activities were expressed as the amount of enzyme required to produce 1 μmol of galacturonic acid per minute with polygalacturonic acid (pectic acid) and pectin used as substrates, respectively. The cellulase activity corresponds to the amount of enzyme required to produce 1 μmol of D-glucose per minute, which was expressed in filter-paper units (FPUs). The activities of *Xac* enzymes were significantly lower when compared to the commercial enzyme activity and 2.4 (pectinase), 8.5 (cellulase), and 20.5 (polygalacturonase) times lower. Knowing that extract of *Xac* proteins is composed from more than 400 proteins, the amounts of pectinase, cellulase, and polygalacturonase are small. Even so, *Xac* enzymes were capable of providing excellent yields of fermentable sugars that were similar to fructose and glucose concentrations obtained with the commercial enzymes used in hydrolysis of the same type of biomass (Awan et al., 2013).

Some of the in silico obtained structures of the two hydrolases from *Xac* are illustrated in Fig. 5.6. The *Xac* enzyme sequence was compared to sequences whose 3D structures were deposited in the Protein Data Bank and resolved by X-ray diffraction. Then, the software Modeler by Accelrys was used to calculate the target enzymes structures. The models shown in Figs. 5.6A and B represent the 3D predicted structures of an endoglucanase (EC 3.2.1.4) with cellobiose and cellotriose as substrates. Figs. 5.6C and D show the model of beta-glucosidase (EC 3.2.1.21) with D-glucose as substrate (Eckert et al., 2009).

Biofuel-related Enzymes From Other Sources

As already stated, many different types of agro-industrial waste can be explored in the bioenergy sector and especially for 2G-ethanol production. And all these wastes, despite having similar composition and addressed as lignocellulosic biomass, have different proportions of their chemical constituents. Therefore, they differ from one another by having different chemical composition and different native biota with specific types and quantities of variety of microorganisms that use the specific biomass as the unique source of organic carbon.

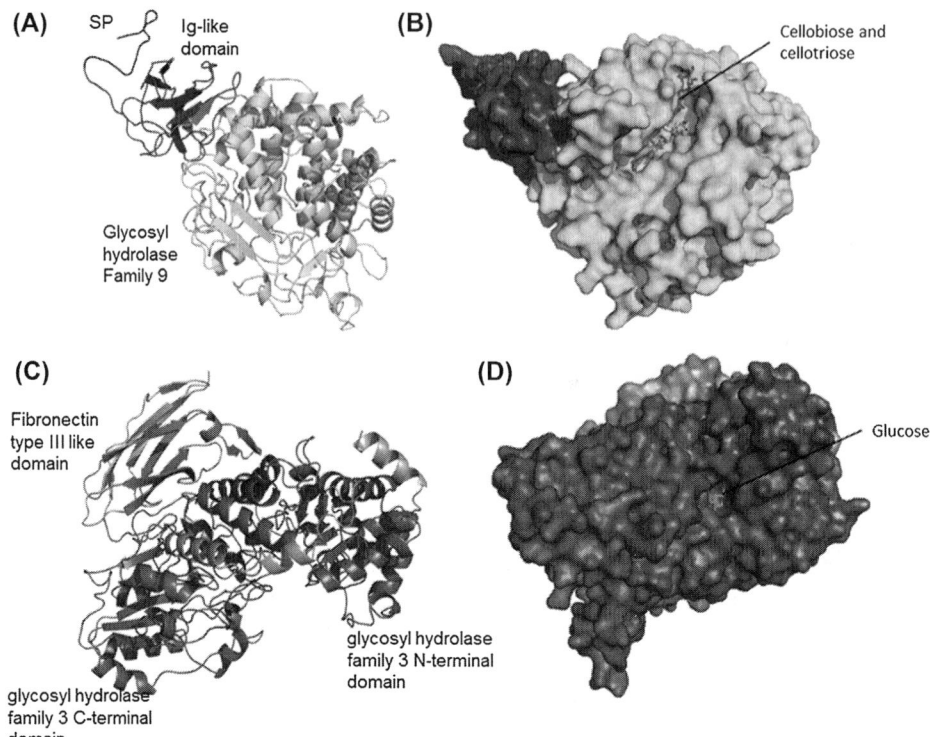

Figure 5.6 (A) Shows the 3D structure of an endoglucanase as predicted by Modeler (Accelrys), the structure presents a signal peptide, Ig-like domain, and glycosyl hydrolase family 9 (Pfam). (B) Shows the surface of the endoglucanase from *Xac* with cellobiose and cellotriose as substrates. (C) Shows the 3D-structure predicted for beta-glucosidase with glycosyl hydrolase family 3 N-terminal domain, glycosyl hydrolase family 3 C-terminal domain and fibronectin type III-like domain (Pfam). (D) The surface of beta-glucosidase from *Xac* with beta-ᴅ-glucose as substrate.

Fungi, bacteria, and actinomycetes among other microorganisms have been reported for cellulase and hemicellulase production using lignocellulosic biomass especially by solid–state fermentation (Saratale et al., 2014). Among the most extensively studied fungi are *Trichoderma, Botrytis, Penicillium,* and *Aspergillus* from agricultural leftovers that have commercially been used for cellulase and hemicellulase production (Xin and Geng, 2010). For example, *Phanerochaete chrysosporium* is also a fungus known for its ability to degrade different agricultural wastes by generating high levels of hydrolytic enzymes under solid–state fermentation (Saratale et al., 2014).

In addition, *Aspergillus fumigates* SK1 isolated from cow dung and later grown on untreated oil palm trunk by solid–state fermentation is known to generate secretome with high levels of cellulases and xylanase along with moderate levels of lignin peroxidase (Ang et al., 2015). Later, crude enzyme mixture obtained from fungal secretomes applied to hydrolyze lemon grass produced maximum quantity of reducing sugars (Ang et al., 2015).

Many bacteria are also recognized for their contributions to biomass utilization. Takasuka et al. (2013) demonstrated the biomass-deconstructing capability of *Streptomyces* sp. SirexAA-E (ActE), an aerobic bacterium. By growing it above the plant biomass, ActE showed enzyme secretions including polysaccharide-monooxygenases, endo- and exocellulases, and hemicellulases. The key enzymes were confirmed through genome analyses and biochemical assays. Many other studies show that the biomass–degrading activity of the enzyme mixture from biomass was comparable to a cellulolytic enzyme cocktail from the fungus *Trichoderma reesei*, showing an example of high cellulolytic capacity of an aerobic bacterium.

Lignin deconstruction for enhancing the release of fermentable sugars from plant cell walls presents a challenge for biofuel production from lignocellulosic substrates. Huang et al. (2013) discovered some unique lignin-degrading enzymes by screening 140 bacterial strains isolated from the soil of a rainforest. Their approach can provide some benefits over fungal enzymes, in terms of enzymes production and relative ease of protein engineering.

Another example of enzymes found in different sources is the hindgut contents of *Holotrichia parallela* that show presence of cellulolytic bacteria, identified as *Pseudomonas* sp. with high endoglucanase activity (Sheng et al., 2012). Studies reveal that the crude endoglucanase enzyme from a novel bacterium (designated as HP207) has high thermostability; approximately 55% of the original activity was maintained after pretreatment at 70°C for 1 h, while the maximum endoglucanase activity obtained was 1.432 UmL^{-1}, higher than those of the most widely studied bacteria and fungi (Sheng et al., 2012).

Unlike rumen bacteria, the rumen fungi such as *Neocallimastix frontalis, Piromyces communis, and Caecomyces communis* are able to degrade lignified secondary walls in lucerne stems (Bootten et al., 2011). These fungi improve the forage utilization in ruminants. Therefore, they have great potential for degrading lignocellulosic biomass in the production of biofuels. Industrial utilization of this data in the bioenergy sector requires further investigations, especially on large pilot scale to allow its suitable application.

Gua et al. (2010) presented an interesting natural synergistic complex enzyme system. They isolated it from a composite native microbiota grown on agricultural and animal wastes that were kept under long-term acclimation. These microbes were capable of degrading lignocellulosic biomass and also produced extracellular xylanase that enabled a more efficient biomass hydrolysis.

Despite growing interest in science and intensive research in the bioenergy sector, still the perfect enzyme cocktail for application in 2G-ethanol production from lignocellulosic biomass cannot be cited, as each agro-industrial waste is processed in a very specific way. Nevertheless, one can say that the best recipe for finding the perfect enzyme cocktail can be started with the screening of the target biomass microbiota, continue with isolation of biomass consortium of the microorganisms, and use this consortium as it is, by exploring enzymes from protein cocktails produced by the microbiota as exoenzymes (secreted) or endoenzymes (minority). The last step can be substituted with techniques of recombinant DNA technology in an attempt to obtain low-cost and efficient catalysts—recombinant enzymes.

Enzyme Recycling

Although hydrolytic processes involving enzymes are considered better options then acid- or base-based hydrolyzes of the biomass, enzymatic hydrolyzes are expensive and contribute significantly to the cost of the overall process of 2G-ethanol production. For instance, the cellulase required for average biomass conversion currently accounts for almost 40% of the total process cost, thus representing the second-highest cost contribution, only less expensive compared to the cost of the raw material (Deswal et al., 2011; Dhillon et al., 2012; Saratale et al., 2014). Thus, it is very important to consider the possibility of recycling of enzymes as the catalysts, once the enzyme reuse can effectively increase the rate and yields of the hydrolysis and lower the costs of an enzyme application (Mes-Hartree et al., 1987; Sun and Cheng, 2002). One strategy to do so is through the immobilization of enzymes onto insoluble supports, which provide many advantages, such as easy separation from reaction mixture along with greater stability and possible modulation of the catalytic properties (Jordan et al., 2011). This gain in stability is because the enzyme is protected, which increases its mechanical stability.

Usually enzymes can be easily recovered from the liquid supernatant, while it is more difficult to recover them from the solid residues. In the cases where the substrate is solid, like in orange bagasse, the immobilization procedure requires an immobilization strategy that diminishes the diffusion limitations of the contact interface and conversion kinetics. Of particular interest is the use of magnetic supports for immobilization (Jordan et al., 2011). Magnetic carrier particles are useful because they confer the desired magnetic properties facilitating their removal from the mixture reaction allowing the enzymes bound to them to be potentially recovered and recycled, prolonging enzyme usage for processing of cellulosic materials (Jordan et al., 2011). Magnetite (Fe_3O_4) nanoparticles are one of the most dominant magnetic materials used; this is because they show low toxicity and strong magnetic properties when used as biocompatible super-paramagnetic materials (Huang et al., 2003; Jordan et al., 2011).

Different approaches are efficient in enzyme immobilization. The general methods to accomplish the enzyme attachment are: physical adsorption to a solid phase, covalent bonding (Fig. 5.7), inclusion in a gel phase, cross-linking with bifunctional reagents, and encapsulation (Garcia et al., 1989).

In order to prevent the free-enzyme loss, using bifunctional reagent has the advantage of interacting with functional groups of the proteins forming covalent bonds, which is important since adsorbed biomolecules are easily desorbed due to changes in the ionic strength, temperature, or the substrate (Torres-Salas et al., 2011). A recent example of a bifunctional binding reagent is carbodiimide, which due to its simple use and high efficiency has gained popularity for enzyme immobilization (Jordan et al., 2011).

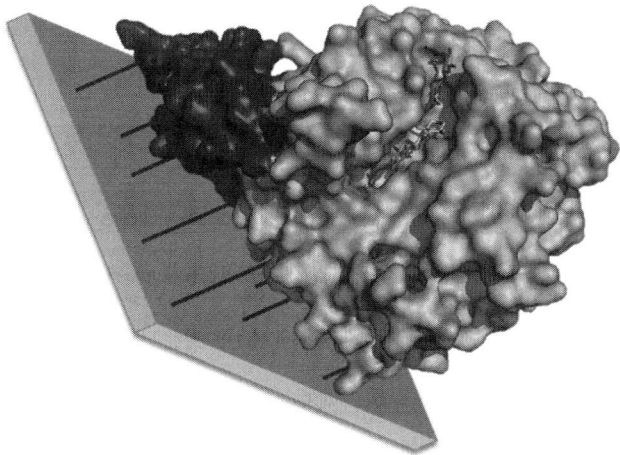

Figure 5.7 Example of a cellulase from *Xac* immobilized on a metal surface (Fe_3O_4) by linker.

NATIVE MICROORGANISMS

In the United States, some efficient processes starting from acid hydrolysates of orange waste are already used for bioethanol production (Grohmann et al., 1994, 1995; Stewart et al., 2006; Lohrasbi et al., 2010). Nevertheless, difficulties concerning fermentation are still unresolved. Currently, the conventional process of bioethanol production is done with yeast cultures of *Saccharomyces cerevisiae*. This microorganism is widely used because it can grow fast, has high yields of ethanol from glucose, can tolerate environmental stresses, like high ethanol concentration and low levels of oxygen, moreover it is very well known in industrial applications and fermentation (Piskur et al., 2006; Alper and Stephanopoulos, 2009).

However, *S. cerevisiae* can be inhibited in some cases (Widmer et al., 2010), for example, by very high sugar content, nutrient deficiency, extreme temperatures, inhibitors, such as limonene, and aromatic components. Those are hostile conditions that usually arise in the industrial bioenergy sector where maximum conversion of lignocellulose to ethanol is the main interest (Pourbafrani et al., 2007; Alper and Stephanopoulos, 2009; Lohrasbi et al. 2010; Tsukamoto et al., 2013b).

Besides the fermentation issue, improvements can be done in the enzymatic hydrolysis process, since there are bacteria and fungi that are naturally able to degrade biomass. These microorganisms have an inherent capacity and potential not only to use recalcitrant substrates but also to survive toxic products. In fact, many native microorganisms hold biochemical pathways for biomass consumption and conversion into products that are similar to biofuels. The main problem of working with these native microorganisms that are quite different from the ones usually used in industrial processes is the lack of genetic and molecular biology tools, consequently any effort to work with them requires significant amount of time and money (Alper and Stephanopoulos, 2009). Therefore, the search for new microorganism strains is still an open field of investigation (Tsukamoto et al., 2013b).

As reported by Alper and Stephanopoulos (2009), numerous advantages in biofuel production can be achieved by application of isolated native microorganisms. Some of these advantages are the ability of native flora to adapt to the biomass, resistance to inhibition provoked by substrates or products, and, therefore, high yield of biofuel is produced. The native microorganisms work for the biofuel production as if it was a native environment. The intrinsic power of all of these native microorganisms cannot be underestimated.

FINAL REMARKS

Nature works perfectly well and adapts to all kinds of different, new, and hostile environmental factors, lower or higher temperatures, among many other circumstances, by modifying the ways of action, using some alternative routes and directions. It sometimes survives even through promoting species differentiation and creating or modulating new enzymes, proteins, and genes. Borrowing or adapting some of the most interesting solutions from nature, we could solve many problems in modern biotechnology and can add remarkable upgrades in the bioenergy sector, such as 2G-ethanol production starting from agro-industrial waste as a target biomass. We can use many combinations of enzymes, cocktails, and conditions similar to the natural degradation processes for the hydrolysis of almost any kind of biomass. Also a great variety of microorganisms isolated from a given biomass can be used for fermentation processes. These isolated colonies can be screened for synergistic effects in either hydrolysis or fermentation processes and selected in the way that can produce the highest yields of the chosen commodity in the shortest time.

ACKNOWLEDGMENTS

We thank FAPESP, CAPES, CNPq/TWAS, and CNPq for financial support and fellowships.

REFERENCES

Alegria, M.C., et al., 2005. Identification of new protein–protein interactions involving the products of the chromosome- and plasmid-encoded type IV secretion loci of the phytopathogen *Xanthomonas axonopodis* pv. citri. Journal of Bacteriology 187 (7), 2315–2325.

Alper, H., Stephanopoulos, G., 2009. Engineering for biofuels: exploiting innate microbial capacity or importing biosynthetic potential? Nature Reviews. Microbiology 7 (10), 715–723.

Ang, S.K., et al., 2015. Isolation, screening, and identification of potential cellulolytic and xylanolytic producers for biodegradation of untreated oil palm trunk and its application in saccharification of lemongrass leaves. Preparative Biochemistry and Biotechnology 45 (3), 279–305.

Awan, A.T., Tsukamoto, J., Tasic, L., 2013. Orange waste as a biomass for 2G-ethanol production using low cost enzymes and co-culture fermentation. RSC Advances 3 (47), 25071–25078.

Awan, A.T., 2013. Orange bagasse as Biomass for 2G-ethanol Production. (Ph.D. thesis, Unicamp).

Bakri, Y., Jacques, P., Thonart, P., 2003. Xylanase production by *Penicillium canescens* 10–10c in solid-state fermentation. Applied Biochemistry and Biotechnology 108 (1–3), 737–748.

Bampidis, V.A., Robinson, P.H., 2006. Citrus by-products as ruminant feeds: a review. Animal Feed Science and Technology 128 (3–4), 175–217.

Baurhoo, B., Ruiz-Feria, C.A., Zhao, X., 2008. Purified lignin: nutritional and health impacts on farm animals—a review. Animal Feed Science and Technology 144 (3–4), 175–184.

Bermejo, A., Llosá, M.J., Cano, A., 2011. Analysis of bioactive compounds in seven citrus cultivars. Food Science and Technology International 17 (1), 55–62.

Booten, T.J., et al., 2011. Degradation of lignified secondary cell walls of Lucerne (*Medicago sativa* L.) by rumen fungi growing in methanogenic co-culture. Journal of Applied Microbiology 111 (5), 1086–1096.

Brunings, A.M., Gabriel, D.W., 2003. *Xanthomonas citri*: breaking the surface. Molecular Plant Pathology 4 (3), 141–157.

Carvalho, R.N.L., de, C., 2009. Dilute Acid and Enzymatic Hydrolysis of Sugarcane Bagasse for Biogas Production (Master's dissertation). Technical University of Lisboa.

Chang, V., Holtzapple, M., 2000. Fundamental factors affecting biomass enzymatic reactivity. In: Finkelstein, M., Davison, B. (Eds.), Twenty-first Symposium on Biotechnology for Fuels and Chemicals SE - 1. Applied Biochemistry and Biotechnology. Humana Press, pp. 5–37.

Choi, I.S., et al., 2013. Bioethanol production from mandarin (*Citrus unshiu*) peel waste using popping pretreatment. Applied Energy 102, 204–210.

Cubero, J., Graham, J.H., Gottwald, T.R., 2001. Quantitative PCR method for diagnosis of citrus bacterial canker. Applied and Environmental Microbiology 67 (6), 2849–2852.

Da Silva, A.C.R., et al., 2002. Comparison of the genomes of two Xanthomonas pathogens with differing host specificities. Nature 417 (6887), 459–463.

Deswal, D., et al., 2011. Optimization of cellulase production by a brown rot fungus *Fomitopsis* sp. RCK2010 under solid state fermentation. Bioresource Technology 102, 6065–6072.

Dhillon, G.S., et al., 2012. Lactoserum as a moistening medium and crude inducer for fungal cellulase and hemicellulase induction through solid-state fermentation of apple pomace. Biomass and Bioenergy 41, 165–174.

Doherty, W.O.S., Mousavioun, P., Fellows, C.M., 2011. Value-adding to cellulosic ethanol: lignin polymers. Industrial Crops and Products 33 (2), 259–276.

Duff, S.J.B., Murray, W.D., 1996. Bioconversion of forest products industry waste cellulosics to fuel ethanol: a review. Bioresource Technology 55 (1), 1–33.

Eckert, K., et al., 2009. Crystal structures of *A. acidocaldarius* endoglucanase Cel9A in complex with cello-oligosaccharides: strong -1 and -2 subsites mimic cellobiohydrolase activity. Journal of Molecular Biology 394 (1), 61–70.

Edwards, M.C., Doran-Peterson, J., 2012. Pectin-rich biomass as feedstock for fuel ethanol production. Applied Microbiology and Biotechnology 95 (3), 565–575.

Fattori, J., et al., 2011. Structural insights on two hypothetical secretion chaperones from *Xanthomonas axonopodis* pv. citri. The Protein Journal 30 (5), 324–333.

Garcia, A., Oh, S., Engler, C.R., 1989. Cellulase immobilization on Fe_3O_4 and characterization. Biotechnology and Bioengineering 33 (3), 321–326.

Gardner, K.H., Blackwell, J., 1974. The structure of native cellulose. Biopolymers 13 (10), 1975–2001.

Graham, J.H., et al., 2004. *Xanthomonas axonopodis* pv. citri: factors affecting successful eradication of citrus canker. Molecular Plant Pathology 5 (1), 1–15.

Grohmann, K., Baldwin, E.A., 1992. Hydrolysis of orange peel with pectinase and cellulase enzymes. Biotechnology Letters 14 (12), 1169–1174.

Grohmann, K., Baldwin, E.A., Buslig, B.S., 1994. Production of ethanol from enzymatically hydrolyzed orange peel by the yeast *Saccharomyces cerevisiae*. Applied Biochemistry and Biotechnology 45–46 (1), 315–327.

Grohmann, K., Cameron, R.G., Buslig, B.S., 1995. Fractionation and pretreatment of orange peel by dilute acid hydrolysis. Bioresource Technology 54 (2), 129–141.

Guo, P., et al., 2010. Functional characteristics and diversity of a novel lignocelluloses degrading composite microbial system with high xylanase activity. Journal of Microbiology and Biotechnology 20 (2), 254–264.

Himmel, M.E., et al., 2007. Biomass recalcitrance: engineering plants and enzymes for biofuels production. Science (New York, NY) 315 (5813), 804–807.

Hon, D.N.-S., Shiraishi, N., 2000. Wood and Cellulosic Chemistry, Second Edition, Revised, and Expanded.

Hsu, T.A., 1996. Handbook on Bioethanol: Production and Utilization. Applied Energy Technology Series. Taylor & Francis, Washington DC, p. 179.

Huang, S.-H., Liao, M.-H., Chen, D.-H., 2003. Direct binding and characterization of lipase onto magnetic nanoparticles. Biotechnology Progress 19 (3), 1095–1100.

Huang, X., et al., 2013. Isolation and characterization of lignin-degrading bacteria from rainforest soils. Biotechnology and Bioengineering 110 (6), 1616–1626.

Jalan, N., et al., 2011. Comparative genomic analysis of Xanthomonas axonopodis pv. citrumelo F1, which causes citrus bacterial spot disease, and related strains provides insights into virulence and host specificity. Journal of Bacteriology 193 (22), 6342–6357.

Jayani, R.S., Saxena, S., Gupta, R., 2005. Microbial pectinolytic enzymes: a review. Process Biochemistry 40 (9), 2931–2944.

Jordan, J., Kumar, C.S.S.R., Theegala, C., 2011. Preparation and characterization of cellulase-bound magnetite nanoparticles. Journal of Molecular Catalysis B: Enzymatic 68 (2), 139–146.

Kale, P.N., Adsule, P.G., 1995. Handbook of Fruit Science and Technology. Production, Composition, Storage and Processing. Marcel Dekker, New York.

Khater, L., et al., 2007. Identification of the flagellar chaperone FlgN in the phytopathogen Xanthomonas axonopodis pathovar citri by its interaction with hook-associated FlgK. Archives of Microbiology 188 (3), 243–250.

Kirk, T.K., 1971. Effects of microorganisms on lignin. Annual Review of Phytopathology 9, 185–210.

Kumar, R., Singh, S., Singh, O.V., 2008. Bioconversion of lignocellulosic biomass: biochemical and molecular perspectives. Journal of Industrial Microbiology & Biotechnology 35 (5), 377–391.

Limayem, A., Ricke, S.C., 2012. Lignocellulosic biomass for bioethanol production: current perspectives, potential issues and future prospects. Progress in Energy and Combustion Science 38 (4), 449–467.

Lin, H.-C., et al., 2010. A pectate lyase homologue PEL1 from Xanthomonas axonopodis pv. citri is associated with the water-soaked margins of canker lesions. Journal of Plant Pathology 92, 149–156.

Liu, L., et al., 2004. Pectin/poly(lactide-co-glycolide) composite matrices for biomedical applications. Biomaterials 25 (16), 3201–3210.

Liu, L.-P., et al., 2012. Construction of EGFP-labeling system for visualizing the infection process of Xanthomonas axonopodis pv. citri in planta. Current Microbiology 65 (3), 304–312.

Lohrasbi, M., et al., 2010. Process design and economic analysis of a citrus waste biorefinery with biofuels and limonene as products. Bioresource Technology 101 (19), 7382–7388.

Mansfield, S., Mooney, C., Saddler, J., 1999. Substrate and enzyme characteristics that limit cellulose hydrolysis. Biotechnology Progress 15 (5), 804–816.

Menon, V., Rao, M., 2012. Trends in bioconversion of lignocellulose: biofuels, platform chemicals & biorefinery concept. Progress in Energy and Combustion Science 38 (4), 522–550.

Mes-Hartree, M., Hogan, C.M., Saddler, J.N., 1987. Recycle of enzymes and substrate following enzymatic hydrolysis of steam-pretreated aspenwood. Biotechnology and Bioengineering 30 (4), 558–564.

Moreira, L.M., et al., 2004. Comparative genomics analyses of citrus-associated bacteria. Annual Review of Phytopathology 42 (1), 163–184.

Mussatto, S.I., Teixeira, J.A., 2010. Lignocellulose as Raw Material in Fermentation Processes. Formatex.

Nielsen, H., et al., 1997. Identification of prokaryotic and eukaryotic signal peptides and prediction of their cleavage sites. Protein Engineering Design and Selection 10 (1–6), 1–15.

Numan, M.T., Bhosle, N.B., 2006. Alpha-L-arabinofuranosidases: the potential applications in biotechnology. Journal of Industrial Microbiology & Biotechnology 33 (4), 247–260.

Oberoi, H.S., et al., 2010. Ethanol production from orange peels: two-stage hydrolysis and fermentation studies using optimized parameters through experimental design. Journal of Agricultural and Food Chemistry 58 (6), 3422–3429.

Piskur, J., et al., 2006. How did Saccharomyces evolve to become a good brewer? Trends in Genetics: TIG 22 (4), 183–186.

Plomion, C., Leprovost, G., Stokes, A., 2001. Wood formation in trees. Plant Physiology 127 (4), 1513–1523.

Pourbafrani, M., et al., 2007. Protective effect of encapsulation in fermentation of limonene-contained media and orange peel hydrolyzate. International Journal of Molecular Sciences 8 (8), 777–787.

Rivas, B., et al., 2008. Submerged citric acid fermentation on orange peel autohydrolysate. Journal of Agricultural and Food Chemistry 56 (7), 2380–2387.

Saha, B.C., et al., 2005. Dilute acid pretreatment, enzymatic saccharification and fermentation of wheat straw to ethanol. Process Biochemistry 40 (12), 3693–3700.

Saratale, G.D., et al., 2014. Cellulolytic enzymes production by utilizing agricultural wastes under solid state fermentation and its application for biohydrogen production. Applied Biochemistry and Biotechnology 174, 2801–2817.

Sarkar, N., et al., 2012. Bioethanol production from agricultural wastes: an overview. Renewable Energy 37 (1), 19–27.

Schaad, N.W., et al., 2006. Emended classification of xanthomonad pathogens on citrus. Systematic and Applied Microbiology 29 (8), 690–695.

Sheng, P., et al., 2012. Isolation, screening, and optimization of the fermentation conditions of highly cellulolytic bacteria from the hindgut of *Holotrichia parallela* larvae (Coleoptera: *Scarabaeidae*). Applied Biochemistry and Biotechnology 167, 270–284.

Smook, G.A., 2002. Handbook for Pulp & Paper Technologists, 3nd ed. Angus Wilde Publications Inc, Vancouver, British Columbia.

Soccol, C.R., et al., 2010. Bioethanol from lignocelluloses: status and perspectives in Brazil. Bioresource technology 101 (13), 4820–4825.

Sternberg, D., 1976. Beta-glucosidase of trichoderma: its biosynthesis and role in saccharification of cellulose. Applied and Environmental Microbiology 31 (5), 648–654.

Stewart, D., et al., 2006. Apparatus Which Serves as Bioreactor Which Ferments Heated Citrus Waste into Ethanol, Acetates, Galacturonic Acid Monomers and Polymers and Five Carbon Sugars. US 20060177916 A1.

Stewart, D., et al., 2008. Ethanol Production from Solid Citrus Processing Waste. US 8372614 B2.

Sticklen, M.B., 2008. Plant genetic engineering for biofuel production: towards affordable cellulosic ethanol. Nature Reviews Genetics 9 (6), 433–443.

Sun, Y., Cheng, J., 2002. Hydrolysis of lignocellulosic materials for ethanol production: a review. Bioresource Technology 83 (1), 1–11.

Takasuka, T.E., et al., 2013. Aerobic deconstruction of cellulosic biomass by an insect-associated *Streptomyces*. Scientific Reports 3:1030.

Tasic, L., et al., 2007. Cloning and characterization of three hypothetical secretion chaperone proteins from *Xanthomonas axonopodis* pv. citri. Protein Expression and Purification 53 (2), 363–369.

Torres-Salas, P., et al., 2011. Immobilized biocatalysts: novel approaches and tools for binding enzymes to supports. Advanced Materials 23 (44), 5275–5282.

Tsukamoto, J., Awan, A.T., Durán, N., Tasic, L., 2013a. 18/12/2013: Processo de obtenção de bioetanol, esperidina e nanocelulose a partir de bagaço de laranja. BR1020130325856.

Tsukamoto, J., Durán, N., Tasic, L., 2013b. Nanocellulose and bioethanol production from orange waste using isolated microorganisms. Journal of the Brazilian Chemical Society 24 (9), 1537–1543. http://dx.doi.org/10.5935/0103-5053.20130195.

Watt, S.A., Wilke, A., Patschkowski, T., Niehaus, K., 2005. Comprehensive analysis of the extracellular proteins from *Xanthomonas campestris* pv. campestris B100. Proteomics 5 (1), 153–167.

Widmer, W., Zhou, W., Grohmann, K., 2010. Pretreatment effects on orange processing waste for making ethanol by simultaneous saccharification and fermentation. Bioresource Technology 101 (14), 5242–5249.

Wilkins, M.R., et al., 2005. Effect of seasonal variation on enzymatic hydrolysis of Valencia orange peel waste. Proceedings of the Florida State Horticultural Society 118, 419–422.

Willats, W.T., et al., 2001. Pectin: cell biology and prospects for functional analysis. In: Carpita, N.C., Campbell, M., Tierney, M. (Eds.), Plant Cell Walls SE - 2. Springer Netherlands, pp. 9–27.

Willats, W.G., Knox, J.P., Mikkelsen, J.D., 2006. Pectin: new insights into an old polymer are starting to gel. Trends in Food Science & Technology 17 (3), 97–104.

Wyman, C., 1996. Handbook on Bioethanol: Production and Utilization Applied Energy Technology Series. Taylor & Francis, Washington DC.

Wyman, C.E., 1999. Biomass ethanol: technical progress, opportunities, and commercial challenges. Annual Review of Energy and the Environment 24 (1), 189–226.

Xin, F., Geng, A., 2010. Horticultural waste as the substrate for cellulase and hemicellulase production by *Trichoderma reesei* under solid-state fermentation. Applied Biochemistry and Biotechnology 162 (1), 295–306. http://dx.doi.org/10.1007/s12010-009-8745-2.

Zhang, Y.-H.P., Lynd, L.R., 2004. Toward an aggregated understanding of enzymatic hydrolysis of cellulose: noncomplexed cellulase systems. Biotechnology and Bioengineering 88 (7), 797–824.

Zhao, X., Zhang, L., Liu, D., 2012. Biomass recalcitrance. Part II: fundamentals of different pre-treatments to increase the enzymatic digestibility of lignocellulose. Biofuels, Bioproducts and Biorefining 6 (5), 561–579.

Zhong, R., Ye, Z.-H., 2009. In: Secondary Cell Walls. eLS. John Wiley & Sons, Ltd.

FURTHER READING

Andorrà, I., et al., 2010. Effect of pure and mixed cultures of the main wine yeast species on grape must fermentations. European Food Research and Technology 231 (2), 215–224.

Andorrà, I., et al., 2012. Effect of mixed culture fermentations on yeast populations and aroma profile. LWT – Food Science and Technology 49 (1), 8–13.

Bansal, N., et al., 2011. A novel strain of *Aspergillus niger* producing a cocktail of hydrolytic depolymerising enzymes for the production of second generation biofuels. BioResources 6 (1), 552–569.

Bradford, M., 1976. A rapid and sensitive method for the quantitation of microgram quantities of protein utilizing the principle of protein-dye binding. Analytical Biochemistry 72 (1–2), 248–254.

Brás, J.L.A., et al., 2011. Structural insights into a unique cellulase fold and mechanism of cellulose hydrolysis. Proceedings of the National Academy of Sciences of the United States of America 108 (13), 5237–5242.

Carrão-Panizzi, M.C., Bordingnon, J.R., 2000. Activity of beta-glucosidase and levels of isoflavone glucosides in soybean cultivars affected by the environment. Pesquisa Agropecuária Brasileira 35 (5), 873–875.

Chen, M., Xia, L., Xue, P., 2007. Enzymatic hydrolysis of corncob and ethanol production from cellulosic hydrolysate. International Biodeterioration & Biodegradation 59 (2), 85–89.

Dinh, T.N., et al., 2008. Adaptation of *Saccharomyces cerevisiae* cells to high ethanol concentration and changes in fatty acid composition of membrane and cell size. PLoS ONE 3 (7), e2623.

Duetz, W.A., et al., 2003. Biotransformation of limonene by bacteria, fungi, yeasts, and plants. Applied Microbiology and Biotechnology 61 (4), 269–277.

Gafner, J., Schütz, M., 1996. Impact of glucose-fructose-ratio on stuck fermentations: practical experiences to restart stuck fermentations. Wein-Wissenschaft 51 (3–4), 214–218.

Garde-Cerdán, T., Ancín-Azpilicueta, C., 2006. Contribution of wild yeasts to the formation of volatile compounds in inoculated wine fermentations. European Food Research and Technology 222 (1–2), 15–25.

Ghose, T.K., 1987. Measurement of cellulase activities. Pure and Applied Chemistry 59 (2), 257–268.

Hari Krishna, S., Chowdary, G.V., 2000. Optimization of simultaneous saccharification and fermentation for the production of ethanol from lignocellulosic biomass. Journal of Agricultural and Food Chemistry 48 (5), 1971–1976.

de Melo, T.C.C., et al., 2012. Hydrous ethanol–gasoline blends – combustion and emission investigations on a flex-fuel engine. Fuel 97, 796–804.

Miller, G.L., 1959. Use of dinitrosalicylic acid reagent for determination of reducing sugar. Analytical Chemistry 31 (3), 426–428.

Moreira, N., et al., 2008. Heavy sulphur compounds, higher alcohols and esters production profile of *Hanseniaspora uvarum* and *Hanseniaspora guilliermondii* grown as pure and mixed cultures in grape must. International Journal of Food Microbiology 124 (3), 231–238.

Pascual, J.M., Carmona, J.F., 1980. Composition of citrus pulp. Animal Feed Science and Technology 5 (1), 1–10.

Perkins, D.N., et al., 1999. Probability-based protein identification by searching sequence databases using mass spectrometry data. Electrophoresis 20 (18), 3551–3567.

Petersen, T.N., et al., 2011. SignalP 4.0: discriminating signal peptides from transmembrane regions. Nature Methods 8 (10), 785–786.

Phutela, U., et al., 2005. Pectinase and polygalacturonase production by a thermophilic *Aspergillus fumigatus* isolated from decomposting orange peels. Brazilian Journal of Microbiology 36 (1), 63–69.

Puri, V.P., 1984. Effect of crystallinity and degree of polymerization of cellulose on enzymatic saccharification. Biotechnology and Bioengineering 26 (10), 1219–1222.

Subba, M.S., Soumithri, T.C., Rao, R.S., 1967. Antimicrobial action of citrus oils. Journal of Food Science 32 (2), 225–227.

Sudhakar, D.V., Maini, S.B., 2000. Isolation and characterization of mango peel pectins. Journal of Food Processing and Preservation 24 (3), 209–227.

Taherzadeh, M.J., Karimi, K., 2008. Pretreatment of lignocellulosic wastes to improve ethanol and biogas production: a review. International Journal of Molecular Sciences 9 (9), 1621–1651.

Warren, R.K., Hill, G.A., Macdonald, D.G., 1990. Improved bioreaction kinetics for the simulation of continuous ethanol fermentation by *Saccharomyces cerevisiae*. Biotechnology Progress 6 (5), 319–325.

Wilkins, M.R., et al., 2007a. Ethanol production by *Saccharomyces cerevisiae* and *Kluyveromyces marxianus* in the presence of orange-peel oil. World Journal of Microbiology and Biotechnology 23 (8), 1161–1168.

Wilkins, M.R., et al., 2007b. Hydrolysis of grapefruit peel waste with cellulase and pectinase enzymes. Bioresource Technology 98 (8), 1596–1601.

Wilkins, M.R., Widmer, W.W., Grohmann, K., 2007c. Simultaneous saccharification and fermentation of citrus peel waste by *Saccharomyces cerevisiae* to produce ethanol. Process Biochemistry 42 (12), 1614–1619.

CHAPTER 6

Role of Enzymes in Environment Cleanup/Remediation

L. Gianfreda[1], M.A. Rao[1], R. Scelza[1], M. de la Luz Mora[2]
[1]University of Naples Federico II, Portici, Italy; [2]University of La Frontera, Temuco, Chile

INTRODUCTION

Nowadays environmental pollution (including water, air, and soil compartments) has reached elevated levels to be of great concern for humans and other living organisms. A large number of organic and inorganic chemicals with severe polluting and hazardous properties have been released into the environment because of natural and anthropogenic activities, including volcanic eruptions and weathering of the soil parent material and the rapid industrialization of agriculture, expansions in the chemical industry, and the need to generate cheap forms of energy that have produced several polluting compounds.

The continuous pollution of the environment has led to the increase of contaminated sites with possible negative, synergistic effects on environmental health. Therefore, decontamination and restoration of polluted sites has become an urgent need to recovering soil health and fertility, detoxifying ground water, and perhaps reutilizing wastewater (particularly in countries with severe water deficiencies) and producing healthy air (Gianfreda et al., 2006a,b).

Several research efforts have been and are dedicated to find effective, eco-friendly, and possibly low-cost tools to mitigate the pollution and to clean up and restore polluted environments. The nature of the contaminant sources and the presence of contaminant cocktails composed of combinations of many organic and inorganic compounds often makes problematic the remediation of polluted sites (Thavamani et al., 2012). Moreover, if the contaminated site is soil, interactions at interfaces between organic and inorganic soil colloids and pollutants, through sorption/desorption mechanisms, may affect the movement of pollutants, their availability for plant or microbial uptake, and their transformation by abiotic or biotic agents.

Two basic approaches have been utilized for cleaning up aquatic or terrestrial polluted environments: physical/chemical or biological strategies. As specified in later discussion physical/chemical methodologies are traditional approaches and often involve high costs and do not imply a final, safe result. Biological strategies involving living

Agro-Industrial Wastes as Feedstock for Enzyme Production
ISBN 978-0-12-802392-1
http://dx.doi.org/10.1016/B978-0-12-802392-1.00006-X

133

organisms (ie, microorganisms, plants, plant–microorganism associations or some of their main components) appear more environmentally friendly, and in many cases they may extensively modify the structure and toxicological properties of the contaminants. The complete conversion of the molecule into innocuous inorganic end products may result. In addition, biological agents may carry out processes for which no efficient chemical transformations have been devised.

Enzymes are the main effectors of metabolic pathways occurring in living organisms. They have either narrow or broad specificity, and their catalytic capability is so high that also recalcitrant (ie, less or not transformable, at all) compounds can be converted into safer or innocuous final products.

Therefore, enzymes constitute an alternative to traditional, non-biological cleanup methodologies (Karam and Nicell, 1997; Nicell, 2001). Enzymes have been used for the transformation of single or complex mixtures of pollutants. Moreover, enzymes also have been used to detect and quantify the status of the environment before, during, and after the remediation process. This chapter provides a brief overview on the current status of enzymes application for pollution removal. Their involvement in the transformation of so-called "emerging pollutants" as well as their use as indicators of polluted/restored environments also will be addressed.

POLLUTION OF THE ENVIRONMENT

Different products from agriculture, industries, and urbanization are continuously spilled in soil, water, and air. Although these compounds have improved our lifestyle, some of them may accumulate in one or all the environmental compartments, with severe effects on human and environmental health. Pollutants can be classified based on their chemical nature (inorganic or organic) and/or on their originating source (natural or anthropogenic activities). Organic pollutants, mostly of synthetic origins, have different structures and different toxicity levels and may persist for long periods being in many cases resistant to any kind of physical, chemical, or biological degradation (Gianfreda and Nannipieri, 2001). Indeed, a synthetic chemical that is not a product of biosynthesis will only be biologically degraded if an enzyme or an enzyme system is able to catalyze its conversion into an intermediate or a substrate able to participate in existing metabolic pathways. Following prolonged exposure, those pollutants have the potential to produce adverse effects in humans and environment.

The most characteristic organic pollutants are pesticides, oil-based compounds, fuels, solvents, alkanes, polycyclic aromatic hydrocarbons (PAHs), nitrogen, and phosphorus compounds. Some compounds, such as phenols, chloro- and nitrophenols, PAHs, and others belong to the class of persistent organic pollutants (POPs) that have high toxicity toward humans, plants, and animals. POPs are highly stable compounds that can last for years or decades before breaking down. For instance, several halogenated organic compounds, with polluting properties, released in the environment as herbicides, insecticides, fungicides, and solvents, present very stable halogen–carbon linkages. The toxicity and harmful effects of these pollutants is strictly related to the number, type, and position of the halogen substituents (Gianfreda and Nannipieri, 2001).

The recalcitrance of POPs to biological transformation into less toxic products results in their accumulation and persistence in the environment in unaltered, less-degradable chemical forms (Jones and de Voogt, 1999; Reid, 2000). As greater is the difference in the structure of the xenobiotic form from natural compounds, the lower is the possibility for significant biodegradation. For example, simple, C_1–C_{15} hydrocarbons, alcohols, phenols, amines, acids, esters, and amides are very easily biodegradable, while C_{20} and higher hydrocarbons, multi-Cl-hydrocarbons, PAHs polychlorinated biphenyls (PCBs), pesticides are moderately or not at all biodegradable. The US Environmental Protection Agency and the European Community have listed PAHs and/or some complex phenols as priority pollutants (ie, whose removal from the environment should be a top priority).

Similarly, inorganic contaminants, principally represented by heavy metals, if present at high concentrations, may become contaminants because of their accumulation in some ecosystems. Preferred sink of these pollutants is soil, sediments, and natural waters. In particular, brown-colored effluents originating from the chlorine-bleaching process of pulp and paper mills contain a large amount of wood resins, chlorinated phenols, and tannins (Rubilar et al., 2008). Similarly, considerable volumes of phenol-polluted waters are produced by the olive oil industry, mainly in the south bank of the Mediterranean where the main olive-producing countries (Italy, Spain, Morocco, Tunisia, and Turkey) lie.

Extensive literature data demonstrate POPs' toxic effects on environmental compartments whereas little information is available about the "emerging pollutants" that are becoming particularly attractive for researchers in recent years. The so-called "emerging pollutants" represent a heterogeneous class of chemical compounds produced from anthropic activities, not subjected to specific regulation, and for some of them, with unknown effects in the mid- or long-term on human health, terrestrial, and aquatic environments (Deblonde et al., 2011; Gavrilescu et al., 2015). They include personal care products, fragrances, plasticizers, steroids and hormones, illicit drugs, gasoline additives, flame retardants, metals, pesticides, phthalates, PAHs, pharmaceuticals, and endocrine-disrupting compounds (EDCs) (Demarche et al., 2012). Their contamination mainly affects aquatic environments, and the problem is the lack of knowledge about their nature, pathways, and fate in wastewaters. Pharmaceutically active compounds (PhACs) constitute a group of emerging pollutants widely present in the environment, mainly wastewater. Analgesics, antibiotics, and hormones, belonging to the pharmaceutical group, need a different approach for monitoring and control because the conventional wastewater treatment plants cannot be not efficient in their removal (Deblonde et al., 2011).

A particular interest is devoted to EDCs because their properties and chemical structure make them particularly active and toxic toward the hormone systems in mammals (Gavrilescu et al., 2015). They are synthetic phenolic compounds, such as bisphenol A 2, 2-bis (4-hydroxyphenyl) propane (BPA), nonylphenol (NP), their alkyl-phenolic derivatives, triclosan, genistein, and others, broadly used in several industrial and residential applications. Due to their chemical structure, they may mimic the effects of hormones (particularly estrogens), and as such, they may have estrogenic (endocrine-disrupting) activity, with adverse effects on reproduction in wildlife and humans. As some of them (hormones, glucocorticoids, and additives in drugs and cosmetics) are daily used as pharmaceuticals and for personal care (Gavrilescu et al., 2015), more research is needed about their toxicological effects. Dyes are another class of synthetic chemicals, widely used in the industry, with a high toxicity toward living organisms (Dhillon et al., 2012).

A synthetic compound that has recently attracted attention as an emerging groundwater contaminant is 1,2,3-trichloropropane (TCP), a toxic industrial waste product largely utilized as paint or varnish remover, cleaning and degreasing agent, and in the production of pesticides. Its degradation is quite difficult and its accumulation in the environment is of public concern.

As suggested by Gavrilescu et al. (2015), an interdisciplinary approach is needed to identify all the emerging pollutants and their interactions with soil, sediment, and water ecosystems. Only such an approach can be successful.

METHODOLOGIES TO CLEANUP POLLUTED ENVIRONMENTS

Natural attenuation is a process occurring without human intervention and to a varying degree at all environmental sites, which results in reducing the concentration of a contaminant (hydrodynamic dispersion, sorption, and volatilization) or the mass of contaminants (biodegradation). If natural attenuation processes are not enough to achieve site-specific remediation objectives within a reasonable and acceptable time, the aid of other remedial measures or the application of enhancers of biological activity should be considered also in conjunction with the natural resilience of the contaminated environment. In recent decades, the approach in remediating environmental sites, differently contaminated by organic and inorganic pollutants, has been based on physical and chemical methodologies. They gather a wide range of in situ and ex situ technologies (Fig. 6.1) specific for soil or groundwater.

The choice of the more suitable technology needs detailed knowledge of all parameters characterizing both sites and contaminants. Besides remediation strategy, in some cases when the contamination levels are high and sites can be only confined, safety measures could be necessary (vitrification, reactive permeable barriers, Fig. 6.1). Traditional methodologies, especially ex situ ones, requiring transport of polluted matrices to specific plants and the use of expensive equipment, considerably increase project costs and, therefore, are less attractive from a practical point of view.

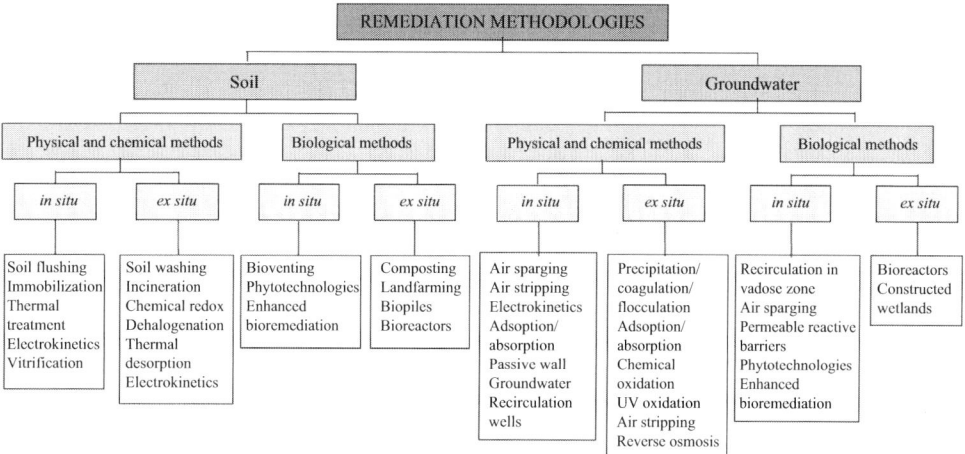

Figure 6.1 Methodologies of soil and groundwater remediation.

A valid alternative with reduced environmental and economic impact are biological methodologies. They consist of the use of biomolecules such as enzymes or directly living organisms such as plants and microorganisms able to degrade contaminants or transform them in less harmful compounds. Ex situ and in situ strategies also belong to this category (Fig. 6.1).

Brief details on some biological technologies, already reported in Fig. 6.1, are reported in the following discussion.

Bioventing is an in situ remediation technology that uses air (or oxygen) flow to enhance indigenous microorganisms to biodegrade organic constituents adsorbed to soils in the unsaturated zone by also adding nutrients, if necessary.

Biosparging is an in situ remediation technology that injects air (or oxygen) and nutrients (if needed) into the saturated zone to increase the biological activity of the indigenous microorganisms.

Biopiles are used to reduce concentrations of petroleum constituents in excavated soils heaped into piles and stimulating aerobic microbial activity within the soils through the aeration and/or addition of minerals, nutrients, and water. *Phytoremediation* is the direct use of green plants and their associated microorganisms to stabilize or reduce contamination (organic and inorganic contaminants) in soils, sludge, sediments, surface water, or groundwater.

Mycoremediation is a biotechnology based on the use of fungi for cleaning up contaminated soils. Principally, white-rot fungi are very effective in degrading a wide range of organic molecules as they release extracellular lignin-modifying enzymes (lignin peroxidase, manganese peroxidase, and laccase) with a low substrate specificity. Carbon sources, such as sawdust, straw, and corn cob can be added to polluted sites to augment degradation processes.

Composting is a treatment where polluted materials (generally organic compounds) are mixed together in piles with a solid organic substance readily degradable, such as straw, wood chips, etc.

Land farming involves spreading excavated contaminated soils in a thin layer on the ground surface and stimulating aerobic microbial activity within the soils through aeration and/or the addition of nutrients and water.

Bioreactors consist of any vessel or container where biological degradation of contaminants in soil and water by existing and/or added microorganisms is isolated and controlled.

Microbial filters are packed-bed bioreactors, where microorganisms are allowed to grow in order to degrade volatile compounds once adsorbed on solid supports (activated carbon, soil, peat, etc.).

The use of biological systems to bioremediation approach requires the characterization of microbial community composition, cellular and molecular activity, especially if complicated by the presence of toxic chemicals that alter the normal behavior of the microbial community. Modern tools of genomics, transcriptomics, proteomics, metabolomics, phenomics, and lipidomics can be applied to realize an integrated research approach, useful to investigate interactions and networks at the molecular, cellular, community, and ecosystem levels. The high cost associated with sample processing equipment, materials and reagents, large amounts of samples required, the need for skilled personnel to process the samples, massive amounts of data generated, and the time-consuming nature in integration and synthesis of the data represent the limits of their application (Chakraborty et al., 2012). It is also true that omic approach needs the choice of the more useful and suitable method mainly depending on the objectives of the research. In fact, tools able to identify RNA and protein (ie, GeoClip, RNAseq, mass spectrometry) should be adopted if the purpose is the knowledge of cellular pathways and the identification of functional genes involved in microbially mediated reactions. Otherwise, more-sophisticated techniques such as nuclear magnetic resonance, matrix-assisted laser desorption (MALDI), or electrospray ionization allow focusing on small molecules produced by microbes. Microbial community proteomics and metabolomics instead contribute to better understanding the microbial cellular functions and gene products interplaying in the environment (Chakraborty et al., 2012).

ENZYMES IMPORTANT TO POLLUTANT REMOVAL

Microorganisms produce enzymes able to react with chemicals different from those utilized as primary carbon and energy sources, thus becoming a potential source of a large array of enzymes useful for the transformation of various xenobiotic compounds. Indeed, microorganisms have unique capabilities absent in species that are more evolved. High adaptability, versatility, and mutability of microorganisms render them potentially active even toward unknown chemicals. The main effectors of these microbial properties are their enzymatic components that can carry out processes for which no efficient chemical transformations have been devised. Both bacterial and fungal enzymes are usually catalysts with either narrow or broad specificity, thus capable of catalyzing the transformation of a large range of different compounds (Whiteley and Lee, 2006). As they are of natural origin, enzymes can be considered eco-friendly catalysts and their application a green technology.

Several studies proposed enzymes as feasible and efficient agents for environment (mainly soil) decontamination (Bollag et al., 1988, 2003; Nannipieri and Bollag, 1991; Bollag, 1992; Gianfreda and Bollag, 1996, 2002; Gianfreda et al., 1999; Ahn et al., 2002). The majority of these reports dealt with the use of isolated, extracellular enzymes, often from microbial origin. Plant enzymes were also considered (Dec and Bollag, 1994; Toscano et al., 2003).

The two main classes of enzymes useful for pollutant transformation are oxidoreductases and hydrolases. Both may perform their catalytic action within their originating cellular sources or outside and far from them (Gianfreda and Rao, 2004). The peculiarity of these two classes of enzymes is their double degradative and protective function. For instance, hydrolases mainly transform complex substrates in partially degraded products that easily enter the metabolic pathways of microbial and plant cells. Moreover, they cooperate with oxidoreductases that besides a degradative action may also have a protective role by catalyzing the polymerization of simple, toxic compounds in polymeric, often insoluble and no more cell-available products.

A list of hydrolases and oxidoreductases involved in the transformation of natural and nonnatural, soluble and insoluble compounds, their preferential polluting substrates and main sources is reported in Table 6.1.

Examples concerning two classes of enzymes, laccases and dehalogenases, representative of oxidoreductases and hydrolases, respectively, will be examined.

Laccases

Laccases are enzymes largely studied from as early as 1883 (Yoshida, 1883) for their capability of transforming a wide range of phenolic and nonphenolic compounds. Numerous papers, book chapters, and reviews are available in literature (Gianfreda and Bollag, 1994; Luterek et al., 1997; Gianfreda et al., 1998, 1999, 2003, 2006; Filazzola et al., 1999; Cameron et al., 2000; Durán and Esposito, 2000; Duran et al., 2002; Gianfreda and Bollag, 2002; Bollag et al., 2003; Claus, 2004; Gianfreda and Rao, 2004, 2008; Verdin et al., 2004; Baldrian, 2006; Rodrıguez Couto and Toca Herrera, 2006; Ruggaber and Talley, 2006; Canfora et al., 2008; Longoria et al., 2008; Rubilar et al., 2008; Iamarino et al., 2009; Kudanga et al., 2011; Strong and Claus, 2011; Dhillon et al., 2012; Fernández-Fernández et al., 2012; Torres et al., 2003; Virk et al., 2012; Viswanath et al., 2014).

Table 6.1 Oxidoreductases and Hydrolases Involved in Pollutant Transformation

Enzyme	Source	Pollutant
Oxidoreductases		
Catechol dioxygenases	*Comamonas testosteroni, Pseudomonas pseudoalcaligenes*	Chlorophenol, diuron, polychlorinated biphenyls (PCB), chloroethanes
Cellobiose dehydrogenase (CDH)	White-rot fungi (WRF), *Phanerochaete chrysosporium*	CCl_4, $CHCl_3$, trinitrotoluene (TNT), polyacrylate
Chloroperoxidase	*Cladariomyces fumago*	Asphaltenes, phenols, PAHs
Laccase	*Pycnoporus sanguineis, P. chrysosporium Trametes versicolor, Pycnoporus cinnabarinus, Trametes hispida, Pyricularia oryzae, Cerrena unicolor*, plant materials	Azo dyes, Bleach plant effluents, PAHs, Polyvinyl alcohol, pentachlorophenol (PCP), dichlorodiphenyltrichloroethane (DDT), PCB, lindane, phenols, urea derivatives, benzopyrene
LDSs	*P. chrysosporium, Inonotus dryophillis, T. versicolor*	PAHs, PCP
Lignin peroxidase (LiP)	White-rot fungi, *P. chrysosporium, Coriolopsis polyzona, Pleurotus ostreatus, Nematolona frowardii, T. versicolor, Chrysonilia sitophila*	Kraft effluent, lignin, CCl_4, CHC3, PCB (1–6 Cl substitutions), TNT, Anthracene, pyrene, estrogenic chemicals, phenols, PAHs
Manganese peroxidase (MnP)	White-rot fungi, *P. chrysosporium, C. polyzona, P. ostreatus, T. versicolor, N. frowardii, Phebia radiata, Lentinula edodes*, unnamed white-rot fungus (+Mn and lactate)	Kraft effluent, lignin, CCl_4, CHCl3, PCB (1–6 Cl substitutions), TNT, estrogenic chemicals, phenols, PAH, nylon, PCP, dyes
Peroxidases	Horseradish, several microorganisms, *Artromyces ramosus*	Phenols, chlorophenols, herbicides. Kraft effluent, black liquor, PAHs
Phenoloxidase-like	*Gloeophyllum trabeum, T. versicolor, P. chrysosporium, Thermoascus aurantiacus*	Kraft effluent, chlorinated compounds
Tyrosinase	*Agaricus bisporus*	Phenols, amines
Hydrolases		
Permethrinase	*Agrobacterium, Pseudomonas sp. Flavobacterium sp.*	Pyrethroids, Parathion, Coumaphos, Diazinon
Parathion hydrolase	*Nocardia sp, Bacillus cereus*	Polyvinyl alcohol
2-4-Pentanedione esterase	*Pseudomonas vericularis*	
Alkaline protease	*Bacillus sp.*	Poly(L-Lactic acid)
Amylase	*Bacillus licheniformis*	Starch materials
Carbamate hydrolase	*Achromobacter sp., Pseudomonas sp.*	Carbofuran, Carbaryl
Cellulase	*Trichoderma reesei, Penicillium funiculosum*	Cellulose materials

Table 6.1 Oxidoreductases and Hydrolases Involved in Pollutant Transformation—cont'd

Enzyme	Source	Pollutant
Chitinase	Actinobacteria	Chitin
Cyanidase, Cyanide hydratase	*Alcaligenes denitrificans*, several fungi	Cyanides
Dehalogenases	Several microorganisms	PCP, DDT, PCB, lindane
Depolymerase	*Amycolatopsis*	Poly(L-Lactic acid)
Esterase	*Curvularia senegalensis* (not WRF), *Corynebacterium*, *Comamonas acidovorans*	Polyurethane
Keratinase	*Chrisosporium keratinophilum*	Keratin
Nitrilase, nitrile hydratase, amidase	*Nocardia* sp., *Rhodococcus* sp., *Fusarium solani*	Nitrile compounds
Protease, Phosphatase	Sulphate-reducing bacteria	Sewage sludge
Xylanase β-Xylosidase	*Streptomyces thermoviolaceus*	Kraft pulp

Along with the lignin peroxidase (LiP) and manganese-dependent peroxidase (MnP), laccase constitutes the lignin-degrading enzyme system (LDS), a very powerful extracellular, fungal oxidative enzymatic system with broad substrate specificity and capable to oxidize several environmental pollutants (Reddy, 1995; Pointing, 2001; Baldrian, 2006; Viswanath et al., 2014). Differently from LiP and MnP, laccase requires only oxygen for catalysis, thus being particularly appropriate for biotechnological applications in the transformation of polluting compounds.

Laccases catalyze the one electron oxidation of substituted phenols, anilines, and aromatic thiols to the corresponding radicals with the concomitant reduction of molecular oxygen to water. These radicals produce polymeric products by self-coupling or cross-coupling with other molecules, and dechlorination, demethoxylation, and decarboxylation during coupling and polymerization of differently substituted substrates may occur (Gianfreda et al., 1999). Less- or non-reactive substances, including also highly recalcitrant compounds, may also be transformed in the presence of highly reactive substances acting as mediators. They can be of both synthetic [2,2′-azino-bis(3-ethylbenzthiazoline-6-sulfonic acid) (ABTS), 3-hydroxyanthranilate, 1-hydroxybenztriazole] or natural (vanillin, acetovanillone, acetosyringone, syringaldehyde, 2,4,6-trimethylphenol, *p*-coumaric acid, ferulic acid, and sinapic acid) origins (Husain and Husain, 2007; Gianfreda et al., 1999; Gianfreda and Rao, 2004). Therefore, these enzymes appear suitable and versatile catalysts, very useful for the application in several biotechnological processes (Rodríguez Couto and Toca Herrera, 2006; Rodríguez Couto, 2009).

As recently well summarized by Viswanath et al. (2014), many aspects of laccases have been investigated. The origin and distribution of these enzymes, the different procedures utilized for their production and isolation, the characteristics at molecular levels of the proteins, the substrate specificity of laccases as well as the basic and applied areas in which laccases have been utilized, have been discussed. In this chapter, few examples among most recent findings will be illustrated. Particular attention has been devoted to the possible use of laccase for transformation of emerging pollutants (Majeau et al., 2010).

Laccase efficiently removed EDCs such as bisphenol A (BPA), nonylphenol (NP), triclosan, and genistein and the addition of mediators greatly improved enzyme activity (Cabana et al., 2007; Husain and Husain, 2007; Tamagawa et al., 2005; Tsutsumi et al., 2001). The efficiency of four different fungal laccases was tested against five different EDCs, also in the presence of both synthetic and natural mediators (Macellaro et al., 2014). Three high redox potential laccases from *Pleurotus ostreatus* (POXC, POXA1b) heterologously expressed in the filamentous fungus *Aspergillus niger*, 1H6C, a POXA1b variant obtained through random mutagenesis and produced in *A. niger* and a commercial laccase, the Novoprime Base 268, were used. The five investigated EDCs were BPA, NP, methylparaben (MTPRB), butylparaben (BTPRB), and dimethylphthalate (DMPTL), and the two mediators were ABTS (synthetic) and acetosyringone (natural). All laccases efficiently transformed EDCs, and their oxidizing activity was substantially increased in the presence of both mediators. Moreover, the enzyme POXC, showing the best catalytic performance, was also able to transform EDC mixtures, although removal rates were different in the mixture if compared to those determined with each EDC alone. Similar results were also observed with mixtures of phenols and chlorinated phenols when incubated with a laccase from *Trametes versicolor* (Canfora et al., 2008; Bollag et al., 2003; Gianfreda et al., 2003).

A better performance in triclosan removal was shown by a fungal laccase when immobilized on vinyl-modified poly(acrylic acid)/SiO_2 nanofibrous membranes (Xu et al., 2014). Immobilized laccase removed triclosan almost two times more than the free laccase, and adsorption of triclosan on the mesostructure of nanofibers was beneficial for its degradation. The immobilized enzyme also showed better storage stability and higher tolerance to pH and temperature changes than the free enzyme, thus providing a new idea for removal of organic pollutants from water environment using enzyme and adsorption technology.

Pezzella et al. (2014) obtained contrasting results. Adsorption mechanisms involved in the immobilization of a laccase from *Pleurotus ostreatus* on perlite, a cheap porous silica material, had detrimental effects on the decolorization of Remazol Brilliant Blue R (RBBR, a reactive dye used as model system) (Pezzella et al., 2014). Results showed that RBBR decolorization was mainly due to enzyme action despite the occurrence of dye adsorption-related enzyme inhibition.

Laccase from *Echinodontium taxodii*, as either free or immobilized enzyme, was able to transform efficiently sulfonamide antibiotics (SAs) belonging to PhACs, as emerging pollutants (Shi et al., 2014). The enzyme was immobilized on concanavalin A-activated Fe_3O_4 nanoparticles, showing higher enzyme loading and activity recovery, compared with conventional covalent binding. Immobilized laccase showed a higher removal rate of SAs compared with the free counterpart, and the presence of siringic acid in the reaction mixture substantially increased the removal of SAs and of SA-type compounds.

Laccase was the only active extracellular lignin-modifying enzyme produced by the white-rot fungi *Trametes hirsute* able to remove multiclass PhACs (17 PhACs) at low and environmentally realistic concentrations (20–500 ng L^{-1}) (Haroune et al., 2014). Results indicated that some of the tested PhACs could act as inducers of laccase production and possibly as mediators of its activity. However, other phenomena, as biosorption and perhaps intracellular enzyme activity, contributed in multiple PhACs removal. As suggested by the authors, further and deeper investigations are needed to understand fully the process (Haroune et al., 2014).

Dehalogenases

Dehalogenases are a family of specific intracellular enzymes produced by terrestrial and marine bacteria (Fetzner and Lingens, 1994; Gianfreda and Bollag, 2002; Zhang et al., 2013). They catalyze the most critical step in the metabolism of halogenated compounds—the cleavage of carbon–halogen bonds, responsible for halogenated pollutants' toxicity. Therefore, dehalogenases appear to be very attractive as detoxification catalysts. Moreover, as claimed by Allpress and Gowland (1998), the study of these enzymes has generated a lot of useful information "about the evolution of catabolic enzymes in general and represents an ideal tool for the investigation and illustration of many key concepts in enzymology, including parallel evolution, convergent evolution, gene transfer, determination of reaction mechanisms and structure–activity relationships."

Dehalogenases differ for their mechanisms of reaction (Fetzner and Lingens, 1994; Gianfreda and Bollag, 2002) and have been classified in different ways, according to dehalogenation mechanisms (eg, hydrolytic dehalogenases, haloalcohol dehalogenases, hydrogen halide lyases, and cofactor-dependent dehalogenases) or to substrate specificity (eg, reductive dehalogenation, oxygenolytic dehalogenation, hydrolytic dehalogenation, thiolytic dehalogenation, intramolecular substitution, dehydrohalogenation, hydration) (Fetzner and Lingens, 1994; Gianfreda and Bollag, 2002). This last classification takes into account the mechanisms by which halogenated compounds may lose their halogen substituents, the specific enzymes involved in the reactions, and the final products of the reaction. Kurihara and Esaki (2008), published an exhaustive review on the occurrence, reaction mechanisms, and applications of bacterial hydrolytic dehalogenases. Structural, biochemical, mutagenesis, and computational studies performed on stereospecific and nonstereospecific dehalogenases have provided detailed insight into different mechanisms and fundamentally different strategies used by the enzymes in the cleavage of carbon–halogen bonds (De Jong and Diikstra, 2003; Huyop and Sudy, 2012; Janssen, 2004). Moreover, intensive research is currently devoted to clarify the structure–function relationships in dehalogenases. Studies have been and are performed by X-ray crystallography, site-directed mutagenesis, kinetic analysis, quantum mechanics and molecular mechanics calculations, statistical analysis, and directed evolution. Findings in these areas have led to rapid progress in this research field (Janssen, 2004).

The list of pollutants partially or totally degraded by these enzymes is very long and includes chloro- and hydroxychloro-benzoate, PCBs, pentachlorophenol, lindane, chloroalkanses, chlorophenylacetate, chlorinated phenoxyacetic acids, haloaromatic compounds, haloalkanes, haloalkanoid acid compounds, bromo- and chloroalcohols, etc. Since some of these compounds are quite novel pollutants, dehalogenases must have been recently engaged to assist in their biodegradation. In addition, the potentialities of these enzymes in the manufacture of chiral intermediates, in the recycling of chlorinated byproducts from chemical manufacturing, and selective treatment of process waste streams have interested several researchers for industrial biocatalysis (Swanson, 1999).

Efficient transformation of TCP was achieved by an immobilized synthetic pathways composed of three enzymes from two different microorganisms: engineered haloalkane dehalogenase from *Rhodococcus rhodochrous* NCIMB 13064, and haloalcohol dehalogenase and epoxide hydrolase from *Agrobacterium radiobacter* AD1 (Dvorak et al., 2014). Cross-linked enzyme aggregates and poly(vinylalcohol) LentiKats lenses were prepared with purified or cell-free enzyme extract and utilized as immobilized biocatalysts. During the operational period, TCP was converted to the intermediates and to the final product (glycerol) by 97% and 78%, respectively. Moreover, immobilized enzymes were suitable to removing TCP from contaminated water up to a 10 mM solubility limit, which is an order of magnitude higher than the concentration tolerated by living microorganisms. Overcoming the restricted biocatalytic use of haloalkane dehalogenases, mainly with pollutants, such as TCP, trichloroethanes and 1,2-dichloropropane, has stimulated several screening, site-directed mutagenesis and directed evolution experiments, aimed at enhancing the catalytic activity and substrate range of this group of enzymes (Janssen, 2004).

A quite novel, potent hydrolyzing enzyme is fluoroacetate dehalogenase responsible for the hydrolysis of the C–F bond, whose dissociation energy is so high that several dehalogenases are practically unavailable to cleave the bond. Fluoroacetate, synthesized by some plants in Australia, Africa, and Central America, is toxic for several organisms and its degradation/removal is of great concern in those lands where fluoroacetate poisoning from plants causes loss of livestock (Kurihara and Esaki, 2008).

ENZYMES AS INDICATORS OF POLLUTED/RESTORED ENVIRONMENTS

Whichever the tool, the process or the organism used to remediate a polluted environment, the successful application of the method has to be validated by at least several further steps. It is necessary to monitor the actual amount of pollutants still present in the restored environment, and to assess the quality of this latter in terms of its safety and regain of initial properties. Enzymes may be helpful in both the scopes. In fact, enzymes may be utilized to reveal the presence of pollutants and quantify their concentration and to evaluate the health and quality status of the polluted/restored target environment. In the first case they behave as biosensors and in the second case as bioindicators.

The comprehensive term *biosensor* denotes a system capable of detecting the presence of a substrate by using biological components, which then provide a quantifiable signal (Rodriguez-Mozaz et al., 2006). According to International Union of Pure and Applied Chemistry (IUPAC), a biosensor is "a self-contained integrated device that is capable of providing specific quantitative or semi-quantitative analytical information using a biological recognition element (biochemical receptor), which is retained in direct spatial contact with a transduction element" (Rodriguez-Mozaz et al., 2006). For an enzymatic biosensor three elements are important: (1) the enzyme is sensitive in such a way to the substance to be determined; (2) the device is acting as physicochemical sensor or transducer to "immobilize" the enzyme and is capable of transforming the change that occurred in the enzyme in a measurable signal (eg, electrochemical, optical, piezoelectrical, or thermal) and; (3) the signal amplifier and the device for data processing. It appears evident that a critical step in the biosensor design is the methodology utilized for immobilizing the enzyme on the transducer. Indeed, it may affect the performance of the biosensor, its sensitivity and selectivity because of the influence on enzyme orientation of the transducer surface and consequent loading, mobility, stability, structure, and biological activity.

As recently summarized by Rao et al. (2014) and Sassolas et al. (2012), several immobilization methods as well as immobilizing supports are available and the choice of each of them has to take into account the properties, such as catalytic activity and stability, achieved by the enzyme after the immobilization process.

Biosensor technology has utilized several hydrolytic and oxidative enzymes (Amine et al., 2006; Sassolas et al., 2012; Rao et al., 2014). Usually, the polluting substance (that is the analyte to be quantified) acts as inhibitor of the enzyme with a resulting decrease of the catalytic activity of the immobilized enzyme (Amine et al., 2006). This change is then transformed in a physical/chemical signal detected by the transducer. Examples that follow still deal with the herein-mentioned laccases and dehalogenases.

Laccases appear suitable for biosensor application having a broad substrate variety, in particular of phenolic nature. Laccase-based sensitive biosensors effectively detected very low amounts of phenolic compounds, such as catechin and/or catechol by optical (Abdullah et al., 2007) or amperometric measurements (Gomes and Rebelo, 2003). A sensitive biosensor for chlorophenol detection was obtained by immobilization of laccase from *T. versicolor* on gold nanoparticles through a complex process and utilizing polyvinyl alcohol and polyethylene oxide–polyoxypropylene–polyethylene oxide (Liu et al., 2011). The enzyme from the same fungus was immobilized in high load with silica spheres on surface of multiwalled carbon nanotubes (MWCNTs)-doped screen-printed electrode (SPE) (Li et al., 2012). It detected dopamine, as model of phenolic compounds, in a rapid, selective, and sensitive way (good linearity in the range 1.3–85.5 µM with a detection limit of 0.42 µM). A novel, sensitive, and stable laccase biosensor, suitable for phenols determination, was designed based on a composite of polydopamine-laccase-nickel nanoparticle-loaded carbon nanofibers (Li et al., 2014). The nickel nanoparticle-loaded carbon nanofibers showed to be very compatible for laccase immobilization and facilitated the direct electron transfer from laccase to the electrode surface. A good sensitivity and stability was observed in the catechol determination even in real water samples (Li et al., 2014).

Biosensors based on dehalogenases utilize protons of halide ions, products of the enzymatic reaction, subsequently detected by appropriate detectors. The shift in the fluorescence intensity of fluorosceinamine provoked by protons, derived from haloalkane dehalogenase catalyzed dehalogenation of ethylene dibromide, was used for a reagent-less enzymatic optical biosensor, able to measure ethylene bromide concentration (Reardon et al., 2009). Both enzyme and layers of fluorosceinamine were immobilized on the tip of an optical fiber. The fluorescence change, subsequent to the enzymatic reaction, was proportional to the concentration of ethylene bromide in a wide range and showed a stable activity within >30 days. The same strategy was utilized by Bidmanova et al. (2010) for the preparation of haloalkane biosensor for the detection not only of ethylene dibromide but also of 3-chloro-2(chloromethyl)-1-propene. Both the biosensors appeared useful tools for continuous in situ monitoring of halogenated pollutants.

A miniaturized and portable dehalogenase microconductometer was designed and characterized for the detection of pesticides (Trnkova et al., 2011). The biosensor was tested with 1-chloroexane. Some experimental conditions as high temperature, closed reaction system (for 1-chloroexane volatility), reaction time, and substrate concentrations were examined to evaluate the detectable range of the compound and to choose the best performance of the biosensor.

A sensor, although not a true biosensor, was obtained by covalently linking a very stable haloacid dehydrogenase from the thermophile *Sulfolobus tokodai*, on an N-hydroxysuccinimidyl Sepharose resin. The immobilized enzyme was packed into disposable columns. The sensor was able to detect 1-2-haloacids by quantification of the chloride ion produced by the enzymatic reaction. The enzyme resin showed a long shelf life, retained more than 70% of its initial activity after six months of storage at 4 °C. It was fully regenerated after continuous operational cycles (Bachas-Daunertet et al., 2009).

The role of enzymes as bioindicators of soil/sediment/water quality and health is documented by several papers and will not be addressed in this chapter. Details on enzyme capabilities to obtaining useful information on pollutants' impact on the quality of polluted environments and on their response to the remediation/restoration process are available in the literature (see review of Dick, 1994, 1997; Gianfreda and Bollag, 1996; Trasar-Cepeda et al., 2000; Nannipieri et al., 2002; Gianfreda and Ruggiero, 2006; Gianfreda and Rao, 2008; Karaca et al., 2011; Burns et al., 2013).

ADVANTAGES AND DRAWBACKS OF ENZYMES AS ENVIRONMENTAL ACTORS

With respect to chemical or microbial processes, enzymes as detoxifying agents may present several advantages. The enzyme-mediated biotransformation produces no toxic side products, as is often the case with chemical and some microbiological processes. If the process occurs in situ, the enzymes may be digested, in situ, by the indigenous microorganisms after the treatment, with obvious advantages mainly for possible side effects produced by the presence of enzymes. If organic cosolvents or surfactants are needed, for instance to enhance the availability of the pollutant to the biological process, the introduction of such additives is much more feasible from an enzymatic point of view than using whole cells. When high amounts of enzymes, with high activity and enhanced stability, are required, new technologies, such as recombinant-DNA technology can potentially satisfy this requirement. In particular, as with respect to microbial processes, enzymes offer additional advantages (Table 6.2).

Indeed, the enzyme usually acts on a single substrate and it is well known which product will be produced by the enzymatic reaction. This aspect can be an advantage when the single reaction step produces a less or no toxic product. However, it becomes a disadvantage when the complete mineralization of the pollutant is needed to obtain the complete removal of its harmful effects. In this case, a multistep process is usually involved, that is more enzymes acting sequentially are needed, and therefore it is only possible to use specific microorganisms. Even in the case of cofactor-requiring enzymes, their use seems problematic unless using a preparation containing both the enzyme and the respective cofactor.

The rapid degradation of cell-free enzymes is another problem in the use of enzymes to detoxify polluted environments. The enzyme can be rapidly degraded by proteases released by microorganisms present in the soil/water environment or can be deactivated under the operational conditions. The incorporation of enzymes in humic-like complexes, their adsorption on clay minerals, or their immobilization in/on synthetic matrices may prevent enzyme deactivation, protect them against proteolytic degradation, stabilize their activity, and prolong the life of the added enzymes (Nannipieri and Bollag, 1991; Ó'Fágáin, 2003; Mateo et al., 2007).

The immobilization of laccase on montmorillonite (Ahn et al., 2002) or of Mn-peroxidase on nanoclays (Acevedo et al., 2010) not only preserved the activity of both enzymes at acceptable transforming levels but also stabilized them against temperature. Both the immobilized enzymes were efficient in the transformation of PAHs also in the presence of soil.

Gassara et al. (2013) demonstrated that laccase encapsulated on polyacrylamide hydrogel and pectin led to 90% transformation of BPA with respect to 26% obtained with the free enzyme. Moreover, the encapsulation stabilized the enzyme and protected it from inactivation by noncompetitive inhibition.

Table 6.2 Advantages and Disadvantages of Enzymes With Respect to Microbial Cells in the Cleanup of Polluted Environments

Microorganisms		Enzymes	
Advantages	Disadvantages	Advantages	Disadvantages
Complete mineralization of pollutants	Cell production and growth	Not inhibited by the substances possibly acting as inhibitors of microbial metabolism	Absence of the complete mineralization of pollutants
Presence of enzymatic sequences	Inhibition of metabolic pathways	Higher reactivity and activity	The single enzymatic reaction can produce a more toxic product than the substrate
	Competition effects with other microorganisms	Active in a large range of experimental conditions; possibly used under extreme conditions limiting microbial activity	Low stability under operational conditions
	Predation effects	Active at low pollutant concentrations and active against a given substrate (whereas microorganisms may prefer more easily degradable compounds than the pollutant)	Need of cofactors (if cofactor-requiring enzymes are needed, their use seems problematic unless to use a preparation containing both the enzyme and the respective cofactor)
	Toxin production	Active in the presence of microbial predators or antagonist and toxins	Susceptibility to protease action
	Cellular movements	Absence of diffusional problems and permeability; more mobile than microorganisms because of their smaller size	High costs and long-term processes for extraction and purification of enzymes
	Need of inductors	Potential elevated mobility in soil Catalytic activity unrelated to cellular growth	

Nanostructures, such as nanoporous media, nanofibers, carbon nanotubes, and nanoparticles, seem to offer several, additional advantages against traditional immobilizing supports (Kim et al., 2006). Enzymes immobilized on nanoparticles often display a high stability to several deactivating factors, are active in a larger range of pH and temperature, and therefore they appear suitable for several applications, including the detoxification and the monitoring of polluted environments (Hu et al., 2007; Garcia et al., 2011). This new research area, called nanobiocatalysis, opens new challenges for enzyme applications, although the interactions occurring between enzymes and nanostructures at the nanoscale require further investigations (Kim et al., 2006, 2008).

CONCLUSIONS AND FUTURE PERSPECTIVES

Overall, experimental evidence so far discussed indicates that enzymes are versatile catalysts and useful tools for the cleanup/remediation of polluted environments. Their capability of transforming several polluting compounds to less hazardous products is attested by a huge amount of findings reported in literature. In particular, examples provided in this chapter clearly suggest that fungal (laccases) and bacterial (dehalogenases) enzymes are particularly suitable for the transformation of compounds considered recalcitrant and listed among emerging pollutants.

However, few examples are available on isolated enzymes efficient in pollutant transformation in natural environments such as soil or water. Indeed, their performance in such harsh conditions can be unsatisfying because of low reaction yields and enzyme stability. Moreover, knowledge of the biochemistry of the degradation processes can be limited. Therefore, the potentiality of enzymes in the process is often unpredictable. In addition, it can be difficult to demonstrate their involvement in the degradation of pollutants in such environments.

Several approaches can be used to improve and increase enzyme efficiency in pollutant removal. They can be aimed at enhancing either the performance of the selected enzyme, by acting on the conditions under which the enzyme displays its catalytic activity, or the intrinsic and inherent catalytic features of the enzyme, by modifying genetically enzyme catalytic characteristics or producing novel and improved enzymatic molecules. Recently, the knowledge of biology of microorganisms (main producers of enzymes useful in pollutant transformation) in several terrestrial and aquatic environments is rapidly increasing by means of modern tools of genomics, transcriptomics, proteomics, and metabolomics. All these approaches may be helpful not only in the understanding of the pollutant biodegradation process, useful for complete removal of the contaminant, but also for the design and possibly production of enzymes designed for a target pollutant.

Indeed, for the degradation of recalcitrant compounds where no natural pathways are known, novel pathways can be generated by combining pathway "cassettes" from various

Table 6.3 Still Open Challenges for Enzymes Efficient in Polluted Environments Detoxification

Methodological Approach	Necessary Steps
Bioprospecting	New enzyme activities among known or previously unknown organisms
Use of nontraditional catalytic properties of enzymes	Ability of enzymes to catalyze reactions quite different from those they catalyze in vivo
	Identification of such unnatural reactions
	Screening of known enzymes for unknown catalytic activities
Genetic engineering techniques	Increased enzyme production by total environmental DNA sampling and DNA recombinant technology
	Improved catalytic performance by site-directed mutagenesis (increased stability, modification of substrate specificity, shift of optimal pH)
Exploration of extreme environments	Enzymes from thermophiles, from psychrophiles, from halophiles (increased stability to pH, increased stability to salinity, shift of optimal pH)
Enzymes in two phase systems, organic solvents and/or micelles	Enzymes with stability in organic solvents
Improvement of existing processes	New immobilizing supports, new immobilization techniques, design of new bioreactors

genetic sources. A relatively new research area, called bioprospecting, is used to search for new enzyme activities among known or previously unknown organisms. It may still employ the quite old culture-dependent methodology for the screening of organisms with specific metabolic characteristics and the identification and subsequent isolation of the involved enzymes. The possible production of such unusual, new specific enzyme activities, by molecular methods or by heterologous expression in industrial host organisms, both well-accepted and widely used technologies, can later provide the catalyst for the target decontamination process. Still open challenges for enzymes efficient in the cleanup of polluted environments are summarized in Table 6.3. The use of methodological approaches as in Table 6.3 may envisage the design and development of a specific enzyme for a given purpose rather than the development of a purpose for an existing enzyme.

LIST OF ABBREVIATIONS

ABTS 2,2′-azino-bis(3-ethylbenzothiazoline-6-sulfonic acid)
BPA Bisphenol A. Synonym: 2,2-bis (4-hydroxyphenyl) propane
EDCs Endocrine-disrupting compounds
LDS Lignin-degrading enzyme system
LiP Lignin peroxidase
MnP Manganese-dependent peroxidase

NP Nonylphenol
PAHs Polycylic aromatic hydrocarbons
PCBs Poly chlorinated biphenyls
PhACs Pharmaceutically active compounds
POPs Persistent organic pollutants
SAs Sulfonamide antibiotics
TCP 1,2,3-Trichloropropane

REFERENCES

Abdullah, J., Ahmad, M., Karuppiah, N., Heng, L.Y., Sidek, H., 2007. An optical biosensor based on immobilization of laccase and MBTH in stacked films for the detection of catechol. Sensors 7, 2238–2250.

Acevedo, F., Pizzul, L., Castillo, M.D., González, M.E., Cea, M., Stenström, J., Gianfreda, L., Diez, M.C., 2010. Degradation of polycyclic aromatic hydrocarbons by free and nanoclay-immobilized manganese peroxidase from *Anthracophyllum discolor*. Chemosphere 80, 271–278.

Ahn, M.Y., Dec, J., Kim, E.Y., Bollag, J.-M., 2002. Treatment of 2,4-dichlorophenol polluted soil with free and immobilized laccase. Journal of Environmental Quality 31, 1509–1515.

AllPress, J.D., Gowland, P.C., 1998. Dehalogenases: environmental defence mechanism and model of enzyme evolution. Biochemical Education 26, 267–276.

Amine, A., Mohammadia, H., Bourais, I., Palleschi, G., 2006. Enzyme inhibition-based biosensors for food safety and environmental monitoring. Biosensensor & Bioelectronics 21, 1405–1423.

Bachas-Daunert, P.G., sellers, Z.P., wei, Y., 2009. Detection of halogenated organic compounds using immobilized thermophilic dehalogenase. Analytical and Bioanalytical Chemistry 395, 1173–1178.

Baldrian, P., 2006. Fungal laccases occurrence and properties. FEMS Microbiology Reviews 30, 215–242.

Bidmanova, S., Chaloupkova, R., Damborsky, J., Prokop, Z., 2010. Development of an enzymatic fiber-optic biosensor for detection of halogenated hydrocarbons. Analytical and Bioanalytical Chemistry 398, 1891–1898.

Bollag, J.-M., Shuttleworth, K.L., Anderson, D.H., 1998. Laccase-mediated detoxification of phenolic compounds. Applied Environmental Microbiology 54, 3086–3091.

Bollag, J.-M., Chu, R., Rao, M.A., Gianfreda, L., 2003. Enzymatic oxidative transformation of chlorophenol mixtures. Journal of Environmental Quality 32, 62–71.

Bollag, J.-M., 1992. Decontaminating soils with enzymes. Environmental Science & Technology 26, 1876–1881.

Burns, R.G., Deforest, J.L., Marxsen, J., Sinsabaugh, R.L., Stromberger, M.E., Wallenstein, M.D., Weintraub, M.N., Zoppini, A., 2013. Soil enzymes in a changing environment: current knowledge and future directions. Soil Biology & Biochemistry 58, 216–234.

Cabana, H., Jiwan, J.L., Rozenberg, R., Elisashvili, V., Penninckx, M., Agathos, S.N., Jones, J.P., 2007. Elimination of endocrine disrupting chemicals nonylphenol and bisphenol A and personal care product ingredient triclosan using enzyme preparation from the white rot fungus *Coriolopsis polyzona*. Chemosphere 67, 770–778.

Cameron, M.D., Timofeevski, S., Aust, S.D., 2000. Enzymology of *Phanerochaete chrysosporium* with respect to the degradation of recalcitrant compounds and xenobiotics. Applied Microbiology and Biotechnology 54, 751–758.

Canfora, L., Iamarino, G., Rao, M.A., Gianfreda, L., 2008. Detoxification of natural and synthetic phenolic mixtures by *Trametes versicolor* laccase. Journal of Agricultural and Food Chemistry 56, 1398–1407.

Chakraborty, R., Wu, C.H., Hazen, T.C., 2012. Systems biology approach to bioremediation. Current Opinion in Biotechnology 23, 483–490.

Claus, H., 2004. Laccases: structure, reactions, distribution. Micron 35, 93–96.

De Jong, R.M., Dijkstra, B., 2003. Structure and mechanism of bacterial dehalogenases: different ways to cleave a carbon–halogen bond. Current Opinion in Structural Biology 13, 722–730.

Deblonde, T., Carole Cossu-Leguille, C., Hartemann, P., 2011. Emerging pollutants in wastewater: a review of the literature. International Journal of Hygiene and Environmental Health 214, 442–448.

Dec, J., Bollag, J.-M., 1994. Use of plant material for the decontamination of water polluted with phenols. Biotechnology & Bioengineering 44, 1132–1139.

Demarche, P., Junghanns, C., Nair, R.R., Agathos, S.N., 2012. Harnessing the power of enzymes for environmental stewardship. Biotechnology Advances 30, 933–953.

Dhillon, G.S., Kaur, S., Brar, S.K., 2012. In-vitro decolorization of recalcitrant dyes through an eco-friendly approach using in-house laccase from *Trametes versicolor* grown on brewer's spent grain. International Biodeterioration and Biodegradation 72, 67–75.

Dick, R.P., 1994. Soil enzymes activities as indicators of soil quality. In: Doran, J.W., Coleman, D.C., Bezdicek, D.F., Stewart, B.A. (Eds.), Defining Soil Quality for a Sustainable Environment. SSSA, Madison, WI, USA. Special Publication no. 35.

Dick, R.P., 1997. Soil enzyme activities as integrative indicators of soil health. In: Pankhurst, C.E., Doube, B.M., Gupta, V.V.S.R. (Eds.), Biological Indicators of Soil Health. CAB International, Wallingford, USA.

Durán, N., Esposito, E., 2000. Potential applications of oxidative enzymes and phenoloxidase-like compounds in wastewater and soil treatment: a review. Journal of Molecular Catalysis B: Enzymatic 28, 83–99.

Durán, N., Rosa, M.A., D'annibale, A., Gianfreda, L., 2002. Applications of laccases and tyrosinases (phenoloxidases) immobilized on different supports: a review. Enzyme Microbiology & Technology 31, 907–931.

Dvorak, P., Bidmanova, S., Damborsky, J., Prokop, Z., 2014. Immobilized synthetic pathway for biodegradation of toxic recalcitrant pollutant 1,2,3-trichloropropane. Environmental Science & Technology 48, 6859–6866.

Fernández-Fernández, M., Sanromán, M.Á., Moldes, D., 2012. Recent developments and applications of immobilized laccase. Biotechnology Advances 31, 1808–1825.

Fetzner, S., Lingens, F., 1994. Bacterial dehalogenases: biochemistry, genetics, and biotechnological applications. Microbiological Reviews 58, 641–685.

Filazzola, M.T., Sannino, F., Rao, M.A., Gianfreda, L., 1999. Effect of various pollutants and soil-like constituents on laccase from *Cerrena unicolor*. Journal of Environmental Quality 28, 1929–1938.

Garcia, J., Zhang, Y., Taylor, H., Cespedes, O., Webb, M.E., Zhou, D., 2011. Multilayer enzyme-coupled magnetic nanoparticles as efficient, reusable biocatalysts and biosensors. Nanoscale 3, 3721–3730.

Gassara, F., Brar, S.K., Verma, M., Tyagi, R.D., 2013. Bisphenol A degradation in water by ligninolytic enzymes. Chemosphere 92, 1356–1360.

Gavrilescu, M., Demnerova, K., Aamand, J., Agathos, S., Fava, F., 2015. Emerging pollutants in the environment: present and future challenges in biomonitoring, ecological risks and bioremediation. New Biotechnology 32, 147–156.

Gianfreda, L., Bollag, J.-M., 1994. Effect of soils on the behaviour of immobilized enzymes. Soil Science Society of America Journal 58, 1672–1681.

Gianfreda, L., Bollag, J.-M., 1996. Influence of natural and anthropogenic factors on enzyme activity in soil. In: Stotzky, G., Bollag, J.-M. (Eds.). Stotzky, G., Bollag, J.-M. (Eds.), Soil Biochemistry, vol. 9. Marcel Dekker, New York.

Gianfreda, L., Bollag, J.-M., 2002. Isolated enzymes for the transformation and detoxification of organic pollutants. In: Burns, R.G., Dick, R. (Eds.), Enzymes in the Environment: Activity, Ecology and Applications. Marcel Dekker, New York.

Gianfreda, L., Nannipieri, P., 2001. Basic principles, agents and feasibility of bioremediation of soil polluted by organic compounds. Minerva Biotecnologica 13, 5–12.

Gianfreda, L., Rao, M.A., 2004. Potential of extra cellular enzymes in remediation of polluted soils: a review. Enzyme Microbiology & Technology 35, 339–354.

Gianfreda, L., Rao, M.A., 2008. Interaction between xenobiotics and microbial and enzymatic soil activity. Critical Review of Environmental Science & Technology 38, 269–310.

Gianfreda, L., Ruggiero, P., 2006. Enzyme activities in soil. In: Nannipieri, P., Smalla, K. (Eds.). Nannipieri, P., Smalla, K. (Eds.), Nucleic Acids and Proteins in Soil. Soil Biology, vol. 8. Springer Verlag, Berlin.

Gianfreda, L., Sannino, F., Filazzola, M.T., Leonowicz, A., 1998. Catalytic behavior and detoxifying ability of a laccase from the fungal strain *Cerrena unicolor*. Journal of Molecular Catalysis B: Enzymatic 14, 13–23.

Gianfreda, L., Xu, X.F., Bollag, J.-M., 1999. Laccases: a useful group of oxidoreductive enzymes. Bioremediation Journal 3, 1–25.

Gianfreda, L., Sannino, F., Rao, M.A., Bollag, J.-M., 2003. Oxidative transformation of phenols in aqueous mixtures. Water Research 37, 3205–3215.

Gianfreda, L., Mora, M., Diez, C.M., 2006a. Restoration of polluted soils by means of microbial and enzymatic processes: a review. Journal of Soil Science and Plant Nutrition 6, 20–40.

Gianfreda, L., Iamarino, G., Scelza, R., Rao, M.A., 2006b. Oxidative catalysts for the transformation of phenolic pollutants: a brief review. Biocatalysis & Biotransformation 24, 177–187.

Gomes, S.A.S.S., Rebelo, M.J.F., 2003. A new laccase biosensor for polyphenol determination. Sensors 3, 166–175.

Haroune, L., Saibi, S., Bellenger, J.-P., Cabana, H., 2014. Evaluation of the efficiency of *Trametes hirsuta* for the removal of multiple pharmaceutical compounds under low concentrations relevant to the environment. Bioresource Technology 171, 199–202.

Hu, X., Zhao, X., Hwang, H., 2007. Comparative study of immobilized *Trametes versicolor* laccase on nanoparticles and kaolinite. Chemosphere 66, 1618–1626.

Husain, M., Husain, Q., 2007. Applications of redox mediators in the treatment of organic pollutants by using oxidoreductive enzymes: a review. Critical Review of Environmental Science & Technology 38, 1–42.

Huyop, F., Sudi, I.Y., 2012. D-specific dehalogenases, a review. Biotechnology and Biotechnological Equipment 26, 2817–2822.

Iamarino, G., Rao, M.A., Gianfreda, L., 2009. Dephenolization and detoxification of olive-mill wastewater (OMW) by purified biotic and abiotic oxidative catalysts. Chemosphere 74, 216–223.

Janssen, D.B., 2004. Evolving haloalkane dehalogenases. Current Opinion in Chemical Biology 8, 150–159.

Jones, K.C., de Voogt, P., 1999. Persistent organic pollutants (POPs): state of the science. Environmental Pollution 100, 209–221.

Karaca, A., Cetin, S.C., Turgay, O.C., Kizilkaya, R., 2011. Soil enzymes as indication of soil quality. In: Shukla, G., Varma, A. (Eds.), Soil Biology. Soil Enzymology, vol. 22. Springer Verlag, Berlin.

Karam, J., Nicell, J.A., 1997. Potential application of enzymes in waste treatment. Journal of Chemical Technology and Biotechnology 69, 141–153.

Kim, J., Grate, J., Wang, P., 2006. Nanostructures for enzyme stabilization. Chemical Engineering Science 61, 1017–1026.

Kim, J., Grate, J., Wang, P., 2008. Nanobiocatalysis and its potential applications. Trends in Biotechnology 26, 639–646.

Kudanga, T., Nyanhongo, G.S., guebitz, G.M., burton, S., 2011. Potential applications of laccase-mediated coupling and grafting reactions: a review. Enzyme and Microbial Technology 48, 195–208.

Kurihara, T., Esaki, N., 2008. Bacterial hydrolytic dehalogenases and related enzymes: occurrences, reaction mechanisms, and applications. The Chemical Record 8, 67–74.

Li, Y., Zhang, L., Li, M., Pan, Z., Li, D., 2012. A disposable biosensor based on immobilization of laccase with silica spheres on the MWCNTs-doped screen-printed electrode. Chemistry Central Journal 6, 103.

Li, D., Luo, L., Pang, Z., Ding, L., Wang, Q., Ke, H., Huang, F., Wei, Q., 2014. Novel phenolic biosensor based on a magnetic polydopamine- laccase-nickel nanoparticle loaded carbon nanofiber composite. Applied Materials & Interfaces 6, 5144–5151.

Liu, J., Niu, J., Yin, L., Jiang, F., 2011. In situ encapsulation of laccase in nanofibers by electrospinning for development of enzyme biosensors for chlorophenol monitoring. Analyst 136, 4802–4808.

Longoria, A., Tinoco, R., Vázquez-Duhalt, R., 2008. Chloroperoxidase-mediated transformation of highly halogenated monoaromatic compounds. Chemosphere 72, 485–490.

Luterek, J., Gianfreda, L., Leonowicz, A., 1997. Screening of wood-rotting fungi for laccase-producing induction by ferulic acid, partial purification and immobilization of laccase from the high laccase-producing strain, *Cerrena unicolor*. Acta Microbiologica Polonica 46, 297–311.

Macellaro, G., Pezzella, C., Cicatiello, P., Sannia, G., Piscitelli, A., 2014. Fungal laccases degradation of endocrine disrupting compounds. BioMed Research International. [Online] Article ID 614038, 8 pages. Available from: http://www.hindawi.com.

Majeau, J., Brar, S.K., Tyagi, R.D., 2010. Laccases for removal of recalcitrant and emerging pollutants. Bioresource Technology 101, 2331–2350.

Mateo, C., Palomo, J.M., Fernandez-Lorente, G., Guisan, J.M., Fernandez-Lafuente, R., 2007. Improvement of enzyme activity, stability and selectivity via immobilization techniques. Enzyme and Microbial Technology 40, 1451–1463.

Nannipieri, P., Bollag, J.-M., 1991. Use of enzymes to detoxify pesticide-contaminated soils and waters. Journal of Environmental Quality 20, 510–517.

Nannipieri, P., Kandeler, E., Ruggiero, P., 2002. Enzyme activities and microbiological and biochemical processes in soil. In: Burns, R.G., Dick, R.P. (Eds.), Enzymes in the Environment: Activity, Ecology and Applications. Marcel Dekker, New York.

Nicell, J.A., 2001. Environmental applications of enzymes. Interdisciplinary Environmental Review 3, 14–41.

Ó'Fágáin, C., 2003. Enzyme stabilization – *recent experimental progress*. Enzyme and Microbial Technology 33, 137–149.

Pezzella, C., Russo, M.E., Marzocchella, A., Salatino, P., Sannia, G., 2014. Immobilization of a *Pleurotus ostreatus* laccase mixture on perlite and its application to dye decolourisation. BioMed Research International. [Online] Article ID 308613, 11 pages. Available from: http://www.hindawi.com.

Pointing, S.B., 2001. Feasibility of bioremediation by white-rot fungi. Applied Microbiology and Biotechnology 57, 20–32.

Rao, M.A., Scelza, R., Acevedo, F., Diez, M.C., Gianfreda, L., 2014. Enzymes as useful tools for environmental purposes. Chemosphere 107, 145–162.

Reardon, K.F., Campbell, D.W., Müller, C., 2009. Optical fiber enzymatic biosensor for reagentless measurement of ethylene dibromide. Engineering in Life Sciences 9, 291–297.

Reddy, C.A., 1995. The potential for white-rot fungi in the treatment of pollutants. Current Opinion in Biotechnology 6, 320–328.

Reid, B.J., Jones, K.C., Semple, K.T., 2000. Bioavailability of persistent organic pollutants in soils and sediment-a perspective mechanism, consequences and assessment. Environmental Pollution 108, 103–112.

Rodríguez Couto, S., Toca Herrera, J.L., 2006. Industrial and biotechnological applications of laccases: a review. Biotechnology Advances 24, 500–513.

Rodríguez Couto, S., 2009. Dye removal by immobilised fungi. Biotechnology Advances 27, 227–235.

Rodriguez-Mozaz, S., Lopez De Alda, M.J., Barceló, D., 2006. Biosensors as useful tools for environmental analysis and monitoring. Analytical and Bioanalytical Chemistry 386, 1025–1041.

Rubilar, O., Diez, M.C., Gianfreda, L., 2008. Transformation of chlorinated phenolic compounds by white rot fungi. Critical Review of Environmental Science & Technology 38, 227–268.

Ruggaber, T.P., Talley, J.W., 2006. Enhancing bioremediation with enzymatic processes: a review. The Practice Periodical of Hazardous Toxic, Radioactive Waste Management 10, 73–85.

Sassolas, A., Blum, L.J., Leca-Bouvier, B.D., 2012. Immobilization strategies to develop enzymatic biosensors. Biotechnology Advances 30, 489–511.

Shi, L., Ma, F., Han, Y., Zhang, X., Yu, H., 2014. Removal of sulfonamide antibiotics by oriented immobilized laccase on Fe_3O_4 nanoparticles with natural mediators. Journal of Hazardous Materials 279, 203–211.

Strong, P.J., Claus, H., 2011. Laccase: a review of its past and its future in bioremediation. Critical Review of Environmental Science & Technology 41, 373–434.

Swanson, P.E., 1999. Dehalogenases applied to industrial-scale biocatalysis. Current Opinion in Biotechnology 10, 365–369.

Tamagawa, Y., Hirai, H., Kawai, S., Nishida, T., 2005. Removal of estrogenic activity of endocrine-disrupting genistein by ligninolytic enzymes from white rot fungi. FEMS Microbiology Letters 244, 93–98.

Thavamani, P., Malik, S., Beer, M., Megharaj, M., Naidu, R., 2012. Microbial activity and diversity in long-term mixed contaminated soils with respect to polyaromatic hydrocarbons and heavy metals. Journal of Environmental Management 99, 10–17.

Torres, E., Bustos-Jaimes, I., La Borgne, S., 2003. Potential use of oxidative enzymes for the detoxification of organic pollutants. Applied Catalysis B: Environmental 46, 1–6.

Toscano, G., Colarieti, M.L., Greco, G.J.R., 2003. Oxidative polymerisation of phenols by a phenol oxidase from green olives. Enzyme Microbiology & Technology 33, 47–54.

Trasar-Cepeda, C., Leirós, M.C., Seoane, S., Gil-Sotres, F., 2000. Limitation of soil enzymes as indicators of soil pollution. Soil Biology & Biochemistry 32, 1867–1875.

Trnkova, L., Hubalek, J., Adam, V., Kizek, R., 2011. An analytical task: a miniaturized and portable μconductometer as a tool for detection of pesticides. In: Stoytcheva, M. (Ed.), Pesticides – Strategies for Pesticides Analysis. InTech. ISBN: 978-953-307-460-3. Available from: http://www.intechopen.com/books/pesticides-strategies-for-pesticides-analysis/an-analytical-task-a-miniaturized-and-portable-conductometer-as-a-tool-for-detection-of-pesticides.

Tsutsumi, Y., Haneda, T., Nishida, T., 2001. Removal of estrogenic activities of bisphenol A and nonylphenol by oxidative enzymes from lignin- degrading basidiomycetes. Chemosphere 42, 271–276.

Verdin, A., Loun, A., Sahraoui, H., Durand, R., 2004. Degradation of benzo[a]pyrene by mitosporic fungi and extracellular oxidative enzymes. International Biodeterioration and Biodegradation 53, 65–70.

Virk, A.P., Sharma, P., Capalash, N., 2012. Use of laccase in pulp and paper industry. Biotechnology Progress 28, 21–32.

Viswanath, B., Rajesh, B., Janardhan, A., Praveen Kumar, A., Narasimha, G., 2014. Fungal laccases and their applications in bioremediation. Enzyme Research 2014. http://dx.doi.org/10.1155/2014/163242 Article ID 163242, 21 pages.

Whiteley, C.G., Lee, D.-J., 2006. Enzyme technology and biological remediation. Review of Enzyme Microbiology and Technology 38, 291–316.

Xu, R., Si, Y., Wu, X., Li, F., Zhang, B., 2014. Triclosan removal by laccase immobilized on mesoporous nanofibers: strong adsorption and efficient degradation. Chemical Engineering Journal 255, 63–70.

Yoshida, H., 1883. Chemistry of lacquer (Urushi), part 1. Journal of Chemical Society 43, 472–486.

Zhang, J., Cao, X., Xin, Y., Xue, S., Zhang, W., 2013. Purification and characterization of a dehalogenase from *Pseudomonas stutzeri* DEH130 isolated from the marine sponge *Hymeniacidon perlevis*. World Journal of Microbiology and Biotechnology 29, 1791–1799.

FURTHER READING

Ansari, S.A., Husain, Q., 2012. Potential applications of enzymes immobilized on/in nano materials: a review. Biotechnology Advances 30, 512–523.

Dai, Y., Yin, L., Niu, J., 2011. Laccase-carrying electrospun fibrous membranes for adsorption and degradation of PAHs in shoal soils. Environmental Science & Technology 45, 10611–10618.

Stockholm Convention on Persistent Organic Pollutants. [Online] Available from: http://pops.int/documents/context/context_en.pdf.

Tuomela, M., Hatakka, A., 2011. Oxidative fungal enzymes for phytoremediation. In: Agathos, A., Moo-Young, M. (Eds.), Comprehensive Biotechnology: Environmental Biotechnology and Safety, vol. 6, second ed. , pp. 183–196.

CHAPTER 7

Enzymes: Applications in Pulp and Paper Industry

G. Singh, N. Capalash, K. Kaur, S. Puri, P. Sharma
Panjab University, Chandigarh, India

INTRODUCTION

Before 2009 the pulp and paper industry was dominated by the United States and Canada, but in 2012 China emerged as the biggest player with the production of 100 million tons out of a total 401 million tons of paper produced worldwide. India has about 515 pulp processing units and is one of the major players for paper production among countries, such as Germany, Italy, Japan, Brazil, and Sweden (http://www.forestindustries.se/; Sadhasivam et al., 2010; Sharma et al., 2014). Indian paper industry contributes about 1.6% of the world's total paper produced. Chemical bleaching is usually performed by chlorine-based agents, such as elemental chlorine, chlorine dioxide, and hypochlorite, which results in the production of various chloro–organic derivatives. Most of them are chlorinated benzenes, epoxystearic acid and dichloromethane, classified as suspected carcinogens and strong mutagens (Central Pollution Control Board Delhi, 2007).

Market demand for chlorine-free bleached pulp has increased during the last decade, mainly due to its suitability toward customer satisfaction (Garcia et al., 2010; Sharma et al., 2014). However, ecofriendly biobleaching and other applications like pitch removal and de-inking of paper waste is feasible by the use of potent biocatalysts, such as xylanases and laccases. In the beginning, use of enzymes for pulp and paper technology was not considered technically and economically feasible because of the lack of ready availability of these biocatalysts. Commercial-scale production and application of these enzymes was a cost-intensive and cumbersome process for the long run. Furthermore, enzymes like laccases require mediators for their bleaching action and also suffer enzyme stability problems (Singh et al., 2008, 2011b, 2011a). Fortunately, research efforts by scientific institutions and enzyme producers have led to the development of biocatalysts that offer significant benefits to the pulp and paper industry (Bajpai, 1999). In the absence of enzymes, available options are (1) bleaching of pulps with oxygen and extended cooking, (2) hydrogen peroxide, and (3) ozone treatment. But most of these methods are highly capital intensive for process change. Thus, an alternative and cost-effective method, that is, use of enzymes, has provided

Agro-Industrial Wastes as Feedstock for Enzyme Production
ISBN 978-0-12-802392-1
http://dx.doi.org/10.1016/B978-0-12-802392-1.00007-1

a very simple and economic way to reduce the use of chlorine and other bleaching chemicals (Bhoria et al., 2012). Xylanase prebleaching technology is now in use in several mills worldwide. This technology has been successfully transferred to full industrial scale in just a few years. The main driving factors have been the economic and environmental advantages that biocatalysts can bring to the pulp-bleaching plant. After xylanases, many research groups have realized laccase as one of the versatile biocatalysts that has the potential to revolutionize the pulping and papermaking processes. It not only plays a role in the delignification and brightening of pulp but has also been recognized for the removal of lipophilic extractives, responsible for pitch deposition, from both wood and nonwood pulps. Laccases are capable of improving physical, chemical as well as mechanical properties of pulp by either forming reactive radicals with lignin or by functionalizing lignocellulosic fibers. Laccases can also target the colored and toxic compounds released as effluents from pulp industries and render them nontoxic through polymerization and depolymerization reactions (Virk et al., 2012).

Such intense demand for the enzyme has pushed enzyme producers to develop an entirely new industry in a remarkably short time (Bajpai, 1999; Singh et al., 2008, 2011a). The objective of this chapter is to review the application of enzymes in the pulp and paper industry.

APPLICATION OF XYLANASES FOR THE DELIGNIFICATION OF PULP

Historically, first-time application of enzymes to modify pulp properties was reported by Paice and Jurasek in 1984, and the later study by Viikari et al. (1986) demonstrated that a similar enzyme treatment reduced the requirement of chemicals to bleach the pulp. Xylanases have been widely tried in the pulp and paper industry, since they showed lignin extractability from kraft pulp by depolymerizing xylan closely associated with lignin in the plant cell wall. Treatment of eucalyptus pulp with commercial xylanases, such as Novozyme 473 and Cartazyme HS-10, reduced chlorine consumption by 31% and increased the final brightness by 2.1–4.9 points (Bajpai et al., 1994). Xylanase P (a commercial enzyme) improved the brightness of kraft pulp by 5.6 points when used at $10\,Ug^{-1}$ pulp and caused 10% reduction in chlorine consumption (Madlala et al., 2001). The utilization of xylanases can reduce 5.0–7.0 kg of chlorine dioxide per ton of kraft pulp and an average decrease of 2.0–4.0 units in the kappa number (KN) of pulp (Polizeli et al., 2005). Xylanase from *Bacillus megaterium* showed 8.12% and 1.16% increase in brightness and viscosity, 13.67% decrease in KN, and 31% decrease in chlorine consumption (Sindhu et al., 2006). Xylanase from *B. stearothermophilus* SDX reduced chlorine consumption up to 15% while its combination with pectinase resulted in 20% reduction (Dhiman et al., 2008). Alkali stable and thermotolerant xylanase from *Bacillus pumilus* SV-85S showed (at pH 9.0 for 2.0 h at 55°C) reduction in KN by 1.6 points and increased brightness by 1.9 points. The pretreatment of pulp with xylanase resulted in 29.16% reduction in chlorine consumption while maintaining the same brightness as in the control (Nagar et al., 2013).

MECHANISM OF DELIGNIFICATION OF PULP WITH XYLANASES

Data collected by Senior et al. (1999) showed that xylanase hydrolyzes the xylan polymer that exists within pulp fibers. Xylans are intimately linked to cellulose and lignin, thus it follows that disruption of the xylan backbone affects their separation during bleaching. Xylanase was also shown to increase fiber wall swelling and in turn increase the speed of diffusion through the walls (Clark et al., 1991). In addition, it was suggested that if lignin, covalently bound to xylan, was made smaller by enzyme use, it would be more easily extracted. Another hypothesis that came from research was that xylanase enzymes catalyze the hydrolysis of xylan that has reprecipitated on the fibers during alkaline pulping. Xylanase acts as a bleaching aid rather than as a true delignification agent, since the enzyme does not directly degrade lignin (Woolridge, 2014).

FACTORS AFFECTING XYLANASE TREATMENT EFFICIENCY

The major factors affecting the xylanase treatment efficiency include pH, temperature, enzyme dosage, pulp consistency, and reaction time (Bajpai, 1999). The optimum pH for xylan treatment varies among enzymes (Sharma et al., 2014). Generally, the xylanases produced from prokaryotes are more potent between pH 6.0 and 9.0, while those derived from fungal cultures act best at 4.0–6.0 pH. The optimum temperature ranges from 35 to 60°C. Xylanase was found completely stable over a broad pH (5–11) range and retained 52% of its activity upon incubation at 70°C for 30 min (Nagar et al., 2011). In addition, pulp consistency must be optimized to obtain effective dispersion of the enzyme used, to improve the efficiency of xylanase treatment. Most of the beneficial effect of bleaching can be obtained after 1.0–2.0 h of treatment (Bajpai, 1999).

COMMERCIAL AVAILABILITY OF XYLANASES

Commercially accessible xylanases in the enzyme market are most often from family 11. This may be attributed to the exclusive specificity of this family of xylanases for substrates containing D-xylose; in contrast, the family 5 xylanases possess carboxymethyl cellulase activity while the family 10 xylanases are active on cellulose (Collins et al., 2005). Viikari et al. (2009) reported that about 20 mills in North America and Scandinavia use enzymes in kraft pulp bleaching and the approximate price of xylanase treatment in 2007 was less than $2.0 per ton of pulp. Xylanases are available that function at both elevated pH and temperature, eg, approaching pH 10 and 90–100°C. There is also continually growing interest in the development of new xylanases and product formulations to enhance their compatibility for industrial applications (Vieira et al., 2009). Xylanases are produced by using microbes that are either naturally free of cellulase-producing ability or have been mutated or genetically engineered to eliminate the cellulases. The most significant advancement is development of xylanases such as Ecopulp TX-200C that function in alkaline pH and at high temperatures. Advances in enzyme production technology have drastically reduced the production cost, making enzymes a sensible economic choice in ecofriendly bleaching (Senior et al., 1999; Woolridge, 2014). There are some important companies that are known for selling commercial xylanases (Bajpai, 1999), including Clariant, UK; Genencor, Finland; Voest Alpine, Austria; Novo Nordisk, Denmark; Biocon, India; Rohn Enzyme OY, Finland; Solvay Interox, USA; Thomas Swan Co., UK; and Iogen Corp., Canada.

APPLICATION OF LACCASES FOR THE DELIGNIFICATION OF PULP (TABLE 7.1)

Laccases are oxidative enzymes that have influenced the pulp and paper industry by their numerous merits over any other bleaching enzyme. Laccases, together with mediators, are able to delignify the pulps by an oxidation chain reaction leading to lignin oxidation without the degradation of cellulose. Nonphenolic lignin compounds and lipophilic units are not oxidized by laccases without mediators (Sharma et al., 2007; Singh et al., 2015). According to reports in the literature, more than 100 compounds as probable laccase mediators have been tested for their ability to oxidize lignin or lignin models through the selective oxidation of their benzylic/hydroxyl groups (Canas and Camarero, 2010; Virk et al., 2012). Synthetic mediators reduce the KN (a measure of the residual lignin content or bleachability of wood pulp) that increases, after L stage (laccase enzyme treatment stage), and decreases in the E stage (alkaline extraction stage). In case of natural mediators, the KN increases and brightness decreases during L stage. In fact, the synthetic mediators probably cause degradation and/or oxidation of carbohydrate chains in cellulose. Natural mediators oxidize carbohydrate chains in cellulose to carbonyl groups during the L stage, thus making the pulp vulnerable to degradation by the strong alkaline medium used in bleaching stage (Fardim and Durána, 2004). Possibilities to expand the laccase range of oxidation through redox mediators offer considerable biotechnological potential for biobleaching of pulps. The use of such compounds has particular merit because once they are oxidized by laccases to stable radicals, these radicals may continue oxidizing other compounds, including those not used directly as substrates of the enzyme.

DELIGNIFICATION OF PULP WITH FUNGAL LACCASES

Trametes villosa laccase–HBT system for biobleaching of eucalyptus pulp resulted in 20–27% decrease in KN. However, an alkaline extraction stage (E) raised delignification to 41–45%, much higher than that obtained in the control without enzymes (16–23%). Treating the pulp with the laccase-HBT system reduced the amount of hydrogen peroxide required for subsequent alkaline bleaching by a factor of three to four relative to control (Moldes and Vidal, 2008). Recombinant laccases were also tested to evaluate

Laccase-Producing Organism	Laccase Ug⁻¹ Pulp	Reaction Time (h)	pH of Process	Reaction Temp.°C	Redox mediator Used and (Concentration mM or %)	Type of Pulp/Consistency (%)	Results and Outcome of the Study	References
Trametes versicolor (ATCC 20869)	5.0^{-1}	2.0	5.0	60°	ABTS (1^a)	Spruce kraft, mixed softwood kraft, mixed hardwood kraft and sulfite pulp/10	Laccase–ABTS treatment delignify first three pulps and sulphite pulp up to 40% and 50% respectively.	Bourbonnais and Paice (1996)
T. versicolor	5.0	2.0	5.0	60	ABTS and HBT (10^b)★	Softwood kraft pulp/10	HBT showed more delignification and less residual laccase activity as comparison to ABTS, 37, 2.0 and 34, 32% respectively.	Bourbonnais et al. (1997)
Coriolus versicolor	10	8.0	4.5	40	HBT and N-hydroxyacetamilide (NHAA)/0.1**	Pine kraft/10	NHAA showed very fast delignification at the beginning of the process as a result of fast formation of the oxidized mediator species.	Balakshin et al. (2001)
Streptomyces cyaneus CECT 3335	10	3.0	5.0	45	ABTS (5)	Eucalyptus kraft/10	Reduction in the kappa number by 2.3 U and increased in brightness by 2.2%.	Arias et al. (2003)
Laccase from *Pycnoporus cinnabarinus* (wild strain) Recombinant laccases were produced in *Aspergillus oryzae* and *Aspergillus niger* hosts by *lacI* gene of wild strain	NA	NA	5.0	NA	HBT/NA	Wheat straw kraft pulp/10	The laccase expressed in *A. niger* has the same efficiency in delignification as the wild-type laccase (close to 50% compared with the control trial without laccase), whereas the laccase expressed in *A. oryzae* showed no delignifying effect.	Sigoillot et al. (2004)

Continued

Table 7.1 Laccases From Fungi and Bacteria, Optimized Conditions (Physical and Chemical) for Biobleaching of Pulp (Singh et al., 2015)—cont'd

Laccase-Producing Organism	Laccase Ug⁻¹ Pulp	Reaction Time (h)	pH of Process	Reaction Temp.°C	Redox mediator Used and (Concentration mM or %)	Type of Pulp/Consistency (%)	Results and Outcome of the Study	References
P. cinnabarinus	20	12	4.0	50	HBT (3^a)	Eucalyptus globulus/3.0	Laccase activity was inhibited 50% and 20% in presence of HBT and pulp+HBT respectively within 4h. Laccase–HBT promote delignification (four points-decrease of KN) and 6% increase brightness.*A. niger A. oryzae*	Ibarra et al. (2006)
Trametes villosa	17	2.0	4.0	50	HBT (1.5^a)★	Eucalyptus kraft/10	The laccase–HBT treatment brought delignification (by 20–27% decrease in KN).	Moldes and Vidal (2008)
Aspergillus fumigatus VkJ2.4.5	10	2.0	6.0	50	HBT (1.5^a)★	Mixed wood/10	KN decreased by 14 and brightness improved by 7%.	Vivekanand et al. (2008)
γ-proteobacterium JB*Aspergillus fumigatus*	20	4.0	8.0	55	ABTS (2)★	Wheat straw/10	Enhanced brightness by 5.89 and reduced KN by 21.1%.	Singh et al. (2008)
T. versicolory	20	2.0	4.0	45	Violuric acid (74)	Eucalyptus globulus and Pinus pinaster kraft pulp/2.5	Reduction in KN and increase in brightness of *E. globulus* and softwood pulp by 49, 10% and 35.9, 11% respectively.	Oudia et al. (2008)
T. villosa	20	4.0	4.0	50	Sinapic acid, ferulic acid, coniferyl aldehyde, and sinapylaldehyde HBT (1.5^a)	Sisal (alkaline pulp from soda–anthraquinone cooking process were)/5.0	HBT inactivated the laccase by 99% and 78% in absence and presence of pulp, respectively. Natural mediators proved less efficient than HBT in facilitating pulp bleaching; rather,	Aracri et al. (2009)

Organism/enzyme					Mediator	Substrate/dose	Observations	Reference
P. cinnabarinus	20	5.0	4.0	50	Acetosyringone, syringaldehyde and p-coumaric acid, which were compared in performance with (HBT). (1.5 or 3[a])★	Flax/3.0	All natural mediators reduced the KN after subsequent alkaline treatment with hydrogen peroxide. HBT and p-coumaric acid were inactivated the laccase in the absence of pulp.	Fillat et al. (2010)
T. villosa	17	2.0	4.0	50	HBT and violuric acid and natural mediator syringaldehyde (SyAl) (1.5[a])p★	Eucalyptus kraft/10	HBT and violuric acid gave high delignification and brightness values (similar to industrial TCF pulp) where as SyAl improved pulp properties in a lower extent.p	Moldes et al. (2010)
Pycnoporus sanguineus	2.4	1.0	3.0	40	Acetosyringone (0.05)★	Eucalyptus kraft/10	Lowers hydrogen peroxide consumption down to 87.4% (94.0% without L) and enhances brightness up to a 59% ISO (51% ISO without L).	Eugenio et al. (2010)
Commercially available enzyme (laccase gene from Myceliophthora thermophila expressed in A. oryzae) Novozyme 51003	22	10	3.5	60	HBT (1.5[a])	Wheat straw/5.0	In presence of HBT laccase was inhibited up to 75% after 12 h.	Dedhia et al. (2014)

[a]Mediator in %.
[b]Mediator in mg/g of pulp.

their efficiency for delignification. *Pycnoporus cinnabarinus* laccase fused to the C-terminal linker and carbohydrate binding module of *Aspergillus niger* cellobiohydrolase B in the presence of HBT was employed for biobleaching of softwood kraft pulp (Ravalason et al., 2009). The first evidence of natural phenols (syringaldehyde, SA; acetosyringone, AS) to mediate delignification of eucalypt pulp with laccase from *P. cinnabarinus* at pH 4.0 and 50°C was shown by Camarero et al. (2007). When sinapic acid, ferulic acid, coniferyl aldehyde, and sinapyl aldehyde were evaluated as laccase mediators, they showed lower bleaching efficiency for sisal pulp as these phenolic compounds tend to bind to pulp fibers (Aracri et al., 2009). Fillat et al. (2010) evaluated SA, AS, and *p*-coumaric acid (PCA) as natural mediators for laccase from *P. cinnabarinus* at pH 4.0 and 50°C to bleach flax fibers. Efficiency of these three was compared to HBT in terms of laccase stability. HBT and PCA were found to inactivate laccase in the absence of pulp. All natural mediators resulted in a reduced KN after the subsequent alkaline treatment with hydrogen peroxide. Generally, the natural mediators were found to increase KN, decrease brightness, and change optical properties of the pulp after the L stage, suggesting that natural mediators tend to couple to fibers during a laccase-mediator treatment (Andreu and Vidal, 2011).

DELIGNIFICATION OF PULP WITH BACTERIAL LACCASES

The majority of laccases evaluated for lignin degradation belong to fungi because bacterial laccases, though widespread in bacterial genomes, are not so commonly characterized (Singh et al., 2011). Bacterial laccases are also able to biobleach different pulps with mediators. The laccase from *Streptomyces cyaneus* in the presence of 2,2'-azino-bis (3-ethylbenzothiazoline-6-sulphonic acid) decreased the KN and an enhanced the brightness by 2.3 U and 2.2%, respectively (Arias et al., 2003). An alkalophilic cellulase-free laccase from γ-proteobacterium JB was applied to wheat straw–rich soda pulp to evaluate its bleaching potential by optimizing the conditions by using response surface methodology based on central composite design. The design was employed by selecting laccase units, ABTS concentration, and pH as model factors. The results of second-order factorial design experiments showed that all three independent variables had a significant effect on brightness and KN of laccase-treated pulp. Optimum conditions for biobleaching of pulp with laccase preparation (specific activity, 65 nkat mg^{-1} protein) were 20 nkat g^{-1} of pulp, 2 mM ABTS and pH 8.0, which enhanced brightness by 5.89% and reduced KN by 21.1% within 4 h of incubation at 55°C without further alkaline extraction of pulp (Singh et al., 2008). Laccases have an advantage over other lignin-oxidizing enzymes, like lignin peroxidases (Lip) in that redox potential of Lip increases with decrease in pH of reaction environment. For laccases such data is not available, as logically alkali-tolerant laccases will be the best choice for pulp bleaching where alkaline conditions are demanded at large (Singh et al., 2009; Canas and Camarero, 2010).

FACTORS AFFECTING THE LACCASE-BASED BIOBLEACHING OF PULP

A search for cost-effective laccase mediator (synthetic or natural) system with increased redox potential of enzyme is required. Except for a few alkali-tolerant laccases (Singh et al., 2008; Eugenio et al., 2011), the rest of the biobleaching studies with these enzymes are based on acidic or neutral pH (3.0–6.0). The pulp and paper industry requires alkali tolerant laccases for delignification purposes because several steps of paper making pass through the alkaline conditions (Singh et al., 2008, 2011b). The next hitch for laccases before biobleaching at industrial scale is that they need high-oxygen concentration for efficient functioning. The majority of laccases have less-specific activities, as a result excessive enzyme units are required before implementing at large-scale biobleaching projects. Less thermal stability of laccases is also a discouraging factor for successful implementation of enzymes at industrial levels. Recently, grafting of biobleached pulp was observed by many workers, when reaction mixture contained natural mediator and laccase. Although, grafting provides strength to pulp fibers but increases the KN and reduces the brightness. Searching for cost-effective methods of enzyme production has always been an important area of enzyme technology. Conventional methods of enzyme production are not appropriate for the large-scale production of laccases from fungi or prokaryotes. Less production of enzyme is a big challenge for continuous supply of laccases to pulp and paper industry at large scale (Singh et al., 2015).

SYNERGISTIC EFFECTS OF ENZYMES INVOLVED IN BIOBLEACHING OF PULP

Delignification mechanisms of xylanases and laccases are different because xylanases help to increase the delignification by making pulp more vulnerable to attack by bleaching chemicals whereas laccases act directly on lignin and cause its extraction from pulps (Virk et al., 2012). The combination of xylanase and laccase-mediator bleaching system in sequential treatments results in enhanced pulp brightness and decreased KN. Such an outcome could improve the competitiveness of enzyme-based, environment friendly processes over the current methods (Kapoor et al., 2007). Mixed-enzyme preparation of xylanase and laccase was evaluated for biobleaching of mixed wood pulp. The enzymes were produced through cocultivation of mutant *Penicillium oxalicum* SAU (E)-3.510 and *Pleurotus ostreatus* MTCC 1804 under solid-state fermentation. Bleaching of pulp with mixed-enzyme preparation resulted in a notable decrease in KN and increased brightness as compared to xylanase alone. An analysis of bleaching conditions denoted that a mixed-enzyme preparation (xylanase:laccase, 22:1) led to enhanced delignification when bleaching was performed at 10% pulp consistency (55°C, pH 9.0) for 3 h (Dwivedi et al., 2009). Xylanase and laccase were produced in a cost-effective manner up to 10 kg substrate level and evaluated in elemental chlorine-free bleaching of eucalyptus kraft pulp.

Compared to the pulp prebleached with xylanase (15%) or laccase (25%) individually, the ClO$_2$ savings were higher with sequential treatment of xylanase followed by laccase (35%) at laboratory scale. The sequential enzyme treatment when applied at pilot scale (50 kg pulp) resulted in improved pulp properties (50% reduced post color number, 15.71% increased tear index) and reduced organochlorine compound (measured as AOX) levels (34%) in bleach effluents (Sharma et al., 2014).

ROLE OF ENZYMES IN PITCH CONTROL

Pitch is constituted of fatty acids, sterols, resin acids, glycerol esters of fatty acids and other fats and is usually considered as the wood component that is soluble in methylene (Allen, 1975). It is less than 10% of the total weight of wood but causes major problems. These deposits have adverse effects, such as altered water absorption by the pulps, hole formation and tearing of the paper due to sticky deposits on dryer rolls, and imparting discoloration and hydrophobic spots on the paper (Hillis and Sumimoto, 1989). Certain types of wood pulps, including sulfite pulps and various mechanical pulps, especially from pines, contained high pitch contents (Gutiérrez et al., 2001). Pitch extraction with enzymes is possible as an efficient biotechnological method (Fischer et al., 1993). Enzymatic pitch control helps to reduce pitch-related problems to a satisfactory level. It reduces defects on paper web as well as the frequency of cleaning pitch deposits on the paper machine (Bajpai, 1999). The laccase-mediator system has been reported to act selectively on lignin but recently its role has been described in the removal of lipophilic compounds, which cause pitch deposition in woody and nonwoody paper pulps (Gutiérrez et al., 2006). A role for lipases has been clearly demonstrated in pitch control, eg, a commercially available lipase, Resinase A 2X (Novo Nordisk AG), which is a recombinant lipase expressed in *A. oryzae*. Resinase hydrolyzed ~95% of the triglycerides in a pine (*Pinus densiflora*) mechanical pulp. In addition, the Resinase treatment reduced the number of deposits, decreased the number of spots and holes from the paper, enabled a reduction in talc dosage to control pitch deposition, and permitted the use of higher amounts of fresh wood. Lipidase 10,000 (American Lab. Inc.) and *Candida* and *Aspergillus* lipases have been investigated for the enzymatic control of pitch (Virk et al., 2012).

DE-INKING OF OLD NEWSPRINT BY ENZYME TREATMENT

Enzymes have revolutionized the de-inking process obtaining brightness levels surpassing conventional de-inking processes. Used paper recycling can mitigate or at least reduce the deforestation and environmental pollution caused by paper mills during traditional papermaking process. Old newsprint (ONP) is one of the main materials reused for papermaking (Virk et al., 2013). De-inking is an important step in fiber recycling

(Lee et al., 2007). However, with each subsequent de-inking cycle the strength of new paper decreases, which further decreases the papermaking potential of recycled fiber (Chen et al., 2010). A comparison of conventional chemical de-inking process with an enzymatic one has gotten attention owing to economic and environmental issues (Pala et al., 2004). Cellulase and hemicellulase alone or in combination have already been routinely applied in many mill practices but the strength of the de-inked pulp is not as good as that of pulp that is chemically de-inked (Zhang et al., 2008). On the other hand, laccase is now considered promising for the de-inking of ONP (Bajpai and Bajpai, 1998). De-inking of ONP with laccase-violuric acid system (LVS) gave 20% and 13% higher tensile and tear strength, respectively. The brightness of the LVS-de-inked pulp was 4.2% ISO more after being bleached with H_2O_2 (Xu et al., 2007). A recent study by Virk et al. (2013) showed that laccase did not require any mediator supplementation for de-inking of ONP pulp with a combination of xylanase and laccase enzymes. When both enzymes were compared for their de-inking potential, the percentage reduction of effective residual ink concentration was higher for the combined xylanase/laccase–de-inked pulp (65.8%) than for xylanase (47%) or laccase (67%) alone. An increase in brightness (21.6%), breaking length (16.5%), burst factor (4.2%), tear factor (6.9%), viscosity (13%), and cellulose crystallinity (10.3%) along with decrease in KN (22%) and chemical consumption (50%) were also observed (Virk et al., 2013).

APPLICATION OF LACCASE FOR GRAFTING OF PULP FIBERS

Laccase-mediated biografting is a versatile method of functionalization to the nonspecific requirements of this enzyme, which allow bonding a wide range of aromatic compounds on fibers. These biografting processes may confer hydrophobicity and may improve mechanical properties (Fardim and Durána, 2004). After using natural mediators on sisal pulp, a decrease in brightness was observed after the L stage. This was due to the coupling of the mediator to the fiber surface. Further work was carried out on grafting of phenols onto flax and sisal pulps (Aracri et al., 2009). Flax pulp and sisal pulp treated with laccase from *P. cinnabarinus* in the presence of syringaldehyde, acetosyringone, and *p*-coumaric acid and with laccase from *T. villosa* in the presence of various syringyl- and guaiacyl-type phenols, respectively, showed an increase in KN and subsequent decrease in brightness even after the acetone washing stage. This suggests the occurrence of cross-linking or cross-coupling reactions of these phenolic compounds onto the fibers. These results suggest that the enzymatic treatment of nonwood pulps with laccases in the presence of simple phenols resulted in their incorporation into the pulps. Flax soda/AQ pulps were treated with *T. villosa* laccase along with lauryl gallate (LG) (Cadena et al., 2011). A major portion of LG became attached to the pulp as revealed by an increase in the KN and further confirmed by thioacidolysis and 1H-NMR analysis of solubilized pulp fractions. It was observed that laccase-histidine-treated fiber surface caused better bonding between fibers in hand sheets and increased paper strength. Xylanase pretreatment enhances fiber properties, such as dry and wet tensile strength. Treating pulp fibers with xylanase followed by laccase provided a collective 25% and 46% increase in dry and wet tensile strength, respectively (Chandra and Ragauskas, 2005).

DECOLORIZATION AND DETOXIFICATION EFFECT OF LACCASE ON WASTEWATERS FROM PULP AND PAPER INDUSTRY

Pulp and paper mills generate large volumes of intensely colored black liquors that contain toxic chlorinated lignin degradation products, like chlorolignins, chlorophenols, and chloroaliphatics (Ali and Sreekrishnan, 2001). These paper mill effluents are highly alkaline and alter the pH of soil and water bodies where they are discharged. Laccase significantly affects the color remediation and toxicity of these samples. Laccase from *T. versicolor* alone caused 82% COD reduction and along with manganese peroxidase caused 99% of pentachlorophenol, 99% of 2,3,4,6-tetrachlorophenol (2,3,4,6-TCP), 98% of 3,4-dichlorophenol (3,4-DCP), and 77% of 4-chlorophenol (4-CP) removal from paper mill effluent (Pedroza et al., 2007). Laccase from *T. versicolor* as pure enzyme decolorized the deep brown effluent from paper and pulp mills to a clear light-yellow solution (Karimi et al., 2010). Similarly, laccase obtained from *T. versicolor* through solid-state fermentation of brewer's spent grain demonstrated significant decolorization (up to 87.7%) of various structurally different industrial dyes (Dhillon et al., 2012).

CONCLUSIONS AND FUTURE PERSPECTIVES

Enzyme-based, green, and eco-friendly technologies for pulp and paper industry are important and fruitful if implemented carefully and strictly adopted by pulp and paper industry. These biotechnologies are designed to use lesser bleaching chemicals that are hazardous and cause environmental pollution, mainly after discharge of paper mill effluents to rivers and ponds (Sharma et al., 2015). Less use of chemicals means less water consumption for the washing of bleaching chemicals from bleached pulp, especially in developing countries where scarcity of water is an acute problem. However, due to high production costs, most of the studies on enzyme production and their application in biobleaching are limited to bench scale, and mill-scale application of enzymes in pulp bleaching is still in the developmental stage (Bajpai, 2012). This indicates the need for the development of efficient and low-cost technologies for enzyme production and applications in pulp industries. Future trends may include the development of more effective systems that use much smaller quantities of chemicals, less water, and less energy to attain maximum product yield and performance. Modern biotechnology can provide novel enzymes with improved properties under diverse physiological conditions, such as broad working pH and temperature range and high-redox potential laccases (Singh et al., 2015).

LIST OF ABBREVIATIONS

ABTS 2,2'-Azino-bis(3-ethylbenzothiazoline-6-sulphonic acid
HBT 1-Hydroxybenzotriazol
NHAA N-Hydroxyacetanilide

ACKNOWLEDGMENTS

The authors thank SERB/DST, Delhi, India, for providing the research funding under the Fast Track Young Scientist Program (SB/FT/LS-315/2012).

REFERENCES

Ali, M., Sreekrishnan, T.R., 2001. Aquatic toxicity from pulp and paper mill effluents: a review. Advances in Environmental Research 5, 175–196.

Allen, L.H., 1975. Pitch in wood pulps. Pulp Paper Canada 76 (5), 70.

Andreu, G., Vidal, T., 2011. Effects of natural laccases mediator systems on kenaf pulp. Bioresource Technology 102 (10), 5932–5937.

Aracri, E., Colom, J.F., Vidal, T., 2009. Applications of laccase natural mediators system to sisal pulp: an effective approach to biobleaching or functionalizing pulp fibres? Bioresource Technology 100, 5911–5916.

Arias, M.E., Arenas, M., Rodrıguez, J., Soliveri, J., Ball, A.S., Hernandez, M., 2003. Kraft pulp biobleaching and mediated oxidation of a nonphenolic substrate by laccase from *Streptomyces cyaneus* CECT 3335. Applied and Environmental Microbiology 69, 1953–1958.

Bajpai, P., Bajpai, P.K., 1998. Deinking with enzymes: a review. Tappi Journal 81, 111–117.

Bajpai, P., Bhardwaj, N.K., Bajpai, P.K., Jauhari, M.B., 1994. The impact of xylanases on bleaching of eucalyptus kraft pulp. Journal of Biotechnology 36 (1), 1–6.

Bajpai, P., 1999. Application of enzymes in the pulp and paper industry. Biotechnology Progress 15, 147–157.

Bajpai, P., 2012. ECF and TCF bleaching. In: Environmentally Benign Approaches for Pulp Bleaching. Elsevier, Waltham, pp. 263–283.

Balakshin, M., Chen, C.L., Gratzl, J.S., Kirkman, A.G., Jakob, H., 2001. Biobleaching of pulp with dioxygen in laccase–mediator system effect of variables on the reaction kinetics. Journal of Molecular Catalysis B: Enzymatic 16, 205–215.

Bhoria, P., Singh, G., Sharma, J.R., Hoondal, G.S., 2012. Biobleaching of wheat straw-rich-soda pulp by the application of alkalophilic and thermophilic mannanase from *Streptomyces* sp. PG-08-3. African Journal of Biotechnology 11 (22), 6111–6116.

Bourbonnais, R., Paice, M.G., 1996. Enzymatic delignification of kraft pulp using laccase and a mediator. Tappi Journal 79, 199–204.

Bourbonnais, R., Paice, M.G., Freiermuth, B., Bodie, E., Borneman, S., 1997. Reactivities of various mediators and laccases with kraft pulp and lignin model compounds. Applied and Environmental Microbiology 63, 4627–4632.

Cadena, E.M., Du, X., Gellerstedt, G., Li, J., Fillat, A., García-Ubasart, J., Vidal, T., Colom, J.F., 2011. On hexenuronic acid (HexA) removal and mediator coupling to pulp fiber in the laccase/mediator treatment. Bioresource Technology 102, 3911–3917.

Camarero, S., Ibarra, D., Martinez, A.T., Romero, J., Gutierrez, A., del Rio, J.C., 2007. Paper pulp delignification using laccase and natural mediators. Enzyme and Microbial Technology 40, 1264–1271.

Canas, A., Camarero, S., 2010. Laccases and their natural mediators: biotechnological tools for sustainable eco-friendly processes. Biotechnology Advances 28, 694–705.

Central Pollution Control Board Delhi, India. Annual Report, 2007.

Chandra, R.P., Ragauskas, A.J., 2005. Modification of high-lignin kraft pulps with laccase. Part 2. Xylanase enhanced strength benefits. Biotechnology Progress 21, 1302–1306.

Chen, Y., Wan, J., Ma, Y., Lv, H., 2010. Modification of properties of old newspaper pulp with biological method. Bioresource Technology 101, 7052–7056.

Clark, T.A., Steward, D., Bruce, M.E., McDonald, A.G., Singh, A.P., Senior, D.J., 1991. In: Proceedings of the 45th Appita Annual General Conference, Melbourne, Australia, vol. 1, p. 193.

Collins, T., Gerday, C., Feller, G., 2005. Xylanases, xylanase families and extremophilic xylanases. FEMS Microbiology Reviews 29, 3–23.

Dedhia, B.S., Vetal, M.D., Rathod, V.K., Levente, C., 2014. Xylanases and laccase aided bio-bleaching of wheat straw pulp. Canadian Journal of Chemical Engineering 92, 131–138.

Dhillon, G.S., Kaur, S., Brar, S.K., 2012. In-vitro decolorization of recalcitrant dyes through an ecofriendly approach using laccase from *Trametes versicolor* grown on brewer's spent grain. International Biodeterioration and Biodegradation 72, 67–75.

Dhiman, S.S., Sharma, J., Battan, B., 2008. Industrial applications and future prospects of microbial xylanases: a review. Bioresource 3, 1377–1402.

Dwivedi, P., Vivekanand, V., Pareek, N., Sharma, A., Singh, R.P., 2009. Bleach enhancement of mixed wood pulp by xylanase-laccase concoction derived through co-culture strategy. Applied Biochemistry and Biotechnology 160, 55–68.

Eugenio, M.E., Santos, S.M., Carbajo, J.M., Martín, J.A., Martín-Sampedro, R., González, A.E., Villar, J.C., 2010. Kraft pulp biobleaching using an extracellular enzymatic fluid produced by *Pycnoporus sanguineus*. Bioresource Technology 101, 1866–1870.

Eugenio, M.E., Hernandez, M., Moya, R., Martin-Stampedro, R., Villar, J.C., Arias, M.E., 2011. Evaluation of new laccase produced by *Streptomyces ipomea* on biobleaching and ageing of kraft pulps. Bioresource 6, 3231–3241.

Fardim, P., Durána, N., 2004. Retention of cellulose, xylan and lignin in kraft pulping of eucalyptus studied by multivariate data analysis: influences on physicochemical and mechanical properties of pulp. Journal of Brazilian Chemical Society 15, 514–522.

Fillat, A., Colom, J.F., Vidal, T., 2010. A new approach to the biobleaching of flax pulp with laccase using natural mediators. Bioresource Technology 101, 4104–4110.

Fischer, K., Puchinger, L., Schloffer, K., Kreiner, W., Messner, K., 1993. Enzymatic pitch reduction of sulfite pulp on pilot scale. Journal of Biotechnology 27, 341–348.

Garcia, J.C., Lopez, F., Perez, A., Pelach, M.A., Mutje, P., Colodette, J.L., 2010. Initiating ECF bleaching sequences of eucalyptus kraft pulps with Z/D and Z/E stages. Holzforschung 64 (1), 1–6.

Gutiérrez, A., del Río, J.C., Martínez, M.J., Martínez, A.T., 2001. The biotechnological control of pitch in paper pulp manufacturing. Trends in Biotechnology 19, 340–348.

Gutiérrez, A., del Río, J.C., Rencoret, J., Ibarra, D., Martínez, A.T., 2006. Main lipophilic extractives in different paper pulp types can be removed using the laccase-mediator system. Applied Microbiology and Biotechnology 72, 845–851.

Hillis, W.E., Sumimoto, M., 1989. Effect of extractives on pulping. In: Natural Products of Woody Plants II. Springer, USA, pp. 880–920.

Ibarra, D., Camarero, S., Romero, J., Martínez, M.J., Martínez, A.T., 2006. Integrating laccase-mediator treatment into an industrial-type sequence for totally chlorine free bleaching eucalypt kraft pulp. Journal of Chemical Technology and Biotechnology 81, 1159–1165.

Kapoor, M., Kapoor, R.K., Kuhad, R.C., 2007. Differential and synergistic effects of xylanase and laccase mediator system (LMS) in bleaching of soda and waste pulps. Journal of Applied Microbiology 103, 305–317.

Karimi, S., Abdulkhani, A., Karimi, A., Ghazali, A.H., Ahmadun, F.L., 2010. The effect of combination enzymatic and advanced oxidation process treatments on the color of pulp and paper mill effluent. Environmental Technology 31, 347–356.

Lee, C.K., Darah, I., Ibrahim, C.O., 2007. Enzymatic deinking of laser printed office waste papers: some governing parameters on deinking efficiency. Bioresource Technology 98, 1684–1689.

Madlala, A.M., Bisson, S., Singh, S., Christov, L., 2001. Xylanase induced reduction of chlorine dioxide consumption during elemental chlorine-free bleaching of different pulp types. Biotechnology Letters 23, 345–351.

Moldes, D., Vidal, T., 2008. Laccase–HBT bleaching of eucalyptus kraft pulp: influence of the operating conditions. Bioresource Technology 99, 8565–8570.

Moldes, D., Cadena, E.M., Vidal, T., 2010. Biobleaching of eucalypt kraft pulp with a two laccase-mediator stages sequence. Bioresource Technology 101, 6924–6929.

Nagar, S., Mittal, A., Kumar, D., Kumar, L., Gupta, V.K., 2011. Hyper production of alkali stable xylanase in lesser duration by *Bacillus pumilus* SV-85S using wheat bran under solid state fermentation. New Biotechnology 28, 581–587.

Nagar, S., Jain, R.K., Thakur, V.V., Gupta, V.K., 2013. Biobleaching application of cellulase poor and alkali stable xylanase from *Bacillus pumilus* SV-85S. Biotech 3, 277–285.

Oudia, A., Queiroz, J., Simões, R., 2008. Potential and limitation of *Trametes versicolor* laccase on biodegradation of eucalyptus globulus and pinus pinaster kraft pulp. Enzyme and Microbial Technology 43, 144–148.

Paice, M.G., Jurasek, L., 1984. Removing hemicellulose from pulps by specific enzymic hydrolysis. Journal of Wood Chemistry and Technology 4, 187–198.

Pala, H., Mota, M., Gama, F.M., 2004. Enzymatic versus chemical deinking of non impact ink printed paper. Journal of Biotechnology 108, 79–89.

Pedroza, A.M., Mosqueda, R., Alonso-Vante, N., Rodríguez-Vázquez, R., 2007. Sequential treatment via *Trametes versicolor* and UV/TiO$_2$/Ru(x)Se(y) to reduce contaminants in waste water resulting from the bleaching process during paper production. Chemosphere 67, 793–801.

Polizeli, M.L., Rizzatti, A.C., Monti, R., Terenzi, H.F., Jorge, J.A., Amorim, D.S., 2005. Xylanases from fungi: properties and industrial applications. Applied Microbiology and Biotechnology 67, 577–591.

Ravalason, H., Herpoël-Gimbert, I., Record, E., Bertaud, F., Grisel, S., deWeert, S., 2009. Fusion of a family 1 carbohydrate binding module of *Aspergillus niger* to the *Pycnoporus cinnabarinus* laccase for efficient softwood kraft pulp biobleaching. Journal of Biotechnology 142, 220–226.

Sadhasivam, S., Savitha, S., Swaminathan, K., 2010. Deployment of *Trichoderma harzianum* WL1 laccase in pulp bleaching and paper industry effluent treatment. Journal of Cleaner Production 18, 799–806.

Senior, D.J., Hamilton, J., Taiplus, P., Torvinen, J., 1999. Enzyme Use Can Lower Bleaching Costs, Aid ECF Conversions. Pulp and Paper International.

Sharma, P., Goel, R., Capalash, N., 2007. Bacterial laccases. World Journal of Microbiology and Biotechnology 23, 823–832.

Sharma, A., Thakur, V.V., Shrivastava, A., Jain, R.K., Mathur, R.M., Gupta, R., Kuhad, R.C., 2014. Xylanase and laccase based enzymatic kraft pulp bleaching reduces adsorbable organic halogen (AOX) in bleach effluents: a pilot scale study. Bioresource Technology 169, 96–102.

Sharma, P., Sood, C., Singh, G., Capalash, N., 2015. An eco-friendly process for biobleaching of eucalyptus kraft pulp with xylanase producing *Bacillus halodurans*. Journal of Cleaner Production 87, 966–970.

Sigoillot, C., Record, E., Belle, V., Robert, J.L., Levasseur, A., Punt, P.J., van den Hondel, Fournel, A., Sigoillot, J.C., Asther, M., 2004. Natural and recombinant fungal laccases for paper pulp bleaching. Applied Microbiology and Biotechnology 64, 346–352.

Sindhu, I., Chhibber, S., Caplash, N., Sharma, P., 2006. Production of cellulase-free xylanase from *Bacillus megaterium* by solid state fermentation for biobleaching of pulp. Current Microbiology 53, 167–172.

Singh, G., Ahuja, N., Batish, M., Capalash, N., Sharma, P., 2008. Biobleaching of wheat straw-rich-soda pulp with alkalophilic laccase from γ-*proteobacterium JB*: optimization of process parameters using response surface methodology. Bioresource Technology 99, 7472–7479.

Singh, G., Sharma, P., Caplash, N., 2009. Performance of an alkalophillic and halotolerant laccase from γ-*proteobacterium JB* in the presence of industrial pollutants. Journal of General and Applied Microbiology 55, 283–289.

Singh, G., Bhalla, A., Kaur, P., Capalash, N., Sharma, P., 2011b. Laccase from prokaryotes: a new source for an old enzyme. Reviews in Environmental Science and Biotechnology 10, 309–326.

Singh, G., Bhalla, A., Ralhn, P.K., 2011a. Extremophiles and extremozymes: importance in current biotechnology. ELBA Bioflux 3, 46–54.

Singh, G., Kavleen, K., Puri, S., Sharma, P., 2015. Critical factors affecting laccase-mediated biobleaching of pulp in paper industry. Applied Microbiology and Biotechnology 99, 155–164.

Vieira, D.S., Degreve, L., Ward, R.J., 2009. Characterization of temperature dependent and substrate-binding cleft movements in *Bacillus circulans* family 11 xylanase: a molecular dynamics investigation. Biochimica et Biophysica Acta 1790, 1301–1306.

Viikari, L., Ranua, M., Kantelinen, A., Sundquist, J., Linko, M., 1986. Bleaching with enzymes. In: Proceedings of the 3rd International Conference of Biotechnology on Pulp and Paper Industry, Stockholm, Sweden, pp. 67–69.

Viikari, L., Suurnäkki, A., Grönqvist, S., Raaska, L., Ragauskas, A., 2009. Forest products: biotechnology in pulp and paper processing. In: Encyclopedia of Microbiology. Academic Press, Oxford, UK, pp. 80–94.

Virk, A.P., Sharma, P., Capalash, N., 2012. Use of laccase in pulp and paper industry. Biotechnology Progress 28, 21–32.

Virk, A.P., Puri, M., Gupta, V., Capalash, N., Sharma, P., 2013. Combined enzymatic and physical deinking methodology for efficient ecofriendly recycling of old newsprint. PLoS One 8 (8), e72346.

Vivekanand, V., Dwivedi, P., Sharma, A., Sabharwal, N., Singh, R.P., 2008. Enhanced delignification of mixed wood pulp by *Aspergillus fumigatus* laccase mediator system. World Journal of Microbiology and Biotechnology 24, 2799–2804.

Woolridge, E.M., 2014. Mixed enzyme systems for delignification of lignocellulosic biomass. Catalysts 4, 1–35.

Xu, Q.H., Fu, Y.J., Qin, M.H., Qiu, H.Y., 2007. Surface properties of old newsprint laccase-violuric acid system deinked pulp. Appita Journal 60, 372–377.

Zhang, X., Renaud, S., Paice, M., 2008. Cellulase deinking of fresh and aged recycled newsprint/magazines (ONP/OMG). Enzyme and Microbial Technology 43, 103–108.

FURTHER READING

Chandra, R., Singh, R., 2012. Decolourisation and detoxification of rayon grade pulp paper mill effluent by mixed bacterial culture isolated from pulp paper mill effluent polluted site. Biochemical Enginering Journal 61, 49–58.

Development of AOX Standards for Large Scale Pulp and Paper Industries., 2007. Central Pollution Control Board, Ministry of Environment and Forests. WWW.cpcb.nic.in.

González, L.F., Sarria, V., Sánchez, O.F., 2010. Degradation of chlorophenols by sequential biological-advance doxidative process using *Trametes pubescens* and TiO2/UV. Bioresource Technology 101, 3493–3499.

Kamali, M., Khodaparast, Z., 2014. Review on recent developments on pulp and paper mill wastewatertreatment. Ecotoxicology and Environmental Safety. http://dx.doi.org/10.1016/j.ecoenv.2014.05.005i14.

Zhang, C., Chen, J., Wen, Z., 2012. Alternative policy assessment for water pollution control in China's pulp and paper industry. Resources, Conservation and Recycling 66, 15–26.

CHAPTER 8

Enzymes in Food Processing

P. Fernandes[1,2], F. Carvalho[2]
[1]Universidade Lusófona de Humanidades e Tecnologias, Lisbon, Portugal; [2]Universidade de Lisboa, Lisbon, Portugal

INTRODUCTION

Enzymes have been used in food production and processing for centuries. Iconic examples include rennet for cheese making, diastase for starch hydrolysis, pectinase for pectin hydrolysis, and amylases and proteases in the production of soy foods (Poulsen and Buchholz, 2003; Gurung et al., 2013). Enzymatic preparations obtained from extracts of plants or animal tissues were commonly used, although currently microorganisms are by far the main providers of commercial enzymes for food applications (van Oort, 2010a; Agarwal and Sahu, 2014). Microorganisms are easier to work with and to manipulate than mammalian or plant cells, and allow for large-scale and cost-effective production of enzymes. Moreover, enzymes from animal origins bring a considerable risk of disease potency both in handling and in manufacturing, which is much more complex and expensive than microbial-based production. The extensive use of enzymes within the scope of the food industry is not only related to their catalytic specificity, ability to operate under mild conditions, and biodegradability but also to their generally regarded as safe (GRAS) label. To achieve the latter status, the candidate enzyme or enzyme formulation must however comply with a complex set of regulatory constraints (Spök, 2006; Agarwal and Sahu, 2014). Currently within the European Union (EU), and on the follow-up of legislation approved in 2009, regulation is being developed in order to harmonize the use of enzymes as food-processing aids in all member states. This will involve safety evaluations of food enzymes, either existing or new, performed by the European Food Safety Authority (EFSA), the European agency for risk assessment in food and feed. Those enzymes considered fit will be sound candidates for inclusion in a list of approved food enzymes by EU decision makers (EC Regulations, 2008; EFSA CEF Panel, 2014).

Once recovered from the culture broth, enzymes are used for the production of a given good, viz glucose from starch; or added to a food preparation, in order to improve a particular feature, viz digestibility, flavor, nutritional value, or texture (Li et al., 2012). Food enzymes correspond to a significant share of the market for industrial enzymes with revenues close to USD 1.3 billion in 2013 and foreseen to reach USD 2.3 billion by 2018 (Markets and Markets, 2013; Adrio and Demain, 2014). Unlike the highly purified enzymes used for analytical or diagnostic purposes, enzymes used within the scope of

Agro-Industrial Wastes as Feedstock for Enzyme Production
ISBN 978-0-12-802392-1
http://dx.doi.org/10.1016/B978-0-12-802392-1.00008-3

the food industry are made available as partly purified or bulk enzymes (van Oort, 2010a; Walsh, 2014). This trend simplifies downstream processing, yet it brings another outcome. Thus, the resulting enzyme preparation typically harbors other enzymes produced by the same (micro)organism, alongside the labeled enzymatic activity. Therefore, and particularly when complex food matrices are processed, care has to be given in order to avoid side-effects, a goal that may require inactivation of the unwanted enzymatic activity (Sieiro et al., 2012).

Actually, enzyme/substrate interactions are often quite intricate when food processing is considered. Food materials, far from pure chemicals, consist of complex structures and/or mixtures, which may convey competing substrates and/or inhibitors or high-salt/metal ion concentrations alongside the desired substrate. Moreover, particulate matter may be present in influent streams to be processed, a feature likely to condition the design of the setup intended to be implemented for processing. Hence, the patterns of enzyme behavior and performance using a pure substrate, required for a basic characterization of the envisaged bioconversion, may not be fully reproduced when a realistic substrate mixture is processed, under food reaction environment. The reliable prediction of the outcome of enzyme action on food material will therefore be increasingly favored as the model system adequately mimics the reaction media and conditions of the envisaged system (Law, 2002).

The features required for a given enzyme are also diverse, depending on the specific application. Thus, for the isomerization of glucose to fructose, the final step in the production of high-fructose corn syrup (HFCS) to be carried out continuously, high operational stability and activity at relatively high temperatures, viz 60°C, is mandatory (DiCosimo et al., 2013). On the other hand, proteases used for meat tenderization should be highly active at room to low temperatures, viz 4°C, and can be easily inactivated during cooking, to avoid overtenderization (Zhao et al., 2012). Given these disparate requirements, there has been a growing interest in microorganisms from extreme environments as sources for enzymes (Trincone, 2011; Nigam, 2013; Gupta et al., 2014).

ENZYMATIC APPLICATIONS IN FOOD PROCESSING

Most of the enzymes used in food processing are hydrolases, but other classes of enzymes are also used. Those of most significance are referred to in Table 8.1. Details on enzyme source and mode of action can be found in online databases, such as BRENDA (http://www.brenda-enzymes.info/oldstart.php), or more specifically for carbohydrates, CAZy (http://www.cazy.org/). Relevant areas for application of those enzymes will be addressed in some detail.

Baked Goods

Flour, a mixture of starch, gluten, lipids, nonstarch polysaccharides, and enzymes, is the raw material in bread making. Mixing flour with water and yeast sets a series of

Continued

Table 8.1 Enzymes in Food Processing: Examples of Key Enzymes

Enzyme	Details	Comments
Acetolactate decarboxylase	EC 4.1.1.5	Promotes the decarboxylation of α–acetolactate to acetoin
Amylases	α–Amylase (EC 3.2.1.1)	These are endoacting α–1,4–glucan glucanohydrolases, that hydrolyze α–1,4–glycosidic linkages in amylose, amylopectin and other starch derivatives, to produce small dextrins with degrees of polymerization (DP) within 2–12. They have activity neither on α–1,6–glycosidic linkages nor on terminal glucose residues. The activity/stability of these enzymes is significantly enhanced by Ca^{2+}
	α–Amylase (EC 3.2.1.133)	These are exacting α–1,4–glucan maltohydrolases that hydrolyze α–1,4–glycosidic linkages in polysaccharides such as amylose, amylopectin, and related glucose polymers, so as to remove successive α–maltose residues from the nonreducing ends of the chains to produce mostly α–maltose
	β–Amylase (EC 3.2.1.2)	These are exacting α–1,4–glucan maltohydrolases that hydrolyze α–1,4–glycosidic linkages but cannot bypass α–1,6–glycosidic linkages, to produce mostly β–maltose, through inversion of the anomeric configuration
	G4 amylase (EC 3.2.1.60)	This exomaltotetraohydrolase produces α–maltotetraose from the nonreducing ends of starch by an exoglycolytic mechanism it produces α–maltotetraose from the nonreducing ends of starch by an exoglycolytic mechanism
	Amyloglucosidase (EC 3.2.1.3)	Also termed glucoamylase, this enzyme promotes the hydrolysis of α–1,4–glycosidic linkages from the nonreducing end of amylose and amylopectin in a stepwise manner. It also acts mildly on α–1,6–glycosidic linkages but at a much lower pace
	Pullulanase (EC 3.2.1.41)	This debranching enzyme promotes the hydrolysis of α–1,6–glycosidic linkages in pullulan and amylopectins
Chymosin	EC 3.4.23.4	An endopeptidase highly specific tor the hydrolysis of peptide bond, highly specific for the hydrolysis of Phe–Met bond
Esterase	EC 3.1.1.1	Promotes the hydrolysis of triacylglycerol to glycerol and fatty acids in aqueous media, while in nonaqueous environment promotes the reverse reaction, producing glycerides from glycerol and fatty acids. Preference for short-chain triacylglycerides/fatty acids as substrates
β–Glucanase	EC 3.2.1.6	Promoted the endohydrolysis of 1,3– or 1,4–linkages in β–D–glucans
Glucansucrases		Promotes the transfer of a hexosyl group
	Alternansucrase (EC 2.4.1.140)	Transfers alternately an α–D–glucosyl residue from sucrose to the 6–position and the 3–position of the nonreducing terminal residue of an α–D –glucan, thus producing a glucan having alternating α–1,6– and α–1,3 –linkages
	Amylosucrase (EC 2.4.1.4)	Transfers α–D–glucosyl units from sucrose to synthesize only α–1,4 linkages
	Dextransucrase (EC 2.4.1.5)	Transfers α–D–glucosyl units from sucrose to synthesize only α–1,6 linkages

Table 8.1 Enzymes in Food Processing: Examples of Key Enzymes—cont'd

Enzyme	Details	Comments
Glucose isomerase	EC 5.3.1.5	Promotes primarily the isomerization of xylose to xylulose, but also displays considerable affinity for glucose allowing its isomerization to fructose
Glucose oxidase	EC 1.1.3.4	In the presence of oxygen, this enzyme promotes the oxidation of β-D-glucose, with formation of H_2O_2 and D-glucono-D-lactone. The latter spontaneously hydrolyzes to gluconic acid
α-glucosidase	EC 3.2.1.20	Promotes the hydrolysis of terminal, nonreducing α-D-1,4-glucose bonds with release of α-D-glucose
β-D-glucosidase	EC 3.2.1.21	Hydrolysis of terminal, nonreducing β-D-glucosyl residues with release of β-D-glucose
Hexose oxidase	EC 1.1.3.5	This enzyme oxidizes D-galactose, D-glucose, D-mannose, maltose, lactose, and cellobiose
Inulinase	Endoinulinase (EC 3.2.1.7) Exoinulinase (EC 3.2.1.80)	β-(2,1)-D-fructanfructanohydrolase that promotes the endohydrolysis of β-(2,1)-D-fructosidic linkages in inulin and related polyfructans Fructan β-fructosidase promotes the hydrolysis of terminal, nonreducing of β-(2,1)-D- and of β-(2,6)-D-fructofuranose residues in inulin and related polyfructans
Invertase	EC 3.2.1.26	A β-D-fructofuranoside fructohydrolase that promotes the hydrolysis of terminal non-reducing β-D-fructofuranoside residues in β-D-fructofuranosides
Laccase	EC 1.10.3.2	Laccases are multicopper, polyphenol oxidases that catalyze the one-electron oxidation of several substrates, viz mono-, di-, and polyphenols, aminophenols, methoxyphenols, aromatic amines, and ascorbate
Lactase	EC 3.2.1.23	Lactase or β-galactosidase promotes the hydrolysis of terminal nonreducing β-D-galactose residues in β-D-galactosides
Lipase	EC 3.1.1.3	Like esterases, but with preference for long-chain triacylglycerides/fatty acids as substrates. Promotes the hydrolysis of triacylglycerol to glycerol and fatty acids in aqueous media, while in nonaqueous environment promotes the reverse reaction, producing glycerides from glycerol and fatty acids
Lipoxygenases	EC 1.13.11.12	Lipoxygenases are nonheme, iron- or manganese-containing enzymes. These enzymes promote the oxidation of unsaturated fatty acids containing a 1-cis, 4-cis-pentadiene structure to fatty acid hydroperoxides. Natural substrates of these enzymes are arachidonic, linoleic, and linolenic acids
Pectin lyase	EC 4.2.2.10	Random cleavage of 1,4-α-D-galacturonan methyl ester to give unsaturated methyloligogalacturonates
Pectin methyl esterase	EC 3.1.1.11	Random hydrolysis of pectin to pectate and methanol

Enzyme (EC number)	Description
Pepsin A (EC 3.4.23.1)	An endopeptidase pepsin that promotes the hydrolysis of peptide bonds involving aromatic amino acids (Phe, Tyr, and Trp)
Peroxidase (EC 1.11.1.7)	In the presence of a phenolic donor and of hydrogen peroxide, the enzyme promotes the formation of a phenoxyl radical of the donor and of water
Polygalacturonase	Endopolygalacturonase (EC 3.2.1.15): Random hydrolysis of $1,4$-α-D-galactosiduronic linkages in pectate and other galacturonans
	Exopolygalacturonase (EC 3.2.1.67): Hydrolysis of $1,4$-α-D-galactosiduronic linkages in pectate and other galacturonans at the terminal bond, leading to monogalacturonates
	Exopolygalacturonase (EC 3.2.1.67): Hydrolysis of $1,4$-α-D-galactosiduronic linkages in pectate and other galacturonans at the penultimate bond, leading to digalacturonates
α-L-rhamnosidase (EC 3.2.1.40)	Hydrolysis of terminal nonreducing α-L-rhamnose residues in α-L-rhamnosides
Transglutaminase (EC 2.3.2.13)	A protein-glutamine γ-glutamyltransferase, the enzyme promotes the cross-linking of proteins through the formation of glutamyllysyl isopeptide bonds
Xylanases	Promoted the depolymerization of hemicellulose to monomer sugars through the synergistic action of multiple enzymes
	Endo-$1,4$-β-xylanase (EC 3.2.1.8): This $1,4$-β-D-xylan xylanohydrolase breaks the glycosidic bonds in the xylan backbone, thus decreasing the degree of polymerization of the substrate
	β-D-xylosidase (EC 3.2.1.37): These are $1,4$-β-D-xylan xylohydrolases, able to hydrolyze xylobiose and higher xylooligosaccharides with decreasing specific affinity, but unable to hydrolyze xylan
	Exo-α-L-arabinofuranosidase (EC 3.2.1.55): These enzymes cleave the arabinose of arabinofuranose side chains of arabinoxylans, allowing the action of endoarabinoses
	Endo-α-L-arabinofuranosidase, (EC 3.2.1.99): These enzymes carry out the hydrolysis of the araban backbone of arabinoxylans. With the combined action of exorabinofuranosidases, branched arabinans are cleaved to L-arabinose
	Acetylxylan esterase (EC 3.1.1.6): Promote the de-esterification of acetyl xylan and acetyl xylooligosaccharides. By cleaving the acetyl side groups, the action of these esterases eases the action of endoxylanases, since acetyl groups could present a steric hindrance to the hydrolysis of the xylan backbone
α-Glucuronidase (EC 3.2.1.131)	These hydrolases remove α-D-glucopyranosyl acid groups are often $1,2$-linked to the xylan backbone

biochemical and biophysical processes in motion, catalyzed by the yeast and by the flour enzymes (the dough phase). After this, and under heat, the dough is transformed into bread (the baking phase). Exogenous enzymes are added to the dough in order to improve several features of bread, improve shelf life, and compensate flour variability. Among these enzymes, amylases stand out.

Starch is of paramount relevance in baked goods, as well as in other areas in the food sector. Starch, the most common storage carbohydrate in plants, is a mixture of two polysaccharides, amylose and amylopectin. Amylose is essentially a linear polymer composed of linked α-(1,4)-D-glucopyranosyl units, with degrees of polymerization (DP) of 100–10,000. The fraction of amylose in nonmutant reserve starch of higher plants varies between 11% and 37%, depending on the species and cultivar of the source, although in genetically modified corn strains up to 50–75% amylose fractions can occur.

Amylopectin is a branched polymer, also composed of α-D-glucopyranosyl units primarily linked by (1,4) bonds, where branches resulting from (1,6) linkages also occur. Branching is relatively random, but on average one branch occurs every 25 glucosyl units. Amylopectin has larger DP, within 10,000–100,000 glucosyl units (Schwartz and Whistler, 2009).

In bread making, α- and β-amylases are used synergistically. Thus, while the former releases low-molecular-chain dextrins from starch, the latter hydrolyzes them to maltose, which is used as fermentable sugars by yeast. Amylase action results in reduced dough viscosity, improved bread volume and crumb texture, as well as in enhanced Maillard reactions for the intensification of browning and flavor of the crust. Fungal α-amylases act mostly on damaged starch, whereas bacterial α-amylases act on gelatinized, amorphous starch. Under the correct dosage, amylases with intermediate thermostability, particularly bacterial amylases, are used to minimize staling (van Oort, 2010b; Miguel et al., 2013). Still, when an antistaling role is aimed at, maltogenic amylases and G4-amylases are clearly favored. Since they are particularly fit to reduce the side chains of amylopectin and release maltooligosaccharides, those enzymes decrease amylopectin retrogradation, allowing crumb softening and increased elasticity, without compromising the amylose network, therefore improving shelf life (Kragh et al., 2011, Fadda et al., 2014). Fungal α-amylases are preferred for flour standardization among other sources given their lower thermostability, as compared to other sources, since the risk of overdextrinization during baking is avoided. Overdextrinization, namely through the formation of an excess of maltodextrins, may also occur in dough if overdosage of α-amylase occurs. This will result in the formation of a sticky or gummy crumb, with an unacceptable mouth feel (Offord, 2008; Popa and Bostan, 2010).

Oxidoreductases, namely lipoxygenase and oxidases, have also been used in the bread-making process. Lipoxygenases from soya bean flour are used to bleach the flour, as the hydroperoxides resulting from enzyme action on lipids react with the yellow carotenoid pigment in wheat flour, leading to the reduction of the yellow color.

Ultimately, a whiter crumb is obtained. Moreover, lipoxygenases have been shown to enhance mixing tolerance and dough-handling properties. This has been related to the oxidation of thiol of gluten proteins, leading to the rearrangement of disulfide bonds and cross-linking of tyrosine residues, ultimately resulting in improved loaf volume. However, the use of lipoxygenases can also result in undesirable flavors in bread, possibly the outcome of the formation of side products during the anaerobic reaction (Addo et al., 1993). Several oxidases have been used in bread making as cross-linking enzymes, in order to enhance dough strength and handling properties, and improve texture and appearance of the baked product. They provide a safe alternative to the use of chemical oxidants, such as potassium bromate or potassium iodate. Often referred to within this type of enzymes is glucose oxidase, which promotes the formation of disulfide bonds in gluten proteins. Improved rheological properties, crumb structure, and a larger volume result from this cross-linking action on proteins. For effective action of glucose oxidase, sufficient amounts of glucose in dough are however needed, a requirement rarely observed in cereal flours (Søe et al., 2004). Hexose oxidases are optionally used, since they have a broader substrate range, namely maltose, but also glucose, lactose, galactose, xylose, arabinose, and cellobiose. These are oxidized to their corresponding lactones and ultimately hydrolyzed to the aldobionic acids. Maltose is present in the freshly prepared dough, and its titer is further increased due to the activity of β-amylase in dough (Søe et al., 2004; Stougaard and Hansen, 2012). Peroxidases also contribute to an improved baked product. These enzymes promote the cross-linking between both arabinoxylans as well as between protein and arabinoxylans, possibly mediated by ferulic acid (Revanappa et al., 2014). Laccases have roughly similar behavior and mode of action as peroxidases in baked goods. Despite their positive role, their use has not yet been considered in the formulation of a commercial product (Osma et al., 2010; van Oort, 2010b). Overdosage of the oxidase can lead to excessive cross-linking, hinder gas retention, and lead to dough difficult to handle and concomitantly to a product of poor quality (Bonet et al., 2006).

Xylanases are used to break down hemicelluloses, particularly the insoluble arabinoxylans, which tamper with the formation of the gluten network. Those nonstarch polysaccharides are widely present in cereals. As a result of enzyme action, flour and water distribution is improved and the dough is softer, more flexible and stable, and easier to handle. During the baking process crumb formation is delayed and the dough is allowed to rise. Moreover, the use of xylanases increases the concentration of arabinoxylooligosaccharides in bread, with a positive impact on human health (Butt et al., 2008; Harris and Ramalingam, 2010; Juturu and Wu, 2012).

Lipases are relatively newcomers to baked goods, their action being on the flour lipids (or added fat, in any case), and essentially improving handling and machinability of dough, enhancing dough strength and stability, and increasing bread oven spring and specific volume (van Oort, 2010b; Pareyt et al., 2011). Moreover, the use of lipase led to a softer crumb in the baked product, suggestive of antistaling effect. This results in prolonged shelf life (Olesen and Qi Si, 1994). Lipases are considered as alternatives to chemical dough strengtheners and emulsifiers, such as diacetyl tartaric esters of mono- and diglycerides (DATEM) or sodium stearoyl-2-lactylate (SSL) (Gerits et al., 2014). Initially lipases were used for the hydrolysis of the ester bond between glycerides and fatty acids in positions one and three of triglycerides, releasing free fatty acids and mono-glycerides. With the emergence of a second generation of lipases, with a broader sub-strate specificity, able to act on polar lipids, and displaying also phospholipase activity, the role of lipase as alternative emulsifiers further strengthened. Moreover, these lipases allowed for a more stable dough and more uniform crumb structure, as compared to the use of the early generation of lipases (Aravindan et al., 2007; Moayedallaie et al., 2010). A third generation of lipases has emerged that has proved more effective in high-mixing and no-time dough processes, and has lower affinity for small-chain fatty acids than their predecessors, hence reducing the risk of off-flavor formation during prolonged storage (Miguel et al., 2013).

Proteases have also been used in baking, namely to decrease mixing time and dough consistency and increase its uniformity. These result in irreversible, proteolytic action on the peptide bonds within the gluten network. This is mostly performed by endopepti-dases, since their endoacting nature has a more direct impact on the gluten network and, hence, on the rheologic properties. Exoacting proteases are also relevant, but the out-come of their action is also noticeable on the flavor and color of the baked product (Goesaert et al., 2005; Hassan et al., 2014). Such effects result from the amino acids released that undergo Maillard-type reaction with different sugars. The use of proteases in dough conditioning has gradually superseded the use of sodium metabisulfite for tampering with the disulfide bonds in the gluten network, since proteases present no toxicity problems and do not react with the heat-stable vitamins (Hassan et al., 2014).

The features presented by these different enzymes have been used in a more advanta-geous way through the selection of adequate enzymatic cocktails, both in terms of com-position and dosage (Keskin et al., 2004; Souppe and Naeye, 2006; Alaunyte et al., 2012; Van Benschop et al., 2012).

Beer Industry

One of the key stages in beer production is the malting process. Malting consists of seed germination, where several endogenous hydrolytic enzymes are combined with hot water to initiate starch, protein, and cell wall polysaccharide degradation (mashing), so that a fermentable extract can later be available for yeast (Jamar et al., 2011). Endogenous proteolytic, amylolytic, and cellulitic enzymes are involved, but their activity may be too scarce, with adverse consequences downstream, namely poor extract yield; slow wort separation and fermentation process, with a too low alcohol titer; impaired filtration of beer; and poor quality and stability of the beer (Kreisz, 2009; Lalor and Goode, 2010; Malomo et al., 2012). Hence, exogenous enzymes are typically added to overcome this drawback. These display pretty much the same type of activity as the endogenous enzymes. Thus, α- and β-amylases, amyloglucosidase, glucanases, and pullulanases are added, so that polysaccharides and proteins are adequately hydrolyzed, leading to a suitable substrate for yeast action, lower viscosity of wort, and improved filterability (Lalor and Goode, 2010; Blanco et al., 2014).

Haze formation in beer involves the interaction of polyphenols with proline-rich proteins, an unwanted feature that can be avoided by the proteolytic action of papain, pepsin, or of specific endoproteases. The latter have proved more adequate, since, unlike papain, they do not tamper with foam stability (Lopez and Edens, 2005; Rehmanji et al., 2005; Steiner et al., 2010; Nguyen et al., 2012). Laccase enzymes from *Trametes versicolor* cultured on brewer's spent grain through solid-state fermentation was also successfully used for haze removal from crude beer samples (Dhillon et al., 2012).

The vicinal diketone diacetyl is a flavor component of beer, but its presence in excess of 0.1 mg/L results in a buttermilk off-flavor. Diacetyl is formed during the exponential growth phase of the yeast during fermentation, and is reduced to acetoin, which is roughly tasteless as compared to diacetyl, by the yeast at late fermentation. This process is however time consuming. By adding α-acetolactate decarboxylase to the fermenting beer, acetolactate is rapidly decarboxylated to acetoin (Fig. 8.1), thus allowing for low levels of diacetyl concentrations during fermentation and shortening of the time span of any maturation (Krogerus and Gibson, 2013).

DAIRY INDUSTRY

The use of enzymes for processing milk and particularly rennet for the production of cheese has a long tradition (Law, 2009). Rennet, as other enzymatic coagulants, are preparations of proteases with a milk-clotting role and indispensable in cheese production.

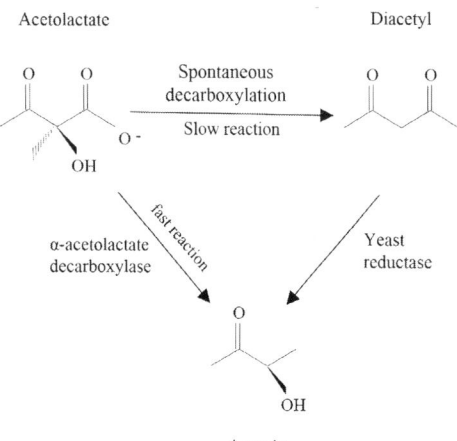

Figure 8.1 Enzymatic decarboxylation of acetolactate to acetoin.

These coagulants hydrolyze k-casein, resulting in the formation of cheese curd (Jacob et al., 2011). More specifically, rennet enzymes bring about the cleavage of Met 105–Phe 106 of k-casein that is present on the surface of casein micelles in milk (Silava and Malcata, 2005; Pereira et al., 2008). There is a sole exception for this trend, since a protease from *Cryphonectria parasitica* cleaves the Ser104–Phe105 bond (Jacob et al., 2011). This hydrolysis brings about the collapse of the micelles, resulting in the separation of milk into solid (curd) and liquid (whey) phases (Pontual et al., 2012; Dalgleish and Corredig, 2012). Although there are many proteases able to coagulate milk, only those that combine a high ratio of specific proteolytic activity on k-casein to overall proteolytic activity under the operational conditional for cheese making with a thermolability that ensures that whey products are devoid of coagulant activity are fit for use. Commercial enzymes that fulfill the role are aspartic proteases (Jacob et al., 2011). These enzymes can be obtained from various sources (Law, 2009). Traditionally, they are from animal origin, preferably from the stomach of young calves, and are known as rennet, a mixture of chymosin and pepsin, where the former tends to be dominant and has a more specific action than pepsin (Theron and Divol, 2014). Chymosin and pepsin convert milk to curd in few minutes (coagulating activity), afterward contributing to proteolysis occurred during cheese ripening (proteolytic activity). Since their specificity is diverse, the ratio of chymosin to pepsin in rennet affects the features of the final product (Jacob et al., 2011; Moschopoulou, 2011). Recombinant versions of ruminant chymosin have been expressed in suitable microbial hosts as to allow for large-scale and fast production of the enzyme under controlled conditions (Mohanty et al., 1999; Mule et al., 2009). Plant extracts are also sources of proteases with milk–clotting activity, viz leaves of Sodom apple, berries of the plant *Solanum dobium*, *Calotropis procera* plant, and cardoon extracts. Cardosins and cyprosins are the aspartic proteases mainly accountable for milk clotting in the latter and are used for the production of cheeses protected by a denomination of origin (POD) (Mazorra-Manzano et al., 2013; Shah et al., 2014). Microbial, nonrecombinant, coagulants are also used, namely from *Rhizomucor miehei*, *Rhizomucor pusillus* (Mucorpepsin, EC 3.4.23.23) and *Cryphonectria parasitica* (endothiapepsin, EC 3.4.23.22), all aspartic proteases. They are easily available and at a lower price than their counterparts (Feijoo-Siota et al., 2014). However, their proteolytic action is less specific than that from animal rennet, and their thermostability tends to lead to excessive proteolysis, resulting in shorter ripening time and bitter cheeses. Moreover, they often display side activities, such as amylase or lipase. They are used in the manufacture of fresh cheeses (Garcia et al., 2012; Abbas et al., 2013).

Lipases are used for cheese ripening and flavoring. Thus, as a result of the lipolytic action of these enzymes, free fatty acids are released, which impart particular flavors on cheese, depending on the side chain. Care has to be taken, however, since excessive lipolysis may result in undesirable odors (El-Hofi et al., 2011). Moreover, lipase action has also been shown to affect the microstructure of cheese, as lipases hydrolyze emulsified substrates such as fat globules, and fat retention in a cheese leads to a softer texture and improved meltability, further water available, and a lower protein density (Karami et al., 2009; Hickey et al., 2015). Besides their use in cheese production, the catalytic activity of lipases over milk fats, most of which (over 95%) are triacylglycerols, have spanned over other products. Hence, lipolysis of milk fats, similar to the action in cheese processing, results in the production of molecules conveying cheese flavor (lipolyzed milk fat) that can be incorporated in other dairy products (viz butter) but also in baked goods and in snacks. On the other hand, (trans)esterification reactions enable the modification of the lipid profile, viz enrichment in conjugated linoleic acid, resulting in healthier foods (Bourlieu et al., 2009; Kontkanen et al., 2011). Moreover, (trans)esterification of lipids from ruminant milk fat has been shown to enable the production of triacylglycerols that mimic those from human milk, and can thus be used in infant formula.

Lysozyme is used to prevent the outgrowth of *Clostridium* spp., namely *Clostridium tyrobutyricum*, *Clostridium beijerinckii*, *Clostridium butyricum*, and *Clostridium sporogenes*. These are known to ferment lactic acid with production of butyric acid, which causes defects in flavor and texture (cracks and slits and irregular holes) of cheese, so-called "late blowing," particularly noticeable in hard and extra-hard nature. The antibacterial role of the enzyme is due to its lytic activity on the $\beta(1,4)$-glycosidic linkages between *N*-acetylmuramic acid and *N*-acetyl-D-glucosamine residues in the peptidoglycan of bacterial cell wall. Food-grade preparations from egg albumin are used as safe alternatives to nitrate and nitrite as food additives, since these two chemicals bring about the formation of potentially carcinogenic nitrosamines (Brasca et al., 2013; Ávila et al., 2014).

In recent years, the trend toward the use of proteases for the production of dairy foods containing bioactive peptides has been gaining increased relevance. These peptides are so termed on account of their biologic activity, which may be of immunomodulatory, antibacterial, or antihypertensive nature. As such these peptides can be relevant ingredients within the scope of the production of dietary supplements, pharmaceutical preparations, or fortified foods (Phelan et al., 2009; Feijoo-Siota et al., 2014). Bioactive peptides are obtained through the controlled hydrolytic action of proteases, viz subtilisin or trypsin, over milk proteins, namely casein. Alongside the use of commercial preparations, cold-active proteases have been looked for, since they allow for efficient operation at low temperatures, viz 4–5°C, hence minimizing the risk of bacterial contaminations (Feijoo-Siota et al., 2014; De Gobba et al., 2014 a,b).

Transglutaminases have a significantly higher affinity toward caseins than to whey proteins, possibly the result of the more open and flexible structure of the former, where lysine and glutamine are more accessible to the enzyme. Hence, the cross-linking action of transglutaminase is predominant in caseins, when milk is considered (Jaros et al., 2006; Jaros and Rohm, 2011). Transglutaminases have been used in the production of yogurt, where their action has resulted in increased gel strength, stability, viscosity, and water-holding capacity of yogurt, as well as decreased acidification during storage and a better distribution of proteins in the gel network, as compared to untreated yogurt (Yüksel and Erdem, 2010; Sanli et al., 2011; Loveday et al., 2013). Improved textural and sensorial characteristics of yogurt were also reported when transglutaminase treatment was combined with high-pressure processing of milk (Tsevdou et al., 2013).

β-galactosidases (lactases) have been used for the hydrolysis of lactose in milk and dairy products for some decades, a trend that was started as lactases became commercially available at the time (Harju et al., 2012). Processing dairy products with lactase enables to overcome lactose intolerance and lactose maldigestion. This phenotypic condition is well disseminated in several areas of the world, including most parts of Africa and Asia, as opposite to Central and Northern Europe, where lactase expression endures to adult age (Leonardi et al., 2012; Harju et al., 2012; Misselwitz, 2014). Furthermore the hydrolysis of lactose in whey leads to the production of a sweet syrup, which can be used in other processes within the dairy industry, but also in baking, confectionary, and soft drinks. Thus, the sweet syrup can be used as an alternative to condensed and skim milk and sugar in several products, viz ice cream, milk desserts, and sauces. Overall, lactose hydrolysis not only significantly widens the market for consumption of milk and derived products but also contributes to tackling the environmental issue presented by whey disposal (Law, 2009; Panesar et al., 2010; Harju et al., 2012). Lactases can be isolated from different microbial sources, but typically those from fungi are most active in acidic environments, whereas those from yeast or bacteria are more active in close to neutral or even in slightly alkaline media. Thus, the former are preferred for the hydrolysis of acid whey, whereas the latter find preferential use in the hydrolysis of sweet whey and of milk. Actually, a commercial scale application for the hydrolysis of lactose in milk involves the entrapment of a neutral-pH lactase in cellulose triacetate fibers. Numerous other methods for lactase immobilization targeting applications in lactose hydrolysis in milk have been implemented at lab scale (Panesar et al., 2010; Elnashar and Hassan, 2014). Some of these have been tested at pilot scale, viz covalent immobilization onto cotton fabric and entrapment in polyvinyl alcohol lenses (Li et al., 2007; Lentikats Bulletin, 2011). Processing of lactose-free milk presents some constraints, due to the increased risk of Maillard reactions between glucose and galactose, more reactive (and more so the latter) than lactose, and proteins, upon heating, such as used in ultra-high temperature treatment of milk. Hence, heat treatment of lactose-free milk requires particular care (Naranjo et al., 2013). Occurrence of those Maillard reactions could be overcome by the aseptic addition of lactase after the heat treatment, but this increases costs and complexity. In any case, Maillard reactions can take place at room temperature during shelf life time, and this has a stronger impact on milk quality than the ultra-high temperature methodology used (Jansson et al., 2014). Other limitations for the implementation of cost–effective processes for lactose-free milk production through lactase hydrolysis may arise from product inhibition, namely from galactose, a feature with a specific nature that depends on the enzyme source (Husein, 2010; Harju et al., 2012). Nevertheless, this drawback has been shown to be mitigated by immobilization and proper choice of reactor and mode of operation (Panesar et al., 2010; Klein et al., 2013). Furthermore, the use of lactases with side activities, viz protease, arylsulphatase, must be either avoided or the side activities removed, not to tamper with the quality of low-lactose or lactose-free milk products (Harju et al., 2012).

Fruit and Vegetable Juices

The turbidity and viscosity of fruit and vegetable juices is mostly due to polysaccharides, such as pectins, starches, and hemicelluloses components—hence, the use of enzymes to tackle those issues and improve extraction and clarification of fruit and vegetable juice. Pectinases are perhaps the primary type of enzymes used in fruit processing, but xylanases, cellulases, and, at a lower scale, amylases are also used. Pectins are polysaccharides with a backbone of galacturonic acid residues linked by $\alpha(1–4)$ linkage and side chains of diverse sugars. They are present in the cell walls of plant cells and may be interconnected with other polysaccharides and proteins. Mechanical crushing of pectin-rich fruits results in a highly viscous, jellified mass, from which juice is difficult to extract by mechanical methods. The hydrolytic action of pectinases breaks different linkages, resulting in the disintegration of the jellified structure, decreased viscosity of the fruit juice drops, improved pressability, the whole allowing for higher recovery yields and clarified juice (Pedrolli et al., 2009; Kuhad et al., 2011; Sieiro et al., 2012; Tapre and Jain, 2014).

The generic term pectinases includes: (1) pectin methyl esterase, which removes methyl esters and makes pectin accessible to polygalacturonases (PGs); (2) PGs, which digest pectin in either a random (endo-PG) or a sequential manner (exo-PG), in either case requiring calcium ions for activity; and (3) pectin lyase, which promotes the random cleavage of pectin to unsaturated methyloligogalacturonates (Rodrigues, 2012; Tapre and Jain, 2014).

Naringinase, an enzyme complex composed of α-rhamnosidase and β-glycosidase, is also used for the debittering of citrus juices. The bitter taste is conveyed by naringin, which is hydrolyzed in two steps: first to prunin, through α-L-rhamnosidase action; then prunin is hydrolyzed to naringenin, through the action of β-D-glycosidase (Puri, 2012).

Meat, Fish, and Seafood

The production of uniform tender meat has been identified by producers as a key element to maintain the confidence of consumers (Bekhit et al., 2014). The use of exogenous proteolytic enzymes is a favored approach to tenderize meat, where the goal is to decrease the amount of connective tissue while avoiding significant degradation of myofibrillar proteins (Marques et al., 2010). Within this scope, bromelain, ficin, and papain were used to accomplish this goal. These cysteine (or thiol) proteases (EC 3.4.22) of plant origin have, however, shown poor selectivity. Moreover, they display unwanted thermal stability, thus they are not fully denatured during cooking. Fungal and particularly bacterial enzymes, viz subtilisin (EC 3.4.21.62) and neutral protease (EC 3.4.24.28) from *Bacillus subtilis,* have proved more promising, since they have mild, if any, action on myofibrillar proteins, whereas they are quite active on elastin and also on collagen, which are part of the connective tissue (Marques et al., 2010; Ha et al., 2013). Moreover, bacterial proteases tend to be active at relatively low temperatures and denaturation occurs upon heating to cooking temperatures (Bekhit et al., 2014). These features are particularly enhanced in cold-active bacterial proteases, which still retain the required specificity (Zhao et al., 2012).

Transglutaminases are used to improve the texture, cohesiveness, and shelf life of meat products, viz sausages, ground beef, or ham, hardening fish protein paste and other fish raw materials, by taking advantage of the ability of the enzyme to promote the formation of isopeptide bonds between proteins, creating a complex network (Jiang and Yin, 2001; Marques et al., 2010; Kieliszek and Misiewicz, 2014).

Diverse proteases have been used for the production of fish hydrolysates, ripening of herring, descaling and deskinning of fish and squid, enzymatic methods being milder than mechanical ones, and for the production of seafood flavors, bases, and stocks for the food industry (Nilsang et al., 2005; Benjakul et al., 2009; Herpandi et al., 2012; Ramakrishnan et al., 2013; Ghaly et al., 2013). Lipases have been used for the enrichment of polyunsaturated fatty acids in fish oil (Carvalho et al., 2009).

Sugars, Sweeteners, and Prebiotics
Starch Hydrolysis

Starch is the raw material for the production of a large variety of sugars and sugar syrups, viz dextrins, dextrose, maltose, fructose/glucose (high-fructose syrup) and fructose syrups; and hydrogenated derivatives, viz hydrogenated starch hydrolysates (HSH, polyglycitol syrups) and sorbitol. In any case, controlled enzymatic hydrolysis of starch is involved. In a typical process, a 30–40% (w/v) starch slurry at pH 5.8–6.2 flows through a jet cooker, operated at 105°C, with a residence time of 5 min in the presence of thermostable α-amylase and calcium ions. The heat treatment results in the gelatinization of starch by disrupting the granular structure, allowing starch molecules to dissolve in water. The gelatinized starch is then cooled to 95°C, and liquefaction is promoted by the α-amylase, where partial hydrolysis of starch and concomitant decrease in viscosity occurs. The liquefied starch, composed mostly of oligosaccharides, but with minute amounts of maltose and glucose, is then further hydrolyzed. Diverse strategies are implemented, depending on the envisaged product and the concomitant dextrose equivalent (DE), which reflects the degree of hydrolysis of starch. It is defined as the direct reducing sugar content (ARD) expressed in percent glucose on a dry basis. The production of maltodextrins, where DE under 20 is required, may rely only on α-amylase action, provided incubation is long enough. For products with higher DE, saccharification of liquefied starch is required, where further hydrolysis releases glucose and maltose. Maltose syrups, with DE within 35–60, require a maltogenic α-amylase or a β-amylase, at 55°C for 48 h. Production of dextrose syrups, with a DE of 97–99, requires the action of a glucoamylase, together with pullulanase, a debranching enzyme, and incubation for about 72 h at 60°C and pH 4.5. HSH and sorbitol are produced by hydrogenation in the presence of a metal catalyst, using as reagents either a low DE hydrolysate or pure dextrose, respectively (Evrendilek, 2012). Dextrose syrup with DE 97–99 can be isomerized to fructose through the catalytic action of glucose (xylose) isomerase. The isomerization of glucose to fructose is a reversible reaction, and in industrial processes it is carried out in a slightly alkaline environment, where commercially available glucose isomerases have pH optima. This limits the temperature of operation to about 60°C, since higher temperatures lead to the formation of browning products. Moreover, the glucose isomerases have limited thermal stability. Thus, the product of the enzymatic reaction is a syrup containing 42% fructose, which undergoes column chromatography, so that a syrup with 55% fructose (used as sucrose replacer in soft drinks) and 90% fructose can be obtained. Efforts have been performed to improve the overall process, namely aiming at amylases not requiring calcium for activity/stability (Chakraborty et al., 2012); and considering the use of amylases active at lower temperature, such as a 4-domain α-amylase containing a starch-binding domain, which, combined with a glucoamylase liquefied starch to DE of 95, at 60°C (Viksø-Nielsen et al., 2006); and thermostable glucose isomerases, displaying high activity in slightly acidic environments (Xu et al., 2014).

Other Sweeteners

Invert sugar syrup is the product of sucrose hydrolysis into fructose and glucose, in a process catalyzed by invertase. Invert sugar syrup is sweeter than sucrose, tends to retain moisture, and is less prone to crystallization. It is widely used in confectionary, baked goods, and pastry (Kotwal and Shankar, 2009).

As an alternative to the multienzymatic process for the production of fructose from starch, the monosaccharide can also be obtained through the hydrolysis of inulin. This is a polyfructan containing β-2,1 linkages and a sucrose unit at the nonreducing end, used by many plants, viz chicory and Jerusalem artichoke, for energy storage (Leenheer and Boot, 1998). Polyfructans from agave are also used as substrates (Partida et al., 1998; Michel-Cuello et al., 2012). The reaction is preferably catalyzed by a mixture of exo- and endoinulinases. The former cleave the β-(2,1) nonreducing end of inulin, releasing fructose, while endoinulinases act randomly on the internal linkages, releasing mostly inulotrioses, inulotetraoses, and inulopentaoses.

Interest in high-potency sweeteners with low-calorie value has attracted attention, as alternatives to high-caloric sugars currently used. Stevioside, a diterpene glycoside isolated from *Stevia rebaudiana* leaves is one of those sweeteners, at least 250 times sweeter than sucrose. It has, however, an aftertaste bitterness that limits its dissemination, a feature that has been shown to be overcome through suitable transglycosylation with suitable glucanotransferases, viz β-cyclodextrin glucanotransferase (Jaitak et al., 2009) (Fig. 8.2).

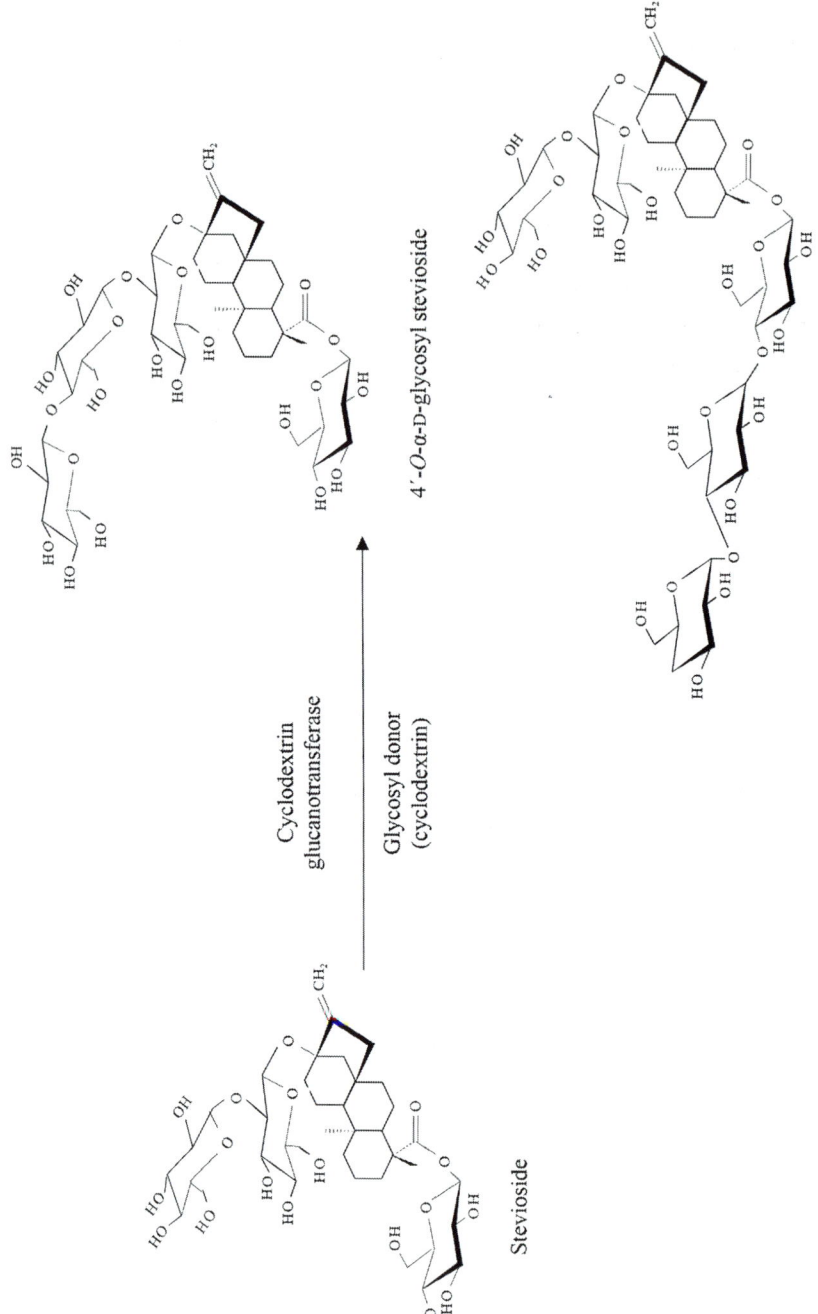

Figure 8.2 Enzymatic transglycosylation of stevioside to glycosyl and maltosyl derivatives, using a cyclodextrin glucanotransferase and cyclodextrin as glycoosyl donor.

Prebiotics

Prebiotics are nondigestible fructo-, galacto-, or glucooligosaccharides that feed the key microbial elements of the intestinal flora. Apart from extraction from plant sources, they can also be obtained through enzyme action.

Fructooligosaccharides

Depending on their relative size, fructooligosaccharides (FOS) can be preferably produced from sucrose elongation or through enzymatic hydrolysis of inulin. In the former case, the FOS mixture has a DP between 2 and 4, whereas in the latter the most common DP is within 2–9. Sucrose elongation is carried out by fructosyltransferases that cleave a sucrose molecule and then transfer the liberated fructose molecule to an acceptor molecule, viz sucrose or another oligosaccharide. Inulin hydrolysis involves the action of endoinulinase. Careful control of dosages and incubation time allow for FOS mixture with a suitable composition. Eventually, an exoinulinase can be used synergistically with the endoinulinase, in which case tighter control of the process is required to prevent full hydrolysis to fructose (Singh and Singh, 2010; Lafraya et al., 2011; Mutanda et al., 2014).

Galactooligosaccharides

Complementary to their hydrolytic role, lactases have also proved effective for the synthesis of galactooligosaccharides, carbohydrates comprising 2–20 molecules of galactose and one molecule of glucose (Otieno, 2010; Torres et al., 2010). They have a prebiotic role since they stimulate the growth of bifidogenic and lactic acid (Patel and Gyal, 2012). Accordingly, they are used as ingredients in functional foods, including as low-calorie sweeteners, viz lactosucrose, and in the fortification of infant formula (Ben et al., 2008; Patel and Gyal, 2012). Synthesis of galactooligosaccharides from lactose is under kinetic control, since it proceeds within a competition between the thermodynamically favored lactose hydrolysis and transgalactosylation. In the latter reaction, the galactosyl moiety of lactose is not transferred to water but instead to another carbohydrate, starting with lactose. The process results in oligosaccharides with an increasing degree of polymerization concomitantly with an increase in reaction time. The process is favored at high-lactose concentration. On the other hand, the products of transgalactosylation are substrates for the hydrolytic role of the enzyme. Overall, the product mixture changes with reaction time, which is therefore a critical parameter to control the galactooligosaccharide composition of the mixture (Husein, 2010; Otieno, 2010; Torres et al., 2010; Díez-Municio et al., 2014). Increased temperature also favors transgalactosylation, yet for a reliable, cost-effective process requires high operational stability, a goal to which enzyme immobilization can contribute (Osman et al., 2014). The use of β-galactosidase as catalyst for the production of lactulose from lactose, as an alternative to the standard chemical transformation, has also been assessed recently. This synthetic ketose disaccharide, composed of fructose and galactose, has a role as prebiotic food additive against constipation

(Wang et al., 2013). Although usually absent in raw, unprocessed milk, it is formed as a result of heat treatment during milk processing (Aluko, 2012). Suggested production processes use whey lactose as substrate and the combined action of β-galactosidase and glucose isomerase; or the sole action of either β-galactosidase or β-glycosidase, in which case fructose addition is required (Fig. 8.3) (Mayer et al., 2010; Song et al., 2013a,b). Synthesis of galactooligosaccharides from lactulose, as well as from lactose, has also been reported (Padilla et al., 2012; Corzo-Martinez et al., 2013).

Glucooligosaccharides

Again, this type of oligosaccharide can be produced through controlled enzymatic hydrolysis or synthesis. Isomaltooligosaccharides, consisting exclusively of D-glucopyranosyl units linked by α-1,6 bonds, can be obtained from the hydrolysis of a starch slurry. The process involves α-amylase, debranching enzymes, such as β-amylase and pullulanase, and α-glucosidase, produced by *Aspergillus niger*. In addition to the hydrolytic role, this glucosidase transfers glucose to C-6 positions of acceptor sugars to form isomaltooligosaccharides (Taniguchi, 2005). Alternatively to the glucosidase, the use of an invertase, also from *A. niger*, with transglucosidase activity, has been implemented (Kwon et al., 2012).

The synthetic approach typically relies on the use of sucrose as D-glucosyl donor and an exogenous acceptor, a wide variety being available, viz sucrose itself, lactose, maltose, gentiobiose, among others. This enables a wide variety of products. Just with sucrose as substrate, and depending on the enzyme, different products can be obtained: dextran, with over 50% of α-1,6 linkages (dextransucrase); alternan, with α-1,3 and α-1,6 linkages (alternansucrase); mutan, with more than 50% α-1,3 linkages (mutansucrase); reuteran, mostly α-1,4 linkages (reuteransucrase) or amylose (amylosucrase). Glucansucrase promotes a transglycosylation reaction where the formation of a covalent β-D-glucosyl-enzyme intermediate is formed, while fructose is released, after which the intermediate is attacked by the hydroxyl group of the acceptor and the glucosyl residue is transferred to an acceptor. Oligosaccharide synthesis takes place as successive transfers of the glucosyl unit onto the oligosaccharide occur (Kralj et al., 2005; Leemhuis et al., 2013; Dudé et al., 2014).

Wine Making

Since wine results from grape maceration and juice extraction, pectinases and xylanases are commonly used to improve the process and improve clarification (Mojsov, 2013). Alongside with these, naringinases are used to enhance the aroma, since they hydrolyze the aglycone moiety of glycosylated secondary metabolites of grapes, and ultimately release the aglycone moiety; urease, to prevent the formation of ethyl carbamate, a potential mild carcinogen resulting from the spontaneous reaction between urea and ethanol; and glucose oxidase, to decrease the alcohol content by reducing the

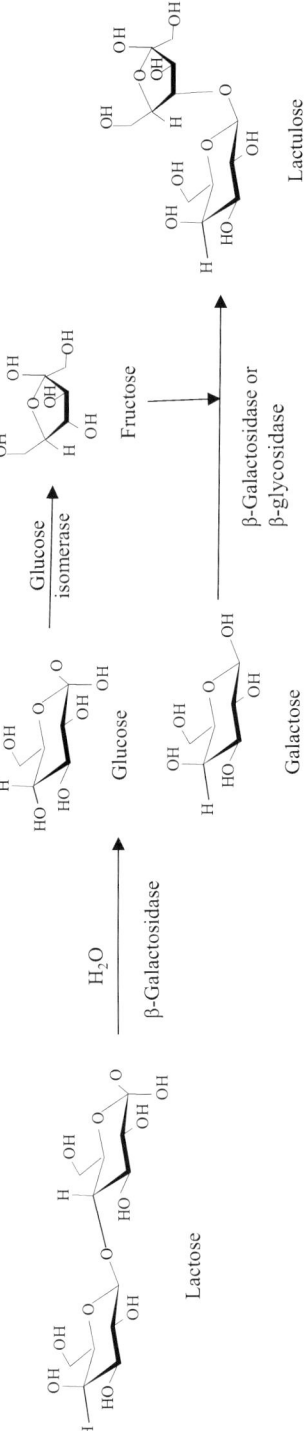

Figure 8.3 Enzymatic synthesis of lactulose.

amount of glucose available for fermentation (Maicas and Mateo, 2005; Biyela et al., 2009; Andrich et al., 2010; González-Pombo et al., 2011; Zhu et al., 2014).

CONCLUSIONS AND FUTURE PERSPECTIVES

The use of enzymes in food production and processing has been for a long time a representative example of the practical applications of biotechnology. The reasons for such enduring success are related to the specificity, fast action, mild operational conditions required, and biodegradable and biocompatible nature of enzymes. These features allow for chemical modifications of substrates to be performed effectively, and for food processing with preservation of nutrients, while complying with public health-and-safety requirements. Most of the enzymes used are from microbial sources, a pattern that can be ascribed to the large array of catalytic activities conveyed, plasticity, and relatively undemanding growth requirements and high productivities conveyed by microorganisms. Moreover, recombinant DNA technology has enabled for enzymes of interest from either higher organisms, or from slow-growing or pathogenic microorganisms, to be expressed in fast-growing microbial strains that comply with health safety regulations. Enzyme applications span from traditional areas, such as baked goods, dairy industries, or sugar industries, to the emerging area of prebiotics, notwithstanding processing of fish, meats, and fruits. As a result of market constraints, which demand processes with increased productivity as well as new products, research efforts are nevertheless constantly required in order to improve the performance of known enzyme catalysts or screen for new ones. In the former case, increased insight on catalytic mechanisms and protein configuration at a molecular level allow for a better understanding of the behavior and stability of enzymes. Together with these, a more rational approach on enzyme formulations, including immobilization methodologies, is sought after. These efforts have a multidisciplinary nature, a trend that is foreseen to be pursued in the future. Screening for new or better performing enzymes has increasingly involved microorganisms from extreme environments, as these are likely to provide enzymes able to operate under operational conditions more suitable for given processes, where the commonly used enzymes, typically from mesophilic organisms, perform poorly. This approach has gained a considerable shot in the arm with recent developments in metagenomics, proteomics, and identification of efficient expression systems. In addition, creating or evolving enzymes with improved properties has become a possibility, given the developments in directed evolution strategies, combined with robust computational methods. Together with findings regarding enzyme promiscuity, these efforts are likely to provide for one-pot multistep reactions using multifunctional catalysts and to the de novo design of enzymes able to perform any envisaged chemical reaction. All these new, exciting developments are likely to be pursued in the near future and therefore contribute to improving and expanding the role played by enzymes in the food industry.

LIST OF ABBREVIATIONS

ARD Direct reducing sugar content
BRENDA BRaunschweig ENzyme DAtabase
CAZy Carbohydrate-Active enZYmes Database
CEF Connecting Europe Facility
DATEM Diacetyl tartaric esters of mono- and diglycerides
DE Dextrose equivalent
DP Degree of polymerization
EC European Commission
EFSA European Food Safety Authority
EU European Union
FOS Fructooligosaccharides
GRAS Generally regarded as safe
HFCS High-fructose corn syrup
HSH Hydrogenated starch hydrolysates
PG Polygalacturonases
POD Protected by a denomination of origin
SSL Sodium stearoyl-2-lactylate
USD United States dollar

ACKNOWLEDGMENTS

Filipe Carvalho acknowledges Fundação para a Ciência e a Tecnologia, Portugal, for the doctoral grant SFRH/BD/74818/2010.

REFERENCES

Abbas, H.M., et al., 2013. Production of white soft cheese using fungal coagulant produced by solid state fermentation technique. World Applied Sciences Journal 25 (6), 939–944.

Addo, K., et al., 1993. Soybean flour lipoxygenase lsozyme mutant effects on bread dough volatiles. Journal of Food Science 58 (3), 583–585.

Adrio, J.L., Demain, A., 2014. Microbial enzymes: tools for biotechnological processes. Biomolecules 4 (1), 117–139.

Agarwal, S., Sahu, S., 2014. Safety and regulatory aspects of food enzymes: an industrial perspective. International Journal of Interdisciplinary and Multidisciplinary Studies 1 (6), 253–267. [Online] Available from: http://www.ijims.com/uploads/0078a9c56ca7560ca162z32.pdf.

Alaunyte, I., et al., 2012. Improving the quality of nutrient-rich Teff (*Eragrostis tef*) breads by combination of enzymes in straight dough and sourdough breadmaking. Journal of Cereal Science 55 (1), 22–30.

Aluko, R.E., 2012. Functional Foods and Nutraceuticals. Springer, New York.

Andrich, L., Esti, M., Moresi, M., 2010. Urea degradation in some white wines by immobilized acid urease in a stirred bioreactor. Journal of Agriculture and Food Chemistry 58 (11), 6747–6753.

Aravindan, R., Anbumathi, P., Viruthagiri, T., 2007. Lipase applications in food industry. Indian Journal of Biotechnology 6, 141–158.

Ávila, M., et al., 2014. Inhibitory activity of reuterin, nisin, lysozyme and nitrite against vegetative cells and spores of dairy-related *Clostridium* species. International Journal of Food Microbiology 172, 70–75.

Bekhit, A.A., et al., 2014. Exogenous proteases for meat tenderization. Critical Reviews in Food Science and Nutrition 54 (8), 1012–1031.

Ben, X.M., et al., 2008. Low level of galacto-oligosaccharide in infant formula stimulates growth of intestinal *Bifidobacteria* and *Lactobacilli*. World Journal of Gastroenterology 14 (42), 6564–6568.

Benjakul, S., et al., 2009. Effects of flavourzyme on yield and some biological activities of Mungoong, an extract paste from the cephalothorax of white shrimp. Journal of Food Science 74 (2), S73–S80.

Biyela, B.N.E., et al., 2009. The production of reduced-alcohol wines using Gluzyme Mono® 10.000 BG-treated grape juice. South African Journal Enology and Viticulture 30 (2), 124–132.

Blanco, C.A., et al., 2014. Innovations in the brewing industry: light beer. International Journal of Food Science and Nutrition 65 (6), 655–660.

Bonet, A., et al., 2006. Glucose oxidase effect on dough rheology and bread quality: a study from macroscopic to molecular level. Food Chemistry 99, 408–415.

Bourlieu, C., Bouhallab, S., Lopez, C., 2009. Biocatalyzed modifications of milk lipids: applications and potentialities. Trends in Food Science & Technology 20, 458–469.

Brasca, M., et al., 2013. Different analytical approaches in assessing antibacterial activity and the purity of commercial lysozyme preparations for dairy application. Molecules 18, 6008–6020.

Butt, M.S., Tahir-Nadeem, M., Ahmad, Z., 2008. Xylanases and their applications in baking industry. Food Technology and Biotechnology 46 (1), 22–31.

Carvalho, P.O., et al., 2009. Enzymatic hydrolysis of salmon oil by native lipases: optimization of process parameter. Journal of the Brazilian Chemical Society 20 (1), 117–124.

Chakraborty, S., et al., 2012. Study on calcium ion independent α-amylase from haloalkaliphilic marine *Streptomyces* strain A3. Indian Journal of Biotechnology 11, 427–437.

Corzo-Martínez, M., et al., 2013. Synthesis of prebiotic carbohydrates derived from cheese whey permeate by a combined process of isomerisation and transgalactosylation. Journal of the Science of Food and Agriculture 93 (7), 1591–1597.

Dalgleish, D.G., Corredig, M., 2012. The structure of the casein micelle of milk and its changes during processing. Annual Review of Food Science and Technology 3, 449–467.

De Gobba, C., et al., 2014a. Antioxidant peptides from goat milk protein fractions hydrolysed by two commercial proteases. International Dairy Journal 39 (1), 28–40.

De Gobba, C., Tomp, G., Otte, J., 2014b. Bioactive peptides from caseins released by cold active proteolytic enzymes from *Arsukibacterium ikkense*. Food Chemistry 165, 205–215.

Dhillon, G.S., et al., 2012. Flocculation and haze removal from crude beer using in-house produced laccase from *Trametes versicolor* cultured on brewer's spent grain. Journal of Agricultural and Food Chemistry 60 (32), 7895–7904.

DiCosimo, R., et al., 2013. Industrial use of immobilized enzymes. Chemical Society Reviews 42, 6437–6474.

Díez-Municio, M., et al., 2014. Synthesis of novel bioactive lactose-derived oligosaccharides by microbial glycoside hydrolases. Microbial Biotechnology 7 (4), 315–331.

Dudé, D., et al., 2014. Successes in engineering glucansucrases to enhance glycodiversification. Carbohydrate Chemistry 40, 624–645.

Regulation (EC) No 1331/2008 Regulation of the European Parliament and of the Council of 16 December 2008, 2008. Available from: http://eur-lex.europa.eu/LexUriServ/LexUriServ.do?uri=OJ:L:2008: 354:0001:0006:EN: PDF.

EFSA CEF Panel (EFSA Panel on Food Contact Materials, Enzymes, Flavourings and Processing Aids, 2014. Scientific Opinion on xylanase from a genetically modified strain of *Aspergillus oryzae* (strain NZYM-FB). EFSA Journal 12 (5), 1–17. http://dx.doi.org/10.2903/j.efsa.2014.3645.

El-Hofi, M., et al., 2011. Industrial application of lipases in cheese making: a review. Internet Journal of Food Safety 13, 293–302.

Elnashar, M.M., Hassan, M.E., 2014. Novel epoxy activated hydrogels for solving lactose intolerance. Biomed Research International 2014, 1–9:817985. [Online] Available from: http://www.hindawi.com/journals/bmri/2014/817985/.

Evrendilek, G.A., 2012. Sugar alcohols (polyols). In: Varzakas, T., Labropoulos, A., Anestis, S. (Eds.), Sweeteners: Nutritional Aspects, Applications, and Production Technology. CRC Press, Boca Raton, Florida, pp. 45–78.

Fadda, C., et al., 2014. Bread staling: updating the view. Comprehensive Reviews in Food Science and Food Safety 13 (4), 473–492.

Feijoo-Siota, L., et al., 2014. Recent patents on microbial proteases for the dairy industry. Recent Advances in DNA and Gene Sequences 8, 44–55.

Garcia, V., et al., 2012. Effect of vegetable coagulant, microbial coagulant and calf rennet on physicochemical, proteolysis, sensory and texture profiles of fresh goats cheese. Dairy Science & Technology 92 (6), 691–707.

Gerits, L.R., et al., 2014. Lipases and their functionality in the production of wheat-based food systems. Comprehensive Reviews in Food Science and Food Safety 13 (5), 978–989.

Ghaly, A.E., et al., 2013. Fish processing wastes as a potential source of proteins, amino acids and oils. Journal of Microbiology and Biochemical Technology 5 (4), 107–129.

Goesaert, H., et al., 2005. Wheat flour constituents: how they impact bread quality, and how to impact their functionality. Trends in Food Science and Technology 16 (1–3), 12–30.

González-Pombo, P., et al., 2011. A novel extracellular β-glucosidase from *Issatchenkia terricola*: isolation, immobilization and application for aroma enhancement of white Muscat wine. Process Biochemistry 46 (1), 385–389.

Gupta, G.N., et al., 2014. Extremophiles: an overview of microorganism from extreme environment. International Journal of Agriculture, Environment & Biotechnology 7 (2), 371–380.

Gurung, N., et al., 2013. A broader view: microbial enzymes and their relevance in industries, medicine, and beyond. BioMed Research International 2013, 1–18:329121 [Online] Available from:http://dx.doi.org/10.1155/2013/329121.

Ha, M., et al., 2013. Comparison of the proteolytic activities of new commercially available bacterial and fungal proteases toward meat proteins. Journal of Food Science 78 (2)), 170–177.

Harju, M., Kallioinen, H., Tossavainen, O., 2012. Lactose hydrolysis and other conversions in dairy products: technological aspects. International Dairy Journal 22 (2), 104–109.

Harris, A.D., Ramalingam, C., 2010. Xylanases and its application in food industry: a review. Journal of Experimental Sciences 1 (7), 1–11.

Hassan, A.A., et al., 2014. Improving dough rheology and cookie quality by protease enzyme. American Journal of Food Science and Nutrition Research 1 (1), 1–7.

Herpandi, H.N., Rosma, A., Wan Nadiah, W.A., 2012. Degree of hydrolysis and free tryptophan content of Skipjack Tuna (*Katsuwonus pelamis*) protein hydrolysates produced with different type of industrial proteases. International Food Research Journal 19 (3), 863–867.

Hickey, C.D., et al., 2015. The influence of cheese manufacture parameters on cheese microstructure, microbial localization and their interactions during ripening: a review. Trends in Food Science & Technology 41 (2)), 135–148.

Husein, Q., 2010. Beta-galactosidases and their potential applications: a review. Critical Reviews in Biotechnology 30 (1), 41–62.

Jacob, M., Jaros, D., Rohm, H., 2011. Recent advances in milk clotting enzymes. International Journal of Dairy Technology 64 (1), 14–33.

Jaitak, V., et al., 2009. Simple and efficient enzymatic transglycosylation of stevioside by β-cyclodextrin glucanotransferase from *Bacillus firmus*. Biotechnology Letters 31, 1415–1420.

Jamar, C., Du-Jardin, P., Fauconnier, M.-L., 2011. Cell wall polysaccharides hydrolysis of malting barley (*Hordeum vulgare* L.): a review. Biotechnology, Agronomy, Society and Environment 15 (2), 301–313.

Jansson, T., et al., 2014. Lactose-hydrolyzed milk is more prone to chemical changes during storage than conventional ultra-high-temperature (UHT) milk. Journal of Agricultural and Food Chemistry 62 (31), 7886–7896.

Jaros, D., Rohm, H., 2011. Enzymes exogenous to milk in dairy technology transglutaminase. In: Fuquay, J.W. (Ed.), Encyclopedia of Dairy Sciences, second ed. Academic Press, San Diego, pp. 297–300.

Jaros, D., et al., 2006. Transglutaminase in dairy products: chemistry, physics, applications. Journal of Texture Studies 37 (2), 113–155.

Jiang, S.-T., Yin, L.-J., 2001. Application of transglutaminase in seafood and meat processings. Journal of the Fisheries Society of Taiwan 28 (3), 151–162.

Juturu, V., Wu, J.C., 2012. Microbial xylanases: engineering, production and industrial applications. Biotechnology Advances 30 (6), 1219–1227.

Karami, M., et al., 2009. Microstructural properties of fat during the accelerated ripening of ultrafiltered-feta cheese. Food Chemistry 113 (2), 424–434.

Keskin, S., Sumnu, G., Sahin, S., 2004. Usage of enzymes in a novel baking process. Nahrung/Food 48 (2), 156–160.

Kieliszek, M., Misiewicz, A., 2014. Microbial transglutaminase and its application in the food industry. A review. Folia Microbiologica (Praha) 59 (3), 241–250.

Klein, M.P., et al., 2013. High stability of immobilized β-d-galactosidase for lactose hydrolysis and galactooligosaccharides synthesis. Carbohydrate Polymers 95 (1), 465–470.

Kontkanen, H., et al., 2011. Enzymatic and physical modification of milk fat: a review. International Dairy Journal 21, 3–13.

Kotwal, S.M., Shankar, V., 2009. Immobilized invertase. Biotechnology Advances 27, 311–322.

Kragh, K.M., et al., 2011. *Pseudomonas saccharophila* G4-amylase Variants and Uses Thereof. United States Patent Application 20110033575 A1.

Kralj, S., et al., 2005. Highly hydrolytic reuteransucrase from probiotic *Lactobacillus reuteri* strain ATCC 55730. Applied and Environmental Microbiology 71 (7), 3942–3950.

Kreisz, S., 2009. Malting. In: Eßlinger, H.M. (Ed.), Handbook of Brewing. Wiley-VCH Verlag GmbH & Co. KGaA, Weinheim, pp. 147–164.

Krogerus, K., Gibson, B.R., 2013. Diacetyl and its control during brewery fermentation. Journal of the Institute of Brewing 119 (3), 86–97.

Kuhad, R.C., Gupta, R., Singh, A., 2011. Microbial cellulases and their industrial applications. Enzyme Research 2011, 1–10:280696. [Online] Available from http://dx.doi.org/10.4061/2011/280696.

Kwon, H.-K., Jeong, H.-S., Lee, J.-H., 2012. Production of Isomaltooligosaccharides and Uses Therefore. European Patent EP 2422630 A1.

Lafraya, Á., et al., 2011. Fructo-oligosaccharide synthesis by mutant versions of *Saccharomyces cerevisiae* invertase. Applied and Environmental Microbiology 77 (17), 6148–6157.

Lalor, E., Goode, D., 2010. Brewing with enzymes. In: Whitehurst, R.J., van Oort, M. (Eds.), Enzymes in Food Technology, second ed. Wiley-Blackwell, Chichester, West Sussex, pp. 163–194.

Law, B.A., 2002. The nature of enzymes and their action in foods. In: Whitehurst, R.J., Law, B.A. (Eds.), Enzymes in Food Technology. Sheffield Academic Press, Sheffield, pp. 1–18.

Law, B.A., 2009. Enzymes in dairy product manufacture. In: Whitehurst, R.J., van Oort, M. (Eds.), Enzymes in Food Technology, second ed. Wiley-Blackwell, Chichester, West Sussex, pp. 88–102.

Leemhuis, H., et al., 2013. Glucansucrases: three-dimensional structures, reactions, mechanism, α-glucan analysis and their implications in biotechnology and food applications. Journal of Biotechnology 163 (2), 250–272.

Leenheer, L., Boot, K., 1998. Process for Producing a Fructose Syrup Rich in Fructose. European Patent Application EP0822262.

Lentikats Bulletin, 2011. Lactose-free Milk Trials. [Online] Available from: http://lentikats.eu/soubory/bulletins/bulletin-11-01.pdf.

Leonardi, M., et al., 2012. The evolution of lactase persistence in Europe. A synthesis of archaeological and genetic evidence. International Dairy Journal 22, 88–97.

Li, X., Zhou, Q.Z.K., Chen, X.D., 2007. Pilot-scale lactose hydrolysis using β-galactosidase immobilized on cotton fabric. Chemical Engineering and Processing: Process Intensification 46 (5), 497–500.

Li, S., et al., 2012. Technology prospecting on enzymes: application, marketing and engineering. Computational and Structural Biotechnology Journal 2 (3), 1–11 [Online] e201209017 Available from: http://csbj.org/articles/e201209017.pdf.

Lopez, M., Edens, L., 2005. Effective prevention of chill-haze in beer using an acid proline-specific endoprotease from *Aspergillus niger*. Journal of Agricultural and Food Chemistry 53 (20), 7944–7949.

Loveday, S.M., Sarkar, A., Singh, H., 2013. Innovative yoghurts: novel processing technologies for improving acid milk gel texture. Trends in Food Science & Technology 33 (1), 5–20.

Maicas, S., Mateo, J.J., 2005. Hydrolysis of terpenyl glycosides in grape juice and other fruit juices: a review. Applied Microbiology and Biotechnology 67 (3), 322–335.

Malomo, O., et al., 2012. Effect of enzymes on the quality of beer/wort developed from proportions of sorghum adjuncts. Advances in Microbiology 2 (4), 447–451.

Markets and Markets, 2013. Food Enzymes Market by Types (Carbohydrase, Protease, Lipase), Applications (Beverages, Dairy, Bakery), Sources (Microorganisms, Plants, Animals) & Geography – Global Trends & Forecasts to 2018. [Online] Available from: http://www.marketsandmarkets.com/Market-Reports/food-enzymes-market-800.html.

Marques, A.C., Maróstica, M.R., Pastore, G.M., 2010. Some nutritional, technological and environmental advances in the use of enzymes in meat products. Enzyme Research 2010, 1–8:480923. [Online] Available from: http://doi:10.4061/2010/480923.

Mayer, J., Kranz, B., Fischer, L., 2010. Continuous production of lactulose by immobilized thermostable β-glycosidase from Pyrococcus furiosus. Journal of Biotechnology 145, 387–393.

Mazorra-Manzano, M.A., et al., 2013. Comparison of the milk-clotting properties of three plant extracts. Food Chemistry 141 (3), 1902–1907.

Michel-Cuello, C., et al., 2012. Study of enzymatic hydrolysis of fructans from agave salmiana characterization and kinetic assessment. The Scientific World Journal 2012, 1–10:863432. [Online] Available from: http://dx.doi.org/10.1100/2012/863432.

Miguel, A.S.M., et al., 2013. Enzymes in bakery: current and future trends. In: Muzzalupo, I. (Ed.), Food Industry. InTech. [Online] Available from: http://www.intechopen.com/books/food-industry/enzymes-in-bakery-current-and-future-trends.

Misselwitz, B., 2014. Lactose intolerance: new insights due to blinded testing? Digestion 90 (1), 72–73.

Moayedallaie, S., Mirzaei, M., Paterson, J., 2010. Bread improvers: comparison of a range of lipases with a traditional emulsifier. Food Chemistry 122 (3), 495–499.

Mohanty, A.K., et al., 1999. Bovine chymosin: production by rDNA technology and application in cheese manufacture. Biotechnology Advances 17, 205–217.

Moschopoulou, E., 2011. Characteristics of rennet and other enzymes from small ruminants used in cheese production. Small Ruminant Research 101 (1–3), 188–195.

Mosjov, K., 2013. Use of enzymes in wine making: a review. International Journal of Marketing and Technology (IJMT) 3 (9), 112–127.

Mule, V.M.R., et al., 2009. Recombinant Calf-chymosin and a Process for Producing the Same. Patent US 7482148 B2.

Mutanda, T., et al., 2014. Microbial enzymatic production and applications of short-chain fructooligosaccharides and inulooligosaccharides: recent advances and current perspectives. Journal of Industrial Microbiology and Biotechnology 41 (6), 893–906.

Naranjo, G.B., et al., 2013. The kinetics of Maillard reaction in lactose-hydrolyzed milk powder and related systems containing carbohydrate mixtures. Food Chemistry 141, 3790–3795.

Nguyen, M.-T., Edens, L., Van Roon, J.L., 2012. Improved Brewing Process. European Patent EP 2402425 A1.

Nigam, P.S., 2013. Microbial enzymes with special characteristics for biotechnological applications. Biomolecules 3 (3), 597–611.

Nilsang, S., et al., 2005. Optimization of enzymatic hydrolysis of fish soluble concentrate by commercial proteases. Journal of Food Engineering 70 (4), 571–578.

Offord, D.A., 2008. Antistaling Agent and Methods. United States Patent Application 20080248178 A1.

Olesen, T., Qi Si, J., 1994. Use of Lipase in Baking. International Patent Application. WO94/0403035.

Osma, J.F., Toca-Herrera, J.L., Rodríguez-Couto, S., 2010. Uses of laccases in the food industry. Enzyme Research 2010, 1–8:918761. [Online] Available from: http://www.ncbi.nlm.nih.gov/pmc/articles/PMC2963825/pdf/ER2010-918761.pdf.

Osman, A., et al., 2014. Synthesis of prebiotic galactooligosaccharides from lactose using bifidobacterial β-galactosidase (BbgIV) immobilised on DEAE-Cellulose, Q-Sepharose and amino-ethyl agarose. Biochemical Engineering Journal 82, 188–199.

Otieno, D.O., 2010. Synthesis of β-galactooligosaccharides from lactose using microbial β-galactosidases. Comprehensive Reviews in Food Science and Food Safety 9, 471–482.

Padilla, B., et al., 2012. Evaluation of oligosaccharide synthesis from lactose and lactulose using β-galactosidases from *Kluyveromyces* isolated from artisanal cheeses. Journal of Agricultural and Food Chemistry 60 (20), 5134–5141.

Panesar, P.S., Kumari, S., Panesar, R., 2010. Potential applications of immobilized β-galactosidase in food processing industries. Enzyme Research 2010, 1–16:473137. [Online] Available from: http://www.ncbi.nlm.nih.gov/pmc/articles/PMC3014700/pdf/ER2010-473137.pdf.

Pareyt, B., et al., 2011. Lipids in bread making: sources, interactions, and impact on bread quality. Journal of Cereal Science 54, 266–279.

Partida, V.Z., Lopez, A.C., Gomez, A.J.M., 1998. Method of Producing Fructose Syrup From Agave Plants. Patent US 5846333 A1.

Patel, S., Gyal, S., 2012. The current trends and future perspectives of prebiotics research: a review. 3 Biotech 2, 115–125.

Pedrolli, D.B., et al., 2009. Pectin and pectinases: production, characterization and industrial application of microbial pectinolytic enzymes. The Open Biotechnology Journal 3, 9–18. [Online] Available from: http://benthamopen.com/tobiotj/articles/V003/9TOBIOTJ.pdf.

Pereira, C., et al., 2008. Proteolysis in model Portuguese cheeses: effects of rennet and starter culture. Food Chemistry 108 (3), 862–868.

Phelan, M., et al., 2009. Casein-derived bioactive peptides: biological effects, industrial uses, safety aspects and regulatory status. International Dairy Journal 19 (11), 643–654.

Pontual, E.V., et al., 2012. Caseinolytic and milk-clotting activities from *Moringa oleifera* flowers. Food Chemistry 135 (3), 1848–1854.

Popa, M., Bostan, R., 2010. Study regarding enzymatic characteristics on the flour. Annales Universitatis Apulensis Series Oeconomica 12 (2), 583–588.

Poulsen, P.B., Buchholz, K., 2003. History of enzymology with emphasis on food production. In: Whitaker, J.R., Voragen, A.G.J., Wong, D.W.S. (Eds.), Handbook of Food Enzymology. Marcel Dekker, New York, pp. 11–20.

Puri, M., 2012. Updates on naringinase: structural and biotechnological aspects. Applied Microbiology and Biotechnology 93, 49–60.

Ramakrishnan, V.V., et al., 2013. Extraction of proteins from mackerel fish processing waste using alcalase enzyme. Journal of Bioprocessing & Biotechniques 3 (130), 1–9. [Online] Available from: http://omicsgroup.org/journals/extraction-of-oil-from-mackerel-fish-processing-waste-using-alcalase-enzyme-2329-6674.1000115.php?aid=19648.

Rehmanji, M., Gopal, C., Mola, A., 2005. Beer stabilization technology – clearly a matter of choice. Master Brewers Association of the Americas Technical Quarterly 42 (4), 332–338.

Revanappa, S.B., Salimath, P.V., Prasada Rao, U.J.S., 2014. Effect of peroxidase on textural quality of dough and arabinoxylan characteristics isolated from whole wheat flour dough. International Journal of Food Properties 17, 2131–2141.

Rodrigues, S., 2012. Enzyme maceration. In: Rodrigues, S., Fernandes, F.A.N. (Eds.), Advances in Fruit Processing Technologies. CRC Press, Boca Raton, Florida, pp. 235–245.

Sanli, T., et al., 2011. Effect of using transglutaminase on physical, chemical and sensory properties of set-type yoghurt. Food Hydrocolloids 25 (6), 1477–1481.

Schwartz, D., Whistler, R.L., 2009. History and future of starch. In: BeMiller, J., Whistler, R. (Eds.), Starch: Chemistry and Technology, third ed. Academic Press, New York, pp. 1–10.

Shah, M.A., et al., 2014. Plant proteases as milk-clotting enzymes in cheesemaking: a review. Dairy Science & Technology 94 (1), 5–16.

Sieiro, C., et al., 2012. Microbial pectic enzymes in the food and wine industry. In: Valdez, B. (Ed.), Food Industrial Processes – Methods and Equipment. InTech. [Online] Available from: http://www.intechopen.com/books/food-industrial-processes-methods-and-equipment/microbial-pectic-enzymes-in-the-food-and-wine-industry.

Silva, S.V., Malcata, F.X., 2005. Studies pertaining to coagulant and proteolytic activities of plant proteases from *Cynara cardunculus*. Food Chemistry 89 (1), 19–26.

Singh, R.S., Singh, R.P., 2010. Production of fructooligosaccharides from inulin by endoinulinases and their prebiotic potential. Food Technology and Biotechnology 48 (4), 435–450.

Søe, J.B., Poulsen, C.H., Høstrup, P.B., 2004. Method of Improving the Properties of a flour Dough, a Flour Dough Improving Composition and Improved Food Products. US6726942 B2.

Song, Y.S., et al., 2013a. Batch and continuous synthesis of lactulose from whey lactose by immobilized β-galactosidase. Food Chemistry 136 (2), 689–694.

Song, Y.S., et al., 2013b. Optimization of lactulose synthesis from whey lactose by immobilized β-galactosidase and glucose isomerase. Carbohydrate Research 369, 1–5.

Souppe, J., Naeye, T.J.-B., 2006. A Novel Enzyme Combination. EP 0796559 B2.

Spök, A., 2006. Safety regulations of food enzymes. Food Technology and Biotechnology 44 (2), 197–209.

Steiner, E., Becker, T., Gastl, M., 2010. Turbidity and haze formation in beer – insights and overview. Journal of the Institute of Brewing 116 (4), 360–368.

Stougaard, P., Hansen, O.C., 2012. Recombinant Hexose Oxidase, a Method of Producing Same and Use of Such Enzyme. US 8338153 B2.

Taniguchi, H., 2005. Carbohydrate active enzymes for the production of oligosaccharides. In: Hou, C.T. (Ed.), Handbook of Industrial Biocatalysis. CRC Press, Boca Raton, Florida, pp. 20.1–20.24.

Tapre, A.R., Jain, R.K., 2014. Pectinases: enzymes for fruit processing industry. International Food Research Journal 21 (2), 447–453.

Theron, L.W., Divol, B., 2014. Microbial aspartic proteases: current and potential applications in industry. Applied Microbiology and Biotechnology 98 (21), 8853–8868.

Torres, D.P.M., et al., 2010. Galacto-oligosaccharides: production, properties, applications, and significance as prebiotics. Comprehensive Reviews in Food Science and Food Safety 9 (5), 438–454.

Trincone, A., 2011. Marine biocatalysts: enzymatic features and applications. Marine Drugs 9 (4), 478–499.

Tsevdou, M.S., Eleftheriou, E.G., Taoukis, P.S., 2013. Transglutaminase treatment of thermally and high pressure processed milk: effects on the properties and storage stability of set yoghurt. Innovative Food Science and Emerging Technologies 17, 144–152.

Van Benschop, C.H.M., Terdu, A.G., Hille, J.D.R., 2012. Baking Enzyme Composition as SSL Replacer. US20120164272A1.

van Oort, M., 2010a. Enzymes in food technology – introduction. In: Whitehurst, R.J., van Oort, M. (Eds.), Enzymes in Food Technology, second ed. Wiley-Blackwell, Chichester, West Sussex, pp. 1–17.

van Oort, M., 2010b. Enzymes in bread making. In: Whitehurst, R.J., van Oort, M. (Eds.), Enzymes in Food Technology, second ed. Wiley-Blackwell, Chichester, West Sussex, pp. 103–143.

Viksø-Nielsen, A., et al., 2006. Development of new α-amylases for raw starch hydrolysis. Biocatalysis and Biotransformation 24 (1–2), 121–127.

Walsh, G., 2014. Proteins: Biochemistry and Biotechnology, second ed. Wiley-Blackwell, Chichester, West Sussex.

Wang, H., et al., 2013. Enzymatic production of lactulose and 1-lactulose: current state and perspectives. Applied Microbiology and Biotechnology 97 (14), 6167–6180.

Xu, H., et al., 2014. Characterization of a mutant glucose isomerase from *Thermoanaerobacterium saccharolyticum*. Journal of Industrial Microbiology & Biotechnology 41 (10), 1581–1589.

Yüksel, Z., Erdem, Y., 2010. The influence of transglutaminase treatment on functional properties of set yoghurt. International Journal of Dairy Technology 63, 86–97.

Zhao, G.Y., et al., 2012. Tenderization effect of cold-adapted collagenolytic protease MCP-01 on beef meat at low temperature and its mechanism. Food Chemistry 134 (4), 1738–1744.

Zhu, F.-M., Du, B., Li, J., 2014. Aroma enhancement and enzymolysis regulation of grape wine using β-glycosidase. Food Science and Nutrition 2 (2), 139–145.

FURTHER READING

Ferreira-Dias, S., Tecelão, C., 2014. Human milk fat substitutes: advances and constraints of enzyme-catalyzed production. Lipid Technology 26 (8), 183–185.

Illanes, A., 2011. Whey upgrading by enzyme biocatalysis. Electronic Journal of Biotechnology 14, 15–42.

Simões, T., et al., 2014. Production of human milk fat substitutes catalyzed by a heterologous rhizopus oryzae lipase and commercial lipases. Journal of the American Chemists Society 91 (3), 411–419.

CHAPTER 9

Seafood Enzymes and Their Application in Food Processing

A.U. Muzaddadi[1], S. Devatkal[1,2], H.S. Oberoi[1,3]
[1]ICAR-Central Institute of Post-Harvest Engineering and Technology, Ludhiana, India; [2]ICAR-National Research Centre for Meat, Hyderabad, India; [3]ICAR-Indian Institute of Horticultural Research, Bengaluru, India

INTRODUCTION

Aquatic organisms are unique and often thrive in extreme environments with high pressure, no light, wide temperature ranges from $-2°C$ to $103°C$ at thermal vents, salinity ranging from 0 to above 50 ppt (parts per thousand), daily fluctuations in dissolved oxygen (DO) availability for tidal organisms, and so on. Therefore, these organisms have unique genotypic characters and hence are the source of useful enzymes. Thus, the application of seafood enzymes becomes an emergent subject for biochemists. Enzyme activity has been potentially used for many exciting applications for the improvement of foods and also as an indicator for quality changes in seafood. Economic factors, such as achievement of optimum yields and efficient recovery of desired protein, are the main deterrents in the use of enzymes (James et al., 1996). New and unique enzymes continue to be developed for use in enzymatic reactions to produce food ingredients by hydrolysis, synthesis, or biocatalysis.

There is a rapid increase in the commercial fish processing industry in the world in recent decades and fish processing units generating large quantities of solid waste and wastewater. According to Awarenet (2004), solid waste, which represents 20–60% of the initial raw material contains various kinds of residues, including whole waste fish, fish heads, viscera, skin, bones, blood, frame liver, gonads, guts, and some muscle tissue, etc. Although these solid wastes are utilized in some developing countries to produce many nonfood items, such as cattle feed, manure, etc., in many other countries these discards are not utilized but incinerated or dumped in sea, causing environmental problems (Bozzano and Sarda, 2002).

There are many traditional disposal methods, such as ensilation and fermentation for the production of high-protein meals for animal feeds as well as composting for manure (Hassan and Heath, 1986; Faid et al., 1997; Liao et al., 1997). Moreover, gelatin or chondroitin sulphate useful in food, cosmetic, and pharmaceutical sectors can be potentially sourced from fish skin or cartilage from some species (Blanco et al., 2006; Karim and Bhat, 2009).

Agro-Industrial Wastes as Feedstock for Enzyme Production
ISBN 978-0-12-802392-1
http://dx.doi.org/10.1016/B978-0-12-802392-1.00009-5

There has been a strong emphasis in recent years on the fact that fish wastes (solid waste and wastewater) are important sources of proteins, lipids, and minerals with high biological value (Kim et al., 2006; Toppe et al., 2007; Kacem et al., 2011). In recent times, different products and processes have emerged that utilize enzymes in a deliberate, controlled fashion (Vilhelmsson, 1997). Fish by-products can be hydrolyzed by applying treatments including heat treatment, enzymatic treatment, and chemical treatment and then improved by enzyme-mediated fermentation (Yamamoto et al., 2005; Yano et al., 2008; Xu et al., 2008). Such hydrolysates are useful in various sectors, such as human nutrition, cosmetology, aquaculture, microbiology, etc., based on their biological and functional properties (Ben Rebah et al., 2008; Liaset et al., 2000; Martone et al., 2005). Recently, Rebah and Miled (2013) prepared and tested fish wastes (heads, viscera, chitinous material, wastewater, etc.) as growth substrates for microbial enzyme production, such as protease, lipase, chitinolytic, and ligninolytic enzymes with a view that this can reduce environmental pollution problems associated with waste disposal and, simultaneously, lower the cost of microbial enzyme production. It is also established that proteolytic hydrolysis produce peptides with interesting medicinal values, such as antihypertensors, immunomodulators, antioxidants, anticoagulants, etc., which are highly useful in the treatment of several diseases, including osteoporosis, arthritis, diabetes, obesity, etc. (Kim and Mendis, 2006).

The enzyme makeup of aquatic organisms is directly related to growth and ultimately to the tissue composition, nutritive value, shelf life, and edible characteristics of seafood (Haard, 2000). Hence, fish-processing wastes and by-products have ample potential to become sources for valuable pharmaceutical and nutraceutical products and thus spearheading a big industry as well as environmentally friendly waste management. This chapter reviews different aspects of seafood-processing waste management, microbial enzyme production, and product development with special reference to enzymes.

SEAFOOD ENZYMES

Nucleotide-Degrading Enzymes

Various enzymes are involved in the breakdown of ATP to the end product, urea.

As described in Fig. 9.1, initially the tissue enzymes (enzymes 1, 2, and 3) are prominent in their action to form hypoxanthine (Hx) from ATP, but at the later stages, bacterial enzymes (enzymes 4 and 5) play a significant role. Thus, it can be considered as a measure of both autolytic deterioration and bacterial spoilage. Hx value progressively increases from near zero in extremely fresh fish to levels as high as 8 mol/g when the fish is considered spoiled.

$$ATP \xrightarrow{\ 1\ } ADP \xrightarrow{\ 2\ } AMP \xrightarrow{\ 3\ } IMP \xrightarrow{\ 4\ } Hx + R.1.P \xrightarrow{\ 5\ } Xa \xrightarrow{\ 5\ } U$$

Figure 9.1 Nucleic acid ATP breakdown process, where *Hx*, hypoxanthine; *R.1.P*, ribose 1 phosphate; *Xa*, Xanthine; *U*, uric acid, *Enzymes 1*, ATPase; *2*, nucleoside monophosphate kinase; *3*, AMP deaminase; *4*, hypoxanthine phosphoribosyl transferase; *5*, xanthine oxidase.

Fish with low xanthine oxidase activity can cause accumulation of hypoxanthine, but the fish might be rated as fresh by sensory analysis. Despite the limitations, hypoxanthine value is one of the best indices so far available for measurement of fish spoilage.

K-value

K-value is considered to be one of the best indices of spoilage in seafood. K-value is calculated from the values of Hx, inosine (I), and total nucleotide levels in fish at the point of measurement. As proposed by Saito et al. (1959), it is defined by the formula given in Fig. 9.2.

The K-value measurement takes into account the role of most enzymes in the ATP breakdown. Hence, it is a more accurate index of loss of fish freshness. Saito et al. (1959) observed that in freshly caught fish, K-value could be as low as zero, 10–20 for moderate-quality fish, and can go up to 90 when the fish is spoiled. Lakshmanan et al. (1996) have shown that extremely fresh fish, when sampled onboard, have K (%) values ranging from 3 to 5 (Table 9.1).

Myosin ATPases

Enzymes like myosin ATPases activity are associated with pre-rigor changes and the onset of rigor mortis in many fish species. This is considered to be an important factor determining the onset of spoilage in freshly caught fish. The reaction catalyzed by the enzyme is

$$K\% = \frac{[I + Hx]}{[ATP + ADP + AMP + IMP + I + Hx]} \times 100.$$

Figure 9.2 The formula for calculating K value, where ATP, adenosine 5′-triphosphate; ADP, adenosine 5′-diphosphate; AMP, adenosine monophosphate; IMP, inosine monophosphate; I, inosine; Hx, hypoxanthine.

Table 9.1 Correlation Between K Values and Sensory Scores

Fish Species	K Value (%)		Storage Time (days)
	Initial	Final	
Silver pomfret (*Pampus argenteus*)	3.5	>50	14
Rainbow sardine (*Dussumieria acuta*)	–	>50	14
Jew fish (*Protonibea diacanthus*)	–	>50	16
Ribbon fish (*Trichiurus lepturus*)	–	>50	16
Squid	–	>50	16
Metapenaeus dobsoni	–	>50	8–10
Parapenaeopsis stylifera	–	>50	8–10
Pearl spot	2.8	55	14
Mullet (*Liza corsula*)	4.2	51	8

Lakshmanan, P.T., Antony, P.D., Gopakumar, K., 1996. Nucleotide degradation and quality changes in mullet (*Liza corsula*) and pearlspot (*Etroplus suratensis*) in ice and at ambient temperatures. Food Control 7 (6), 277–283.

Deterioration of fish muscle during processing and storage is greatly influenced by denaturation of myosin (Fig. 9.3). Recent literature suggests loss of ATPase activity as a potential index for evaluating fish quality.

Lactate Dehydrogenase

Lactate dehydrogenase (LDH) is a key glycolytic enzyme, catalyzing the reduction of pyruvate to lactate in the presence of nicotinamide adenine dinucleotide (NADH) (Table 9.2). In species that are mentioned in Table 9.2, accumulation of lactic acid is faster, and the amount of lactic acid formed is a measure of the enzyme activity. After postmortem fish muscle, in the absence of oxygen, the normal pathway of glucose degradation involves a step in which pyruvate is reduced by LDH to lactic acid. Significant loss in LDH activity occurs under both iced and frozen storage conditions.

Lysosomal and Mitochondrial Enzymes

It has been observed that rapid freezing and thawing of fish tissues result in irreversible damage to mitochondria and lysosomes, releasing a number of enzymes into the cellular fluid. Some of the lysosomal enzymes, such as α–glucosidases, β–N-glucosaminidase, and acid phosphatase play a significant role in fish muscle damage. The mitochondrial enzymes that are significant in quality assessment are glutamate oxaloacetate transaminase, malate dehydrogenase and cytochromes oxidase, lipoamide reductase, and 5′ AMP deaminase.

Histidine Decarboxylase

There are several endogenous and exogenous (mainly bacterial) decarboxylases responsible for fish spoilage. They produce biogenic amines including histamine, putrescine, cadaverine, and also volatile bases, such as ammonia, trimethylamine, dimethylamine, among others. Most

$$M + ATP \leftrightarrow M1ATP \rightarrow M2ATP \leftrightarrow M.ADP.Pi \rightarrow M.ADP + Pi \rightarrow M + ADP + Pi$$

Figure 9.3 Myosin denaturation where, *M*, myosin, *ATP*, adenosine 5′-triphosphate; *ADP*, adenosine – 5′-diphosphate; *Pi*, inorganic phosphate.

Table 9.2 Specific Lactate Dehydrogenase Activity Values

Fish/Shrimp	LDH Activity µmole/min mg protein
Mrigal	545
Mullet	384.7
Pearl spot	312.1
Tilapia	513
Milk fish	933.3
Pindicus	21.5

Nambudiri, D.D., 1987. Enzyme Reactions as Freshness of Fish and Shellfish (Ph.D. thesis). Cochin University of Science and Technology, Cochin, India.

mesophilic bacteria, particularly *Pseudomonas, Bacillus, Clostridium,* and *Aeromonas* reported in tropical fish, contain the enzyme histidine decarboxylase. Visceral proteinases and other proteolytic enzymes accelerate the quantity of histidine production by releasing more histidine into the pool of free amino acids. Though decarboxylases act on a variety of amino acids in fish muscle, decarboxylase activity is measured by estimating the amount of histamine released from histidine and is expressed as μ mole histamine/mg protein.

Most pelagic and scombroid fish contain a good amount of histidine in free state as well as with proteins. Histamine production occurs in fresh fish 40–50 h after death when fish is not chilled properly. It causes food poisoning known as scombroid poisoning as it is linked with eating tuna, mackerel, and other species of the Scombroidea family.

Urease

Ureases decompose urea to form ammonia, a volatile base. Total volatile bases are expressed as nitrogen content, and values reflect both trimethylamine nitrogen and other metabolites that result from either bacterial or enzymatic breakdown of fish flesh.

Trimethylamine Oxide Degradation by Enzymes

Trimethylamine oxide (TMAO) is found largely in most marine fish; in contrast, its presence is negligible or nil in freshwater fish. Two major enzymes including oxidoreductase and TMAO demethylase reduce TMAO to trimethylamine (TMA) during postmortem changes and associated biochemical and microbial reactions in fish muscle. TMAO is further degraded into dimethylamine and formaldehyde by enzymatic action.

TMAO Oxidoreductase

The odorless TMAO is reduced to trimethylamine, a volatile base.

TMAO is reduced partly by intrinsic and mostly by bacterial enzymes of the group reductases (Fig. 9.4). The enzyme is 86-kDa monomer and a molybdoenzyme. Most spoilage organisms belonging to Enterobacteriaceae and the genera *Alteromonas, Photobacteria,* and *Vibrio* reduce TMAO, which serve as a terminal electron acceptor during anaerobic growth. Trimethylamine is associated with fatty substance and is responsible for the fishy smell of spoiled fish. Production of TMA is exponential, slow initially and increasing rapidly after a few days of chilled storage.

TMAO Demethylase

The methyl group of TMAO is removed to form dimethylamine (DMA) and formaldehyde by the enzymes TMAO demethylase in the absence of oxygen (Fig. 9.5).

$$TMAO + 2H^+ + 2e \xrightarrow{\quad TMAOOxydoreductase \quad} TMA + H_2O$$

Figure 9.4 Breakdown of TMAO to TMA by the action of TMAO oxydoreductase enzyme.

$$TMAO \xrightarrow{\textit{TMAO demethylase}} DMA + FA$$

Figure 9.5 Breakdown of TMAO to DMA by the action of TMAO demethylase enzyme.

This enzyme is abundant in kidney and other tissues of viscera. In many marine fishes, the endogenous activity of this enzyme is important in determining the quality deterioration, chiefly textural deterioration of frozen stored fish. Formaldehyde production is responsible for the increase in the firmness of the fish muscle under frozen storage. The amount of DMA and formaldehyde can be related to the freshness of fish.

TMA-specific gas sensor technology is now available for routine rapid assay. For a good quality fish, TMA nitrogen value of 1.25–2.00 mg% is recommended, and levels of 10–15 mg% can be considered as the safety limit beyond which most chilled fish become spoiled.

Lipoxygenases

Lipoxygenase (LOX) catalyzes the incorporation of oxygen at 5, 12, or 15 position of various eicosanoic acids to produce hydroperoxide derivatives. LOX in eggs is considered important in oocyte, maturation, or regulating membrane permeability. LOX activity of fish skin and gill was proposed to be responsible for flavor formation of fresh and processed fish.

LOX present in fish skin generates hydroperoxyarachidonic acid (12-HPETE) and 14-hydroperoxydocosahexaenoic acid (14-HPDHE). The α-cleavage at either side of the carbon atom bearing the hydroperoxy groups can result in the formation of volatile compounds. These compounds are responsible for the typical oxidative fishy odor, such as 2-nonenal from 12-HPETE and 3,6 nonadienal from 14-HPDHE.

Lipases and Phospholipases

Lipases are able to degrade phospholipids. They are grouped into two categories: acylhydrolases and phosphodiesterases. The phospholipases attack the phospholipids. Phospholipase A1 splits the sn–1 fatty–acyl ester bond; phospholipase A2 splits the sn–2 fatty–acyl ester bond; phospholipase B splits both or the remaining ester in the lysophospholipid; phospholipase C attacks the diglyceride–phosphate link; and phospholipase D, the choline–phosphate link lipases (Fig. 9.6).

Lysophosphatidylcholine (LPC) and free fatty acids (FFA) are the products of the hydrolysis of the acid moiety from the sn-2 position of the diacylglycerophospholipid catalyzed by phospholipase A2. Phospholipase A2 hydrolyzes essential dietary phospholipids in marine teleosts. Information about the organs that produce and secrete digestive phospholipase A2 and the ontogeny of the phospholipase A2 cells would help to improve the composition and quality of diets used in cultured fish (Uematsu et al., 1992).

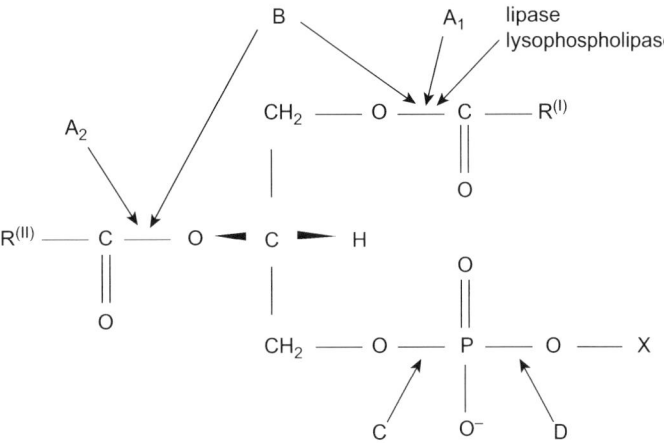

Figure 9.6 Attack sites of phospholipases. R(I), R(II) are alkyl chains; X denotes any of the moieties (choline, serine, ethanolamine, etc.) present in 3-sn-phosphoglycerides (Gurr and Harwood, 1991).

Lipoprotein lipase (LPL), hepatic lipase (HL), and pancreatic lipase (PL) are derived from a common ancestral gene of vertebrates (Hide et al., 1992). In addition to PL, the pancreas synthesizes another lipolytic enzyme that catalyzes not only the hydrolysis of acylglycerols but also phospholipid, cholesterol esters, and vitamin esters. This enzyme has various other names, such as nonspecific lipase, lipase A, lysophospholipase, cholesterol esterase, carboxylesterase, carboxyl ester hydrolase, carboxyl ester lipase, bile salt–stimulated lipase, bile salt–dependent lipase, and bile salt–activated lipase (BAL) (Amaya and Marngoni, 2000). BALs are very important because they are the predominant, if not the sole, lipolytic enzymes present in the pancreas of marine fish. This means that the role of BALs involves not only hydrolysis of cholesterol esters or vitamin esters but also digestion of triacylglyceride and wax esters.

Transglutaminases

Transglutaminase (TGase) is a transferase enzyme, also known as protein–glutamine γ-glutamyltransferase (EC 2.3.2.13). It catalyzes the acyl transfer reaction between γ-carboxyamide groups of glutamine residues in proteins, peptides, and various primary amines. The reactions are shown in Fig. 9.7.

TGase (protein-glutamyltransferase, EC 2.3.2.13) catalyzes an acyl–transfer reaction between the γ-carboxyamide group of peptide-bound glutamine residues (acyl donors) and a variety of primary amines (acyl acceptors), including the γ-amino group of lysine residues in certain proteins. These reactions can lead to cross-linking of proteins (Motoki and Seguro, 1998). The extent of cross-linking depends upon the nature of the actomyosin, which differs among fish species. The contribution of intrinsic TGase in cross-linking offish proteins resulting in hardening of fish pastes at low

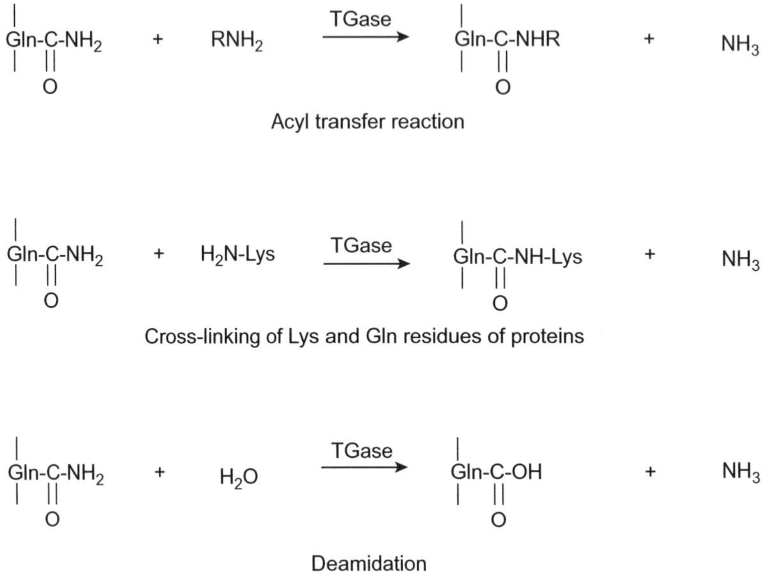

Figure 9.7 Reactions catalyzed by TGases (Ashie and Lanier, 2000).

temperature has been demonstrated (Seki et al., 1990; Nielsen, 1995; Yasumaga et al., 1996; Lee et al., 1997). This enzyme is also present in microorganisms (Motoki and Seguro, 1998), crustaceans (Maruyama et al., 1995), and finfish species including those from temperate waters (Neilsen, 1995; Zhu et al., 1995). Surimi is a preparation of washed fish meat consisting essentially of myofibrillar proteins. It is widely used in many countries, particularly Japan, for development of textured seafood analogues, such as crab legs, shrimp, etc. TGase is unique in nature as it can modify protein functionality by covalent cross-linking. Studies on surimi derived from some fish species (*Theragra chalcogramma*, *Sardinops melanostictus*, and *Micropogan undulatus*) led to the discovery that an endogenous transglutaminase is responsible for the spontaneous gelation of surimi pastes at low temperature (5–40°C) (Seki et al., 1990; Kamath et al., 1992; Tsukamasa et al., 1993). This discovery has led to further studies of the content and nature of TGase endogenous to seafood, as well as its evaluation from other sources in seafood applications as a processing aid for quality improvement and product development. The endogenous TGase activity of fish muscle has long been utilized to build strong texture in surimi-based gelled foods by allowing the enzyme to cross-link myosin during a lower temperature preincubation (setting treatment) prior to cooking.

TGase has been identified and purified from different seafood (Table 9.3). The primary applications of TGase in seafood processing have been for cold restructuring, cold gelation of pastes, or gel-strength enhancement through myosin cross-linking. The enzyme may be used in the future to modify the functionality of marine-derived proteins for food, pharmaceutical, or industrial use. TGase activity may be potentially

Table 9.3 Sources and Characteristics of Some Transglutaminases

Source	MW (kDa)	Optimum Temp. (°C)	Optimum pH	References
Red sea bream liver	78	55	9.0–9.5	Yasueda et al. (1994)
Carp muscle	80			Kishi et al. (1991)
Walleye pollack liver	77	50	9.0	Kumazawa et al. (1996)
Lobster muscle	200			Myhrman and Lorand (1970)
Japanese oyster	84/90	40/25	8.0	Kumajawa et al. (1997)
Limulus haemocyte	86			Tokunaga et al. (1993)
Scallop	80			Nozawa et al. (1997)
Botan shrimp	80			Nozawa et al. (1997)
Squid	80			Nozawa et al. (1997)
Rainbow trout	80			Nozawa et al. (1997)
Atka mackerel	80			Nozawa et al. (1997)

utilized to enrich the nutritional value of various foods by covalently cross-linking proteins containing complementary limiting essential amino acids.

Digestive Proteinases

Proteinases, also known as proteolytic enzymes or proteases, are hydrolytic in their action and catalyze the cleavage of peptide bonds with participation of water molecules as reactants. They are used to improve product-handling characteristics and texture of cereals and baked goods, enhance the drying and quality of egg products, tenderize meat, recover proteins/peptides from bones, hydrolyze blood proteins, for the production of protein hydrolysates, reduction of stickwater viscosity, roe (eggs of fish) processing, to make pulses and rennet puddings, in cheese making/cheese ripening, for biomedical applications to reduce tissue inflammation, dissolve blood clots, promote wound healing, activate hormones, diagnose candidiasis, and to aid or facilitate digestion (Chaplin and Bucke, 1990; Whitaker, 1994; Haard and Simpson, 1994).

Proteases may be classified on the basis of their pH sensitivities as acidic, neutral, or alkaline proteinases. They may also be described based on their substrate specificities, response to inhibitors, or by their mode of catalysis proposed by the Enzyme Commission (EC) of the International Union of Biochemists. Proteinases are broadly classified into four groups: (1) acid/aspartyl proteinases: active and stable at acidic pH, eg, pepsin and pepsin-like proteinases, chymosins (formerly known as rennin), gastricsin; (2) serine proteinases: with a serine residue at their catalytic site, eg, tripsins, chymotripsins, elastase, and collagenases; (3) thiol/cysteine proteinases: endoproteinases with cysteine and histidine residues as the essential components in their catalytic sites, eg, cathepsin B; and (4) metalloproteinases: their activity depends on the presence of bound divalent cations, not common in marine animals (Simpson, 2000).

Based on the EC system, all the acid/aspartyl proteinases from marine animals have the first three digits in common: EC 3.4.23. The three common aspartyl proteinases that have been isolated and characterized from the stomach of marine animals are pepsin, chymosin, and gastricsin. Serine proteinases have three common digits: EC 3.4.21. For example, trypsin has its code as EC 3.4.21.4, chymotrypsin is coded as EC 3.4.21.1, while elastase is coded as EC 3.4.21.11. The first three digits common to thiol proteinases are EC 3.4.22. and cathepsin B is designated as EC 3.4.22.1. Few metalloproteinases have been characterized from marine animals (eg, rockfish, carp, and squid mantle) and they are designated by common first three digits, EC 3.4.24 (Simpson, 2000).

Purification of these enzymes is based on the solubility differences (eg, precipitation with neutral salts or organic solvents); size differences (eg, size exclusion chromatography or dialysis); charge differences (eg, ion exchange chromatography or electrophoresis); and binding to specific ligands (eg, affinity chromatography or hydrophobic interaction chromatography).

Trypsin Isozymes

Being a serine proteinase trypsin, trypsin isozymes consist of a single protein chain with a molecular weight of 24 kDa. The pancreas secretes it as an inactive precursor, trypsinogen, from the pancreatic acinar cells together with chymotrypsinogen and proelastase, which had been demonstrated in chum salmon, *Oncorhynchus keta* (Godfrey and Reichelt, 1983), catfish, *Parasilurus asotus* (Haard and Simpson, 1994), and in eel, *Anguilla japonica* (Haard, 1992). The association between fish size and different trypsin isozyme patterns in the pyloric cecal tissues was established in Atlantic salmon fry (Rehbein, 1988). The enzyme was thus used as genetic marker in salmonids, which is an easy and simple method for selection of salmon imparting improved production in both freshwater and seawater periods.

Polyphenoloxidases

The most important color reactions that affect seafoods, especially crustaceans, is enzymatic browning, caused by the enzyme polyphenol oxidase (1,2 benzenediol; oxygen oxidoreductase, EC 1.10.3.1) including phenoloxidase, phenolase, monophenol and diphenol oxidase, and tyrosinase (Kim and Marchall, 2000). The discoloration caused is known as melanosis, and phenoloxidase is responsible for it in crustacean species, such as lobster, shrimp, and crab. This is considered to be a postmortem spoilage in shrimp, also called black spot formation. The scales of melanosis are presented in Table 9.4.

The chemical reactions involve conversion of monophenol oxidase to *o*-diphenols (Fig. 9.8), polyphenol oxidase to aromatic amines, and *o*-aminophenols (Fig 9.9) and catechol to quinines (Fig. 9.10), and finally to the formation of melanin. The formation of melanin from tyrosine is shown in Fig. 9.11.

The aesthetic value of fish is determined by many attributes, and color is the most important attribute for any food. Many compounds can influence color of seafood and

Table 9.4 Scales of Melanosis in Pink Shrimp

Melanosis Scale	Description
0	Absent
2	Slight, noticeable on some shrimp
4	Slight, noticeable on most shrimp
6	Moderate, noticeable on most shrimp
8	Heavy, noticeable on most shrimp
10	Heavy, totally unacceptable

Otwell, W.S., Marshall, M.R., 1986. Studies on the Use of Sulfites to Control Shrimp Melanosis (Blackspot): Screen Alternatives to Sulfiting Agents to Control Shrimp Melanosis (Florida seagrant technical paper no. 46), pp. 1–10.

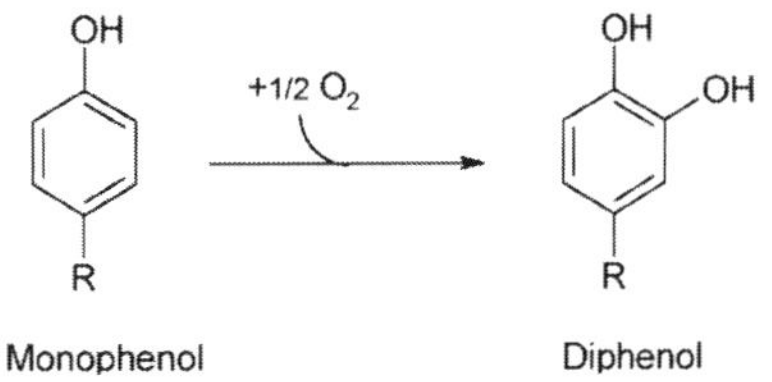

Figure 9.8 Monophenol oxidase pathway producing the diphenol.

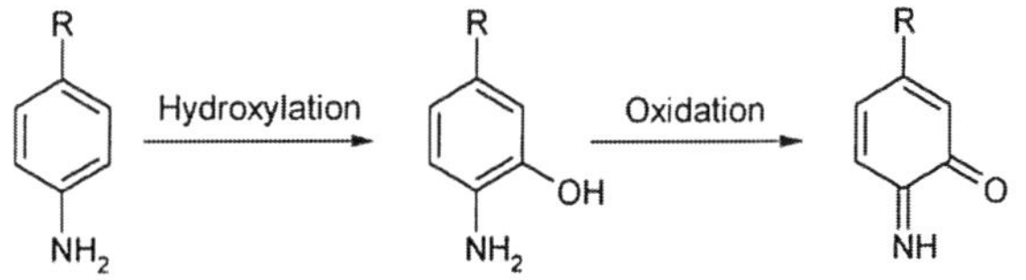

Figure 9.9 Polyphenol oxidase activity for aromatic amines and *o*-aminophenol substrates.

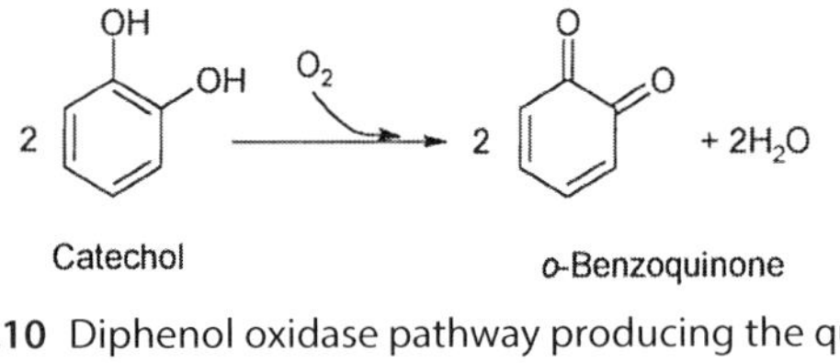

Figure 9.10 Diphenol oxidase pathway producing the quinones.

the naturally occurring pigments or colors formed through enzymatic and nonenzymatic reactions. One of the most important color reactions affecting many fruits, vegetables, and seafood, especially crustaceans, is enzymatic browning caused by the enzyme polyphenol oxidase. Other names by which this enzyme is known are phenoloxidase, phenolase, monophenol and diphenol oxidase, and tyrosinase.

Phenoloxidase is responsible for discoloration called melanosis or blackspot in crustacean species, such as lobster, shrimp, and crab. The enzyme is responsible for catalyzing two reactions: the hydroxylation of the *o*-position adjacent to an existing hydroxyl group in tyrosine, and the oxidation of the diphenol or dihydroxydiphenylalanine (dopa) to

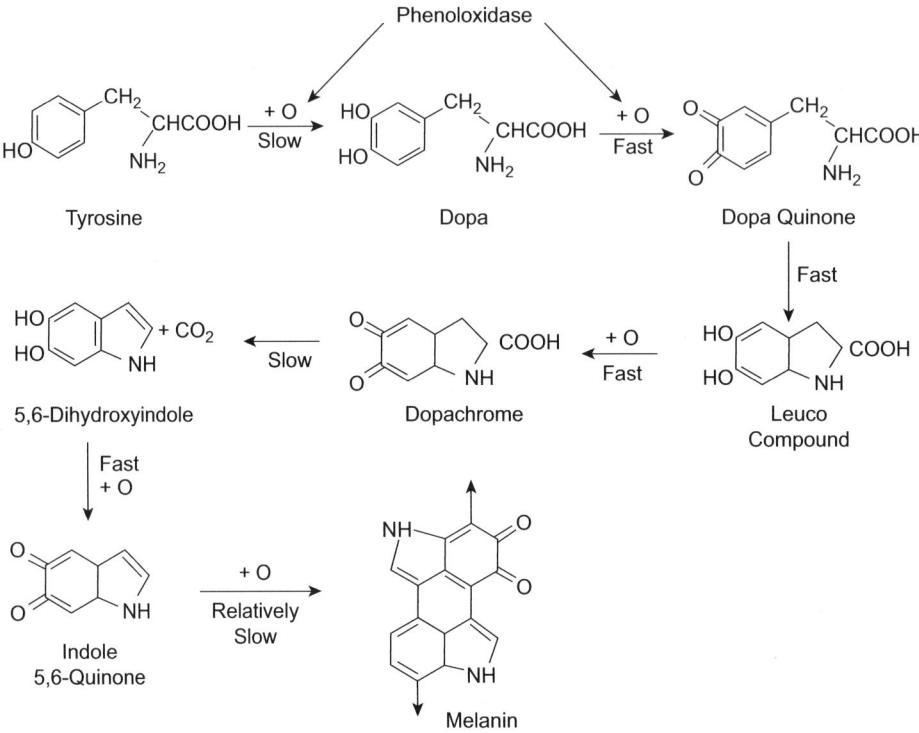

Figure 9.11 Formation of melanin from tyrosine (Lerner, 1953).

o-benzoquinones. The latter then can polymerize spontaneously to form high-molecular-weight compounds or brown pigments called melanins or react with amino acids and proteins that enhance the brown color produced. The whole process is known as melanosis. Melanosis indicates spoilage, and consumers are averse to such discolored products.

Enzymatic browning is controlled or prevented in many ways. The methods are based on the principle leading to elimination of one or more of its essential components from the reaction. Essential components are oxygen, enzyme, copper, or substrate in the reaction. Processing procedures like heating, refrigeration, freezing, dehydration by partial drying or by adding water binding agents, such as polyols, sugars, and salts, high-pressure treatment, and irradiation cause denaturation of various enzymes involved in spoilage of fish. Maintaining supercritical fluids in the food system helps in controlling such blackening. Supercritical fluid is a fluid at conditions above its critical temperature and pressure, and a pure substance exhibits gas-like and liquid-like properties. Supercritical carbon dioxide reduces the number of microorganisms and inactivates unwanted enzymes in food. Phenoloxidase is reported to be in high concentrations in shrimp heads. Hence beheading and use of effective enzyme inhibitors or reducing agents (Table 9.5) can also control blackening.

Table 9.5 Representative Inhibitors of Enzymatic Browning

Group of Inhibitors	Inhibitors
Reducing agents	Sodium bisulfite, ascorbic acid and analogues, cysteine Glutathione
Chelating agents	Phosphates, ethylenediaminetetraacetic acid (EDTA), Kojic acid (a-pyrone derivative, a fungal metabolite produced by *Aspergillus* and *Penicillium* species).
Acidulants	Citric acid, phosphoric acid
Enzyme inhibitors	Aromatic carboxylic acids, aliphatic alcohol, anions, peptides, substituted resorcinols
Enzyme treatments	Oxygenases, *O*-methyl transferase, proteases
Complexing agents	Cydodextrins

Tryptophanase

Tryptophanase converts free tryptophan present in the fish and shellfish muscle to indole (Fig. 9.12) that imparts a foul smell particularly to shrimp.

Formation of excess indole in shrimp is hazardous to human health and limits are prescribed by importing countries. This enzyme originates from bacteria, chiefly from *Escherichia coli* and some *Pseudomonas* sp. The maximum permissible level of indole in shrimp is 25 mg/100 g. Frozen shrimp containing such a higher concentration are considered decomposed.

Chitinolytic Enzymes

Chitinolytic enzymes can degrade chitin and are produced by organisms such as bacteria, fungi, insects, higher plants, and animals (Park et al., 1997). Many chitinase/chitosanase-producing microorganisms were screened out from Taiwan soil (Wang et al., 2011). These enzymes are potentially used in preparation of chitooligosaccharides and *N*-acetyl D-glucosamine, which are known to have various biological activities, such as antimicrobial, antifungal, immunoenhancing, antitumor, pharmaceutical, etc. (Tsai et al., 2000; Shen et al., 2009; Wen et al., 2002). Moreover, the control of pathogenic fungi in agriculture could be done by the use of such enzymes (Dahiya et al., 2005). The degradation of crustacean chitinous waste in seafood industry could be enhanced by these enzymes (Dhillon et al., 2013). They are also useful for the preparation of single-cell protein and also for the isolation of protoplasts from fungi and yeast, etc. (Dahiya et al., 2006).

Various chitinous materials from marine sources such as shrimp shell powder (SSP), squid pen powder (SPP), and shrimp and crab shell powder (SCSP) serve as substrates for chitinolytic enzyme production. Silage obtained by lactic acid fermentation of shrimp head wastes was also used as substrate and inducer for β-*N*-acetylhexosaminidase that is produced by *Verticillium lecanii* in submerged fermentations (SmF) and solid-state fermentation (SSF) processes (Shirai et al., 2001).

$$C_{11}H_{12}N_2O_2 \xrightarrow{\text{tryptophanase}} \text{Indole}$$

Figure 9.12 Breakdown of tryptophan to indole by tryptophanase enzyme.

Ligninolytic Enzymes

Ligninolytic enzymes act upon lignins, the most abundant natural aromatic polymer on earth. The ligninolytic enzymes degrade the recalcitrant aromatic polymer from natural processes in plants, animals, fungi, and bacteria (Kirk and Farrell, 1987). Ligninolytic enzymes have capacities to remove xenobiotic substances (such as hydrocarbons, phenols, perchloroethylene, azo dyes, carbon tetrachloride aromatics, pesticides, lignin, humic substances, etc.) which are introduced into the environment by numerous industrial activities. The production of ligninolytic enzymes using fishery waste showed promising results (Gassara et al., 2010). The production of lignin peroxidase, manganese peroxidase, and laccase by *Phanerochaete chrysosporium* BKM-F-1767 was carried out using fishery residues procured from Saummom Inc., Montreal, Canada.

Lipoxygenases

Lipoxygenases (LOX; EC 1.13.11.12) catalyze the incorporation of oxygen into the moiety of *cis*, *cis*-1,4 pentadiene in polyunsaturated fatty acids, such as arachidonic acid to form the hydroperoxide derivatives at a specific position (Pan and Kuo, 2000). Many researchers have found LOX activity in finfishes and shellfishes including rainbow trout, ayu, sardine, grey mullet, lake herring, tilapia, menhaden, shrimp, crab, mussel, starfish, sea urchin, etc., and also in algae. It is established that LOX activities have functions in ovarian development and immune regulations. LOX present in fish skin tissue generates hydroperoxy arachidonic acid (12-HPETE) and 14-hydroperoxydocosahexaenoic acid (14-HPDHE) (German and Kinsella, 1985; Hsieh and Kinsella, 1989). The α-cleavage at either side of the carbon atoms bearing the hydroperoxy groups can result in the volatile compounds responsible for the typical oxidative fishy odor, such as 2-nonenal from 12-HPETE and 3,6-nonadienal from 14-HPDHE (Josephson and Lindsay, 1986).

The pleasant green, seaweedy and melon-like aromas of freshly harvested fish are derived from long-chain polyunsaturated fatty acid (PUFA) reactions catalyzed by LOX yielding volatile alcohols and aldehydes of C_6, C_8, and C_9 (Josephson and Lindsay, 1986). These compounds originate from PUFA by the action of endogenous lipoxygenases such as 12-and l5-lipoxygenases and hydroperoxide lyase. Fresh shrimp flavor, characterized by total volatiles and 1–octen-3-ol, increases with increase in LOX activities from either endogenous or exogenous sources when added to shrimp homogenate (Kuo and Pan, 1991; Pan and Kuo, 1994; Pan et al., 1997). Measurement of the products of lipid oxidation is now being considered as accepted way of assessing fish quality. Two common indices used for lipid oxidation are the peroxide value (PV) and thiobarbituric acid number (TBA).

APPLICATIONS OF ENZYMES

Most of the enzymes are important tools of biotechnology, as they carry out biotransformation processes (Lopez and Carreno, 2000). Enzymes are important components of food and feed for many reasons as they have important roles to play in growth, maturation, production, processing, storage, and spoilage; consumer preferences and selection; safety and control of predators; food intake, digestion, and assimilation and disease; and as analytical tools (Reed, 1993). Enzymes have been applied traditionally for the production of fish protein hydrolysate, fish sauces or paste, semipaste fermented fish products of Southeast Asian countries wherein mostly endogenous proteases in the fish are used (Haard, 1992).

Fish and shellfish enzymes nowadays are being used exogenously to improve the traditional applications to accelerate the process in production of other products including PUFA-enriched fish oils and selective removal of skin, fish scale removal, or riddling process in cured roe production (Haraldsson, 1990; Gildberg, 1993). Many enzymes are being used for producing products either for human consumption or for other uses. A few of such applications are discussed in the following sections.

Specialty Products

Specialty products are those products that invariably warrant the use of enzymes in the processes leading to development of value-added products. Lopez and Carreno (2000) defined specialty product as "a product which has been manufactured, directly or indirectly, by the use of enzymes." The conventional products are manufactured without the application of enzymes, which include fresh, frozen, or canned fish or shellfish, which however cannot be categorized as specialty products. Some products that have industrial importance are discussed briefly here.

Mince Products

These products are mainly based on surimi and crab leg analogues and various types of Japanese heat-gelled products, such as *itatsuki*, *kamaboko*, *chikuwa*, *hampen*, and Satsuma-age, etc. As discussed previously, transglutaminase-catalyzed reactions are used to modify the functional properties of food proteins, including surimi production. Transglutaminase has been used to catalyze the cross-linking of a number of proteins, such as whey proteins, meat proteins (myosin and actomyosin), soybean proteins, and gluten. Such modification of protein techniques is further used to produce textured products, such as hamburger, meatballs, canned meat, frozen meat, molded meat, surimi (fish paste), krill paste, baked foods, protein powders from plants, etc. This process also helps to protect lysine in food proteins from various chemical reactions, encapsulate lipids and lipid-soluble materials, avoid heat treatment for gelation, improve elasticity and water-holding capacity, modify solubility and functional properties, and produce food proteins of higher nutritive value by cross-linking of different proteins containing complementary limiting essential amino acids (Kitabatake and Doi, 1993; Motoki and Seguro, 1994).

PUFA-Enriched Fish Oils

PUFAs are essential components in human nutrition, and the *n*–3 and *n*–6 fatty acids have distinct and sometimes opposing roles in human metabolism (Bjerve et al., 1992). Because the conversion of *n*–3 and *n*–6 fatty acids is not reversible in humans, they must be present in the diet. An appropriate ratio of *n*–3/*n*–6 PUFAs is 1:4. The dietary requirement of *n*–3 fatty acids for humans is about 1.0 g per day (Bjerve et al., 1992), however, fish and fish dishes currently provide an average intake of 0.2 g *n*–3 PUFAs per day, mainly as EPA and DHA according to the Food and Agriculture Organization of the United Nations. PUFA-enriched fish oil thus is of great commercial interest. It is possible to prepare triacylglycerides containing up to 30% EPA and DHA directly from fish oils without splitting the fat by using enzymes and processes, such as winterization, molecular distillation, and solvent crystallization. The PUFA concentrate can then be resynthesized into triacylglycerols using lipases from microorganisms, such as yeast, fungi, and bacteria, under conditions that favor the reesterification process for triacylglycerol synthesis (Bjerve et al., 1992). These oils are further treated with enzymes such as lipase (triacylglycerol acylhydrolase, EC 3.1.1.3) for increasing heat resistance.

Caviar and Roe Production

Caviar refers only to the riddled and cured roe (fish egg) of the sturgeon (*Acipenser* spp. and *Huso huso*), which are very expensive. Less-expensive caviar is made from the eggs of several fish including cod, catfish (*Parasilurus astus*), herring (*Clupea harengus*), capelin (*Mallotus villosus*), lumpfish (*Protopterus aethiopicus*), and some freshwater species (Lanier, 1994). The riddling process (manually or mechanically separating the eggs from ovarian connective tissue) is a laborious task and involves destruction of a large amount of the roe, sometimes yielding intact roe as low as 50% (Raa, 1986). Solving this problem, a US patent describes a method to release salmon roe from the connective tissue by the use of proteinases under acidic, neutral, or alkaline conditions (Sugihara et al., 1973).

Cured Fish Products

The cured products are made by a maturing or ripening process of raw materials that includes salting and enzymatic changes in proteins, lipids, and carbohydrates (Venugopal and Shahidi, 1998). In traditional cured fish products, a partial proteolysis occurs, catalyzed by endogenous fish enzymes including muscle proteases and proteases from the gastrointestinal tract. The proteases from intestine play a decisive role in the maturing or ripening process of fish meat, which, in turn produces the characteristic texture and flavor (Haard, 1992). Trypsin-type enzymes from the pyloric caecum are the major contributors to endopeptidase activity in the ripening process. For accelerating the fermentation or ripening process, enzymes are added in the form of crude extract obtained from pyloric cecum of cod or squid (Lee et al., 1982; Simpson and Haard, 1984a,b).

Protein Hydrolysates

Fish protein hydrolysates (FPHs) are prepared by proteolytic enzymes (endogenous and exogenous) at the optimal temperature and pH, digesting whole fish or other aquatic animals or parts thereof. Since the process of hydrolysis breaks down the protein into smaller peptides, FPH appears like a water-soluble product consisting of proteinaceous fragments (Venugopal and Shahidi, 1995). Proteases include mostly pepsin, chymotrypsin, and trypsin and are obtained from various sources, such as mammals, bromelain, papain, and ficin from plants, microbial proteases, and also from fish (Haard, 1992).

FPH is used in dietetic foods, such as soufflés, meringues, macaroni, or bread, and for the preparation of fish soup, fish paste, and shellfish analogues as flavoring compounds as a source of small peptides and amino acids. The major feed applications of FPH are as milk replacers for calves and weaning pigs and as proteins and attractants in fish feed (Gildberg, 1993). The combination of 14% hydrolysate and 0.2% salt achieved nearly the same level of water retention as the standard industry practice of adding nearly 2% salt. This could mean that processors could reduce current levels of salt addition by 90% without compromising process performance, color, or texture in the final product (Ibarra et al., 2013).

Seafood Flavorings

Seafood flavorings from fish and shellfish or their by-products can be produced by enzymatic hydrolysis. Exogenous and endogenous proteolytic enzymes and also some other types of enzymes are being used extensively. Certain fresh seafood is rich in 5′-AMP (adenosine-5′-monophosphate), which may be converted to 5′-IMP (inosine-5′-monophosphate) by the action of endogenous enzymes released during processing, which are potential meat flavorings (Diaz-Lopez and Garcia-Carreno, 2000). Various kinds of by-products are used to produce marketable seafood flavorings. Seafood flavors are in high demand for use as additives in products, such as kamaboko, artificial crab, and fish sausage, and cereal-based extrusion products, such as shrimp chips (Kawai, 1996). Proteolytic enzymes can aid the extraction of flavor compounds from shells and other materials. This allows separation of flavors from bones and shells, which can then be concentrated up to 50–60% dry matter. The taste and volatile components of shrimp heads were recovered by enzymatic digestion using COROLASE N and koji or bacterial strains with high proteolysis rates followed by concentration and spray-drying (Haard and Simpson, 1994; Anon, 1992).

Fish Sauces and Fermented Products

Fish sauces are liquid products made by storing heavily salt-preserved fish material at tropical temperatures until it is solubilized by endogenous enzymes. Many other products are fermented from fish or shellfish by endogenous enzymes to make solid, paste, semipaste, and semiliquid products that are popular in Southeast Asian countries. Some typical examples of fish sauces are the *nuoc-mam* produced in Vietnam and Cambodia, *nam-pla* in Thailand, *patis* in the Philippines, *uwoshoyu* in Japan, and *ngapi* in Myanmar. Semisolid and paste products include *budu* in Malaysia, kapi of the Philippines (Saisithi, 1994), and shidal of Northeast India (Muzaddadi, 2015). Fish digestive enzymes play a major role in the fermentation of such products. In several Southeast Asian countries, small, commercially unimportant fish are allowed to autolyze in the presence of high (>25%) concentration of salt, which takes about three months to two years. The autolytic, trypsin-like enzymes present in the muscle slowly degrade the fish tissue. Because of the high salt content, bacteria are not able to thrive. The final product is a clear liquid, which is obtained by decantation or filtration and contains a high level of soluble nitrogenous compounds and a high concentration of salt. Acceleration of fish sauce production can be achieved by the action of externally added pepsin under controlled conditions. In this case, the pH of the fish slurry has to be lowered to about 4.0 and the salt concentration is also reduced for optimal action of pepsin (Beddows, 1985; Owens and Mendoza, 1985; Raa and Gildberg, 1982). Capelin is one of the low-cost fish abundantly available in the Atlantic ocean, which has been examined for sauce production. Addition of proteolytic enzymes, such as fungal protease, pronase, trypsin, chymotrypsin, squid protease, or squid hepatopancreas to minced and salted capelin increased the rate of protein solubilization during the first month of fermentation. Previous study showed that the product supplemented with squid hepatopancrease was highly acceptable and preferred as a product in the Philippines (Raksakulthai et al., 1986). Fermentation of squid meat using proteolytic enzymes has also been reported (Lee et al., 1982).

Isolation of Pigments

Shellfish contain significant amounts of carotenoids bound to their exoskeletons. An enzymatic method has been developed to recover the carotenoid along with the protein in its native carotenoprotein state from shrimp shell waste using proteolytic enzymes (Simpson and Haard, 1985) and also from crab (Manu-Tawiah and Haard, 1987). About 80% of the protein and 90% of the astaxanthin pigment from shrimp process waste can be recovered as an aqueous dispersion after trypsin hydrolysis. The extracted pigment has been found to be a good agent to enhance the color of cultured fish species, such as salmon. The protein–pigment complex is more resistant to oxidation and is deposited easily in the muscle of cultured fish. The color of the flesh of salmonids is caused by astaxanthin, which belongs to a large group of compounds named carotenoids. In salmonid-farmed production, carotenoid pigments are used in the feed for improving the attractive pinkish red color of the fish meat. Astaxanthin is the major carotenoid pigment in shrimp and lobster. Proteolytic enzyme treatment of shrimp, snow crab shell, and shellfish (Ya et al., 1991) wastes allows the recovery of the carotenoid pigment along with the protein. The carotenoid pigment is recovered in the form of a protein–carotenoid complex, which is more resistant to oxidation and gives better results than free astaxanthin in the coloring of farmed rainbow trout (Haard, 1992).

Ripening of Salted Fish

Salting has been one of the traditional methods of processing fish. Salted fresh herring (*C. harengus*) is a traditional product available in all countries in the North Atlantic and North Sea regions. The salted fish (whole or gutted) during storage develops a characteristic flavor and soft texture. This is due to the action of endogenous proteases and exopeptidases, which hydrolyze the proteins, thus causing an increase in soluble nitrogenous compounds, such as peptides and amino acids with associated changes in the tissue texture (Voskrensensky, 1965). The final salt concentration can vary from approximately 4–18%, depending on the recipe used. The ripening time, which normally lasts up to a year in heavily salted herring, can be reduced by incorporation of exogenous proteases and lipases, although the actual mechanism of flavor improvement is not clear (Borresen, 1992). The enzyme mixtures for ripening purposes are available commercially. Curd formation in canned salmon can also be avoided by protease treatment (Yamahoto and Mackey, 1981).

Meat Tenderization

Many consumers carry a negative opinion of the eating quality of squid due to its characteristic tough texture. Removal of squid skin required a pretreatment in 5% salt solution at 45°C for 10 min, which also activates endogenous enzymes causing a softening of the skin (Raa et al., 1985). At low temperature (3°C), a combination of 5% salt and an enzyme preparation from papaya latex could be used (Borrcscn, 1992). Melendo et al. (1997, 1998) tenderized squid meat by treatment with commercial bromelain and a crude extract from bovine spleen. Squid mantles were transversally cut as rings and treated with one of the enzymes at a pH of 7.0 and temperature of 37°C for 30 min. The tenderization was demonstrated by sheer force measurement as well as sensory evaluation.

Reduction in the Viscosity of Stickwater

Stickwater is a by-product obtained during the production of fish meal. It is high in protein and has excellent nutritional value. It can be sold as a feedstuff labeled as condensed or dried fish solubles. The protein content of stickwater presents a problem during evaporation. As the dissolved solids are increased beyond 25%, viscosity increases. This change causes inefficient water removal and may even lead to clogging of the evaporator. The problem can be solved by reducing the viscosity of the stickwater by protease treatment. Commercial proteases, such as alcalase 0.6 L or neutrase 0.5 L at 0.2% could be added to the stickwater at neutral pH at a temperature of 50°C for significantly reducing viscosity, enabling further evaporative concentration (Jacobsen and Rasmussen, 1984).

Extension of Shelf Life

The enzyme treatment is beneficial for onboard extension of the shelf life of shrimp. In this case, the catch is treated with a solution of the enzymes containing glucose at 0–2°C (Kantt et al., 1993). Lysozyme, which is found in almost all human and animal cells, is known to have antibacterial properties. The preservative effect of the enzyme is due to its action on the mucopeptide structure of bacterial cell walls. Igarashj and Zama (1972) observed that the enzyme could be used for seafood preservation. Ramesh and Lewis (1980) reported that the enzyme at concentrations up to 150 pg mL^{-1} retarded the growth of microorganisms isolated from shrimp. The amount of the enzyme could be reduced by incorporation of traces of disodium ethylenediaminetetraacetic acid. At 4°C, the enzyme exhibited about 55% of its activity compared to that at 37°C, suggesting its potency at lower temperatures (Myrnes and Johansen, 1994). Lysozyme from clamshell (*Chlamys islandica*) has been examined as a bacteriostatic agent in specialty food and feed items (Sikorski et al., 1995). Oxidative deterioration is a major quality loss in flesh foods. The potential of glutathione peroxidase to prevent oxidative deterioration in fish muscle during handling, storage, and processing has been suggested. Other enzymes that may be of use to control endogenous enzyme activity in seafood include superoxide dismutase, protocatechuate-3,4-dioxygenase, and *o*-methyltransferase (Ashie et al., 1996).

Removal of Off-Odor and Fishy Taste

Jeffreys and Krell (1965) observed that treating fish portions with enzymes derived from *Aspergillus* could eliminate much of the off-odor and fishy taste. Meat of Elasmobranch species, ie, sharks and rays, contains significant amounts of urea, which adversely influences the consumer acceptance of these products. Application of urease is useful, in addition to its use in biosensor construction. The enzyme has been purified from *Cajanus cajan* seeds (Pandey and Pandey, 1991).

Enzymes as Processing Aids

Enzymes are used as gentle knives for selective removal or modification of certain tissue structures.

Deskinning

Enzymatic deskinning can be done by cold-adapted fish pepsin at low-reaction temperatures. Proteases are usually mixed with carbohydrases to facilitate skin removal (Stefánsson and Steingrímsdóttir, 1990). In addition, fish pepsin at low pH breaks down fish skin rapidly but breaks down muscle protein relatively slowly. In Iceland, experiments have been carried out on enzymatic deskinning of skate wings, ocean perch, and haddock (Anon, 1992; Stefansson and Steingrímsdóttir, 1990; Stefansson, 1988). The skin is first denatured by warm water and subsequently treated with an enzyme mixture containing proteolytic and glycolytic enzymes for 4 h at 25°C. A patented process for the deskinning of tuna consists of preheating the fish to 60°C with steam followed by digestion of the skin with proteolytic enzymes at about 50°C (Fehmerling, 1970). Herring skin has been successfully removed by pepsin treatment (Joakinsson, 1984; Stefansson, 1988). Starry ray (*Raja radiata*) skin contains a high amount of collagen, which can be degraded by enzymes. The skinning is accomplished by a gentle heat denaturation of the skin collagen, followed by incubating the skate wings at low temperatures (0–10°C) in the enzyme bath for a few hours. The skin is slowly solubilized during the incubation, which is removed from the wings by rinsing. At 0°C the process can be accomplished by overnight incubation (Stefansson, 1988). Fehmerling (1970) has developed a process for loosening the shells of shrimp and also for removal of visceral mass in clam processing by the use of a mixture of carbohydrases and cellulases derived from *Aspergillus niger*.

Descaling

Fish processing generates valuable by-products that are high in proteins and lipids, such as viscera, skin, tails, heads, and frames. The disposal of these nutrient-rich byproducts can often increase costs. By using enzymes to valorize byproducts, processors can potentially obtain additional revenue from nutritional supplements, ingredients for pet food, biodiesels, fertilizers, agricultural products, and animal and fish feeds. Many processors do not capture the full value of these by-products. This may be due to inefficient secondary processing and overreliance on mechanical, heat, or chemical methods. Enzymes can enhance the quality and value of fish processing by-products, optimizing time, energy, and resources while also increasing revenue by the following:

- Increase yield
- Improve extraction
- Convert by-products into valuable ingredients

In some fish species, such as redfish (*Sebastes marinus*) or haddock (*Melanogrammus aeglefinus*), the mechanical means of descaling, which are harsh and tend to damage the skin and lower fillet yield, could be replaced/addressed by enzyme treatment. The scales can be gently removed without affecting the skin or flesh after incubating the fish in an enzyme solution at 0°C followed by spraying with water. Descaling of large fish is a major fish processing operation, which is done manually or by mechanical scrubbing and scraping methods (Venugopal et al., 2000). The process becomes all the more cumbersome and tedious for fish having large numbers of scales. There are several concerns with the process, namely, incomplete removal of scales, partial damage of the skin associated with loss of color and shiny appearance, damage to muscle texture, etc. Scaling of species like ocean perch and haddock by mechanical means is harsh and results in tearing of skin and a lower fillet yield. The demand for scaleless fish fillets of species such as salmon, perch, ocean bream, silver carp, and tuna has resulted in the development of a new enzymatic method to remove scales from fish. The process includes three steps: denaturation and loosening of the mucous layer and outer protein structures of the skin, followed by enzymatic degradation of the outer skin structures, and finally washing off the scales with water jets. In the process, the fish is incubated in a specially designed incubation tank filled with a slightly acidified water solution of the descaling enzyme, a collagenase.

Membrane Removal

Membrane removal is a problem during the production of canned cod liver. To avoid seal-worm infestation in cod liver oil and canned cod, the membrane has to be removed before liver oil processing and canning. Since no suitable machinery is available for its removal, it is presently removed by hand but only with moderate success and considerable labor cost. An enzymatic method can be a solution for this as dissolution of the cod liver membrane could be achieved using fish proteases (Stefánsson and Steingrímsdóttir, 1990).

As mentioned previously as well, Fehmerling (1970) developed a process for loosening the shells of shrimp and for removal of visceral mass in clam processing by the use of a mixture of carbohydrases and cellulases derived from *A. niger*. Cod frames can be treated with a crude proteinase from tuna pyloric caeca for a period of 12 h at 50°C to recover up to 80% of the protein (Kim et al., 1997). Ferreira and Hultin (1994) used an acidic fungal protease (Newlase A from *Rhizopus niveus*) to recover protein from codfish frames. The enzyme hydrolyzed the fish proteins at 40–50°C at a pH of 3.5 within 75 min. The shellfish waste is a good source of chitin, which has numerous industrial and laboratory applications (Venugopal, 1994; Kaur and Dhillon, 2015). Recovery of chitin from these wastes necessitates removal of adhering proteinaceous matter. Traditionally, preparation of chitin involves demineralization and deproteinization using strong acids or bases (Simpson et al., 1994; Dhillon et al., 2013). However, these chemicals may cause a partial deacetylation of chitin and hydrolysis of the polymer resulting in a final product with lower molecular weight and inconsistent physiological properties. Chemical treatment may also degrade tryptophan residues in the protein. Such shell wastes could be treated with proteases for efficient protein removal. Wang and Chio (1998) reported that a microbial protease from *Pseudomonas aeruginosa* can be employed for the process. The authors further reported that fermentation for a period of seven days removed proteins from shrimp, crab shell powder, shrimp shells, and shrimp heads up to 55%, 48% and 61%, respectively. Solid fermentation of the materials for a period of 5–10 days removed proteins in the range of 46–81%. A process for removal of membranes surrounding cod liver is also available (Stefansson, 1988).

Application in Other Foods and Feeds

The seafood enzymes are also used in many other foods including dairy technology, meat tenderizing, feed products, and the enzymatic clarification of fruit juice.

Seafood Enzymes in Dairy Technology

The major consumers of seafood enzymes are the dairy product processing plants. The dairy industry has been potentially using these enzymes in (1) enzymatic milk coagulation, (2) prevention of oxidized flavor in milk, and (3) preparation of infant milk. Kim and Dewapriya (2014) have comprehensively described the applications of seafood enzymes in dairy industry. Seafood enzymes are used in cheese making for coagulating milk. Such milk-clotting enzymes are termed as "rennet" and made from the fourth stomach of bovine animals, sheep or goats, or microbial origin. These rennet substitutes are significantly cheaper when extracted from seafood and their by-products. The proteases with excessive general proteolysis ability lead to lower curd yield and bring undesirable changes in the cheese aging. So, selective enzymes only can give high yields. Cold-adapted gastric proteases from fish and chymosin-like enzymes from marine mammals have several characteristics that make them suitable for both milk-clotting phases and that avoid the major problems with rennet substitutes (Haard, 1992). Tuna (Atlantic tuna, *Thunnus obesus*) gastric enzyme was proposed as milk-coagulating enzyme (Tavares et al., 1997). A semipurified aminopeptidase preparation from squid hepatopancreas has potential to reduce bitterness and enhance the flavor of cheese effectively (García-Carreño et al., 1997). The rennet substitutes from seafood have several advantages including that they have a lower temperature coefficient (so lower Arrhenius activation energy) for milk clotting, require lower enzyme concentrations (thereby conserving rennet), have lower temperature optimum for hydrolysis, helping avoid softening problems and off-flavors with much lower levels of free amino acids and bitter peptides during ripening (Simpson and Haard, 1987). The milk-clotting enzymes from fish stomach mucosa or shellfish hepatopancreas for cheese manufacture have great potential as an inexpensive alternative to rennet, and could become a new food-related industry.

Peroxidase (EC 1.11.1.7), which is stable at pasteurization temperatures, catalyzes oxidative reactions in milk and dairy products. Xanthine oxidase (EC 1.2.3.2), which catalyzes oxidation of xanthine or hypoxanthine to uric acid and hydrogen peroxide, also causes oxidative flavors (Brown, 1993). Such a problem is common in dairy products and is normally prevented by using bovine trypsin. The residual effect of trypsin in the pack again deteriorates milk quality. However, trypsin from cold-adapted fish (eg, Greenland cod) has a lower free energy of activation at reaction temperatures below 30°C and more thermal instability than bovine trypsin. Thus, the prevention of oxidized flavors in milk by Greenland cod trypsin has been demonstrated (Simpson and Haard, 1984a,b).

Lysozyme is used to make cow's milk suitable for infants. Lysozyme purified from Arctic scallop (*Chlamys islandica*) has low activation energy at low temperature with specific activity 300% higher than egg white lysozyme (Myrnes and Johansen, 1994) and thereby it has immense potential in infant milk production.

Seafood Enzymes in Feed Products
Fish Silage
Fish silage is a liquid product made from whole fish (bycatch fish) or fish waste plus acid, or, less frequently, alkali. Liquefaction is caused by the action of enzymes naturally present in the fish and is accelerated by the acid, creating the right conditions for the enzymes that are active at low pH to hydrolyze quickly most of the protein. This process yields an aqueous solution and limits the growth of spoilage bacteria (Arason, 1994). Commercial enzymes like bromelain were used in fermented silage production to accelerate the liquefaction process, and addition of bromelain (0.7–0.9%, w/w) and *Lactobacillus plantarum* to minced whole fish with 15% molasses increased the proteolysis rate and decreased the liquefaction time from 15 days to 12 h (Tome et al., 1995). The digestive enzymes from fish may also have similar or even higher effect in liquefaction process.

Cold-active Enzymes

The most extensively studied enzymes from the marine environment are trypsin, chymotrypsin, elastase, collagenase, and alkaline phosphatase that have been isolated from Atlantic cod viscera (stomach, pyloric caeca, and intestine). Atlantic lives in relatively frigid waters, so it seems reasonable to surmise that its enzymes are cold active. Several of these enzymes have indeed been demonstrated to be cold active, including elastase, collagen, and chymotrypsin. The enzyme activity at low temperatures could render them useful in various food-processing applications, for example, in instances where proteolysis must be carried out at low temperatures (Vilhelmsson, 1997). Examples from fish processing include caviar production and the extraction of carotenoprotein, for use as a colorant or flavoring in food or feed, from crustacean processing offal (Table 9.6). An added benefit is that these enzymes are typically quite heat labile making thermal inactivation possible with only a modest increase in temperature. The cod enzymes have also been shown to have a higher catalytic efficiency than corresponding mammalian enzyme. This is in line with numerous other observations that indicate that alterations in enzyme activity are of central importance in cold adaptations.

Table 9.6 Summary of Some of the Enzymatic Applications in Seafood Processing

Problem	Solution	Enzyme	Benefits
Poor oil yield	FoodPro alkaline protease	Protease	Improved oil separation; lower temperatures; higher oil yields; energy savings
Low solids from stickwater evaporation	FoodPro alkaline protease	Protease	Reduced water binding of proteins; increased water removal in evaporators; lower need for water removal in spray-drying; energy savings; less evaporator downtime
Unrealized value from fish protein hydrolysates	FoodPro alkaline protease FoodPro PNL FoodPro 51FP	Protease	Improved protein value, flavor and digestibility; enhanced protein functionality: emulsifying properties, solubility, fat- and water-binding; generation of high-value peptides and proteins
Poor fish sauce flavor and fermentation	FoodPro alkaline protease FoodPro PNL FoodPro 51FP	Protease	Improved fish sauce flavor; faster fermentation; reduced fermentation cost; increased fish sauce yield
Remaining protein in mollusk shells	FoodPro alkaline protease FoodPro PXT	Protease	Efficient removal of residual meat; "Clean" shells for use as calcium source; increased recovery of hydrolyzed mollusk protein
Fish and seafood spoilage	FoodPro GOL	Glucose oxidase	Removal of oxygen; reduced potential for oxidative rancidity; preservation of color
Residual hydrogen peroxide after removing discoloration from fish roe	FoodPro CAT	Catalase	Fast, efficient removal of hydrogen peroxide

CONCLUSIONS

The production of seafood enzymes from fish and shellfish has disadvantages, such as variable and limited availability of raw materials and expensive production processes because of low enzyme concentration in the raw materials. However, many seafood enzymes have unique properties that are not present in cheaper enzymes from animals, plants, and microbes, and this aspect renders the seafood enzymes invaluable, despite being expensive. In future, such enzymes may be produced by using advanced biotechnological tools including recombinant DNA technology, site-directed mutagenesis, etc. The enzymes from fish and shellfish collected from extreme environments of the ocean can become useful for industrial applications. Cold-adapted enzymes from fish living at temperatures at about the freezing point of seawater and thermoresistant enzymes from organisms, including crustaceans living in the hydrothermal vents, are some of the examples. Enzymes working better at 2–4 M NaCl from organisms living in salt mines are already in use. In case of enzymes being used in seafood processing, the functional characteristics with enhanced enzyme concentration in crude filtrate is extremely important. Therefore, a combination of technologies or the use of genomics, proteomics, and metabolomics in isolation or again in conjunction may help in improving the enzyme production potential of the microbes for the desired enzyme. In addition, the enzyme-producing microbial strains thus developed must produce only the desired enzyme(s) so as to make the downstream process simpler and convenient. Each time an enzyme with special kinetic abilities is discovered, a new potential application for food technology arises. An aggressive approach is needed to open new opportunities for enzyme applications that can benefit the food industry. There is still, however, a long way to go in realizing the actual potential of seafood enzymes.

LIST OF ABBREVIATIONS

ADP Adenosine 5′-diphosphate
AMP Adenosine monophosphate
ATP Adenosine 5′-triphosphate
BAL Bile salt-activated lipase
DHA Docosahexaenoic acid
DMA Dimethylamine
EC Enzyme Commission
EDTA Ethylenediaminetetraacetic acid
EPA Eicosapentaenoic acid
FA Formaldehyde
FAO Food and Agriculture Organization of the United Nations
FFA Free fatty acids
FPH Fish protein hydrolysate
HL Hepatic lipase
HPDTE Hydroperoxydocosahexaenoic
HPETE Hydroperoxyarachidonic acid/Hydroperoxyeicosatetraenoic acid
Hx Hypoxanthine
I Inosine
IMP Inosine monophosphate
IUB International Union of Biochemists
kDa Kilo Dalton
LDH Lactate dehydrogenase
LOX Lipoxygenase
LPC Lysophosphatidylcholine
LPL Lipoprotein lipase
M Myosine

NADH Nicotinamide adenine dinucleotide
Pi Inorganic phosphate
PL Pancreatic lipase
PUFA Polyunsaturated fatty acid
R.l.P Ribose 1 phosphate
TGase Transglutaminase
TMA Trimethylamine
TMAO Trimethylamine oxide
U Uric acid
Xa Xanthine

REFERENCES

Amaya, C.L., Marngoni, A.G., 2000. Lipases. In: Haard, N.F., Simpson, B.K. (Eds.), Seafood Enzymes-Utilization and Influence on Postharvest Seafood Quality. Marcel Dekker, Inc., New York, pp. 120–146.

Anon., 1992. Marine enzymes in fish processing. Infofish International 6, 53.

Arason, S., 1994. Production of fish silage. In: Martin, A.M. (Ed.), Fisheries Processing: Biotechnological Applications. Chapman & Hall, London, pp. 244–272.

Ashie, I.N.A., Lanier, T.C., 2000. Transglutaminases in seafood processing. In: Haard, N.F., Simpson, B.K. (Eds.), Seafood Enzymes. Marcel Dekker Inc., New York, pp. 147–190.

Ashie, I.N.A., Simson, B.K., Smith, J.P., 1996. Spoilage and shelf life extension of fresh fish and shellfish. Critical Reviews in Food Science and Nutrition 36, 87–121.

Awarenet, 2004. Handbook for the Prevention and Minimization of Waste and Valorization of By-products in European Agro-Food Industries. Agro-Food Waste Minimization and Reduction Network (AWARENET). Grow Programme, European Commission, pp. 1–7.

Beddows, C.G., 1985. Fermented fish and fishery products. In: Wood, B.J.B. (Ed.), Microbiology of Fermented Foods, vol. 2. Elsevier Applied Science, London, pp. 1–34.

Ben Rebah, F., Frikha, F., Kammoun, W., Belbahri, L., Gargouri, Y., Miled, N., 2008. Culture of *Staphylococcus xylosus* in fish processing by-product-based media for lipase production. Letters in Applied Microbiology 47, 549–554. http://dx.doi.org/10.1111/j.1472-765X.2008.02465.x.

Bjerve, K.S., Thoresen, L., Bønaa, K., Vik, T., Johnsen, H., Brubakk, A.M., 1992. Clinical studies with α-linolenic and long chain n-3 fatty acids. Nutrition 8, 130–135.

Blanco, M., Sotelo, C.G., Chapela, M.J., Perez-Martin, R.I., 2006. Towards sustainable and efficient use of fishery resources: present and future trends. Trends in Food Science & Technology 18, 29–36. http://dx.doi.org/10.1016/j.tifs.2006.07.015.

Borresen, T., 1992. Biotechnology, by-products and aquaculture. In: Seafood Science and Technology. Fishing News Books, pp. 278–287.

Bozzano, A., Sarda, F., 2002. Fishery discard consumption rate and scavenging activity in the northwestern Mediterranean Sea. ICES Journal of Marine Science 59, 15–28. http://dx.doi.org/10.1006/jmsc.2001.1142.

Brown, R.J., 1993. Dairy products. In: Nagodawithana, T., Reed, G. (Eds.), Enzymes in Food Processing, third ed. Academic Press, San Diego, pp. 347–361.

Chaplin, M.F., Bucke, C., 1990. Enzyme Technology. Cambridge University Press, Cambridge (UK), pp. 40–79.

Dahiya, N., Tewari, R., Tiwari, R.P., Hoondal, G.S., 2005. Production of an antifungal chitinase from *Enterobacter* sp. NRG4 and its application in protoplast production. World Journal of Microbiology and Biotechnology 21, 1611–1616.

Dahiya, N., Tewari, R., Hoondal, G.S., 2006. Biotechnological aspects of chitinolytic enzymes: a review. Applied Microbiology and Biotechnology 71, 773–782.

Dhillon, G.S., Kaur, S., Brar, S.K., Verma, M., 2013. Green synthesis approach: extraction of chitosan from fungus mycelia. Critical Reviews in Biotechnology 3 (4). http://dx.doi.org/10.3109/07388551.2012.717217.

Diaz-Lopez, M., Garcia-Carreno, F.L., 2000. Applications of fish and shelf life enzymes in food and feed products. In: Haard, N.F., Simpson, B.K. (Eds.), Seafood Enzymes. Marcel-Dekker, Inc., New York, pp. 571–618.

Faid, M., Zouiten, A., Elmarrakchi, A., Achkari-Begdouri, A., 1997. Biotransformation of fish waste into a stable feed ingredient. Food Chemistry 60, 13–18. http://dx.doi.org/10.1016/S0308-8146(96)00291-9.

Fehmerling, G.B, 1970. U.S. Patent 3513071.

Ferreira, N.G., Hultin, H.O., 1994. Liquefying cod fish frames under acidic conditions with fungal enzyme. Journal of Food Processing and Preservation 18, 87–101.

García-Carreño, F.L., Raksakulthai, R., Haard, N.F., 1997. Processing wastes. Exopeptidases from shellfish. In: Bremmer, A., Davis, C., Austin, B. (Eds.), Making the Most of the Catch. AUSEAS, Hamilton, Queensland, pp. 37–43.

Gassara, F., Brar, S.K., Tyagi, R.D., Verma, M., Surampalli, R.Y., 2010. Screening of agro-industrial wastes to produce ligninolytic enzymes by Phanerochaete chrysosporium. Biochemical Engineering Journal 49, 388–394.

German, J.B., Kinsella, J.E., 1985. Lipid oxidation in fish tissue. Enzymatic initiation via lipoxygenase. Journal of Agricultural and Food Chemistry 33, 680–683.

Gildberg, A., 1993. Enzymic processing of marine raw materials. Process Biochemistry 28, 1–15.

Godfrey, T., Reichelt, J., 1983. Industrial Enzymology. The Nature Press, Surrey, UK, p. 582.

Gurr, M., Harwood, J.L., 1991. Lipid Biochemistry: An Introduction, fourth ed. Chapman and Hall, New York, pp. 307–314.

Haard, N.F., Simpson, B.K., 1994. Proteases from aquatic organisms and their uses in the seafood industry. In: Martin, A.M. (Ed.), Fish Process: Biotechnological Applications. Chapman & Hall, London, UK, pp. 133–154.

Haard, N.F., 1992. A review of proteolytic enzymes from marine organisms and their application in the food industry. Journal of Aquatic Food Product Technology 1, 17–35.

Haard, N.F., 2000. Seafood enzymes: the role of adaptation and other intra-specific factors. In: Haard, N.F., Simpson, B.K. (Eds.), Seafood Enzymes : Utilization and Influence on Postharvest Seafood Quality. Mercel & Dekker, New York, p. 689.

Haraldsson, G.G., 1990. The applications of lipases for modification of facts and oils, including marine oils. In: Voigt, M.N., Botta, J.R. (Eds.), Advances in Fisheries Technology and Biotechnology for Increased Profitability. Technomic Publishing, Lancaster, PA, pp. 337–357.

Hassan, T.E., Heath, J.L., 1986. Biological fermentation of fish waste for potential use in animal and poultry feeds. Agricultural Wastes 15, 1–15. http://dx.doi.org/10.1016/0141-4607(86)90122-8.

Hide, W.A., Chan, L., Li, W.-H., 1992. Structure and evolution of the lipase superfamily. Journal of Lipid Research 33, 167–178.

Hsieh, R.J., Kinsella, J.E., 1989. Lipoxygenase generation of specific volatile flavor carbonyl compounds in fish tissues. Journal of Agricultural and Food Chemistry 37 (2), 280–286.

Ibarra, P., Teixeira, A., Simpson, R., Valencia, P., Pinto, M., Almonacid, 2013. Addition of fish protein hydrolysate for enhanced water retention in sous vide processing of Salmon. Journal of Food Processing & Technology 4, 241. http://dx.doi.org/10.4172/2157-7110.1000241.

Igarashi, H., Zama, K., 1972. Preservation of edible seafoods. Japan Patent 71: 19, 576 (Cl.A23b).

Jacobsen, F., Rasmussen, O.L., 1984. Energy savings through enzymatic treatment of stickwater in the fish meal industry. Process Biochemistry 19, 165–169.

James, J., Simpson, B.K., Marshall, M.R., 1996. Application of enzymes in food processing. Critical Reviews in Food Science and Nutrition 36 (5), 437–463.

Jeffreys, G.A., Krell, A.J., 1965. U.S. Patent 3170794.

Joakinsson, K.G., 1984. Enzymatic Deskinning of Herring (*Clupea Harengus*) (thesis). Institute of Fisheries, University of Tromso, Norway.

Josephson, D.B., Lindsay, R.C., 1986. Enzymatic generation of volatile aroma compounds from fresh fish. In: Parliment, T.H., Croteau, R. (Eds.), Biogeneration of Aromas. American Chem. Soc., Washington, DC, pp. 201–221.

Kacem, M., Sellami, M., Kammoun, W., Frikha, F., Miled, N., Ben Rebah, F., 2011. Seasonal variations of chemical composition and fatty acid profiles of viscera of three marine species from the Tunisian coast. Journal of Aquatic Food Product Technology 20, 233–246. http://dx.doi.org/10.1080/10498850.2011.560365.

Kamath, G.G., Lanier, T.C., Foegeding, E.A., Hamann, D.D., 1992. Nondisulfide covalent crosslinking of myosin heavy chain in setting of Alaska pollock and Atlantic croaker surimi. Journal of Food Biochemistry 16, 151–172.

Kantt, C.A., Bouzas, J., Dondero, M., Torres, J.A., 1993. Glucose oxidase/catalase solution for on-board control of shrimp microbial spoilage: model studies. Journal of Food Science 58, 104–109.

Karim, A.A., Bhat, R., 2009. Fish gelatin: properties, challenges, and prospects as an alternative to mammalian gelatins. Food Hydrocolloid 23, 563–576. http://dx.doi.org/10.1016/j.foodhyd.2008.07.002.

Kaur, S., Dhillon, G.S., 2015. Recent trends in biological extraction of chitin from marine shell wastes: a review. Critical Reviews in Biotechnology 35 (1), 44–61.

Kawai, T., 1996. Fish flavor. Critical Reviews in Food Science and Nutrition 36, 257–298.

Kim, S.-K., Dewapriya, P., 2014. Enzymes from fish processing waste materials and their commercial applications. In: Kim, S.-K. (Ed.), Seafood Processing By-products: Trends and Applications. Springer, London, pp. 183–196.

Kim, J., Marshall, M.R., 2000. Polyphenoloxidase. In: Haard, N.F., Simpson, B.K. (Eds.), Seafood Enzymes-Utilization and Influence on Postharvest Seafood Quality. Marcel Dekker, Inc., New York, pp. 271–315.

Kim, S.K., Mendis, E., 2006. Bioactive compounds from marine processing by-products—a review. Food Research International 39, 383–393. http://dx.doi.org/10.1016/j.foodres.2005.10.010.

Kim, S.K., Jeon, Y.J., Byenn, H.G., Kim, Y.T., Lee, C.K., 1997. Enzymatic recovery of cod frame proteins with crude proteinase from tuna pyloric caeca. Fish Science (Tokyo) 63, 421–427.

Kim, Y.J., Kim, H.J., No, J.K., Chung, H.Y., Fernandes, G., 2006. Anti-inflammatory action of dietary fish oil and calorie restriction. Life Sciences 78, 2523–2532. http://dx.doi.org/10.1016/j.lfs.2005.10.034.

Kirk, T.K., Farrell, R.L., 1987. Enzymatic "combustion": the microbial degradation of lignin. Annual Review of Microbiology 41, 465–501.

Kishi, H., Nozawa, H., Seki, N., 1991. Reactivity of muscle transglutaminase on carp myofibrils and myosin B. Nippon Suisan Gakkaishi 57, 1203–1210.

Kitabatake, N., Doi, E., 1993. Improvement of protein gel by physical and enzymatic treatment. Food Reviews International 9, 445–471.

Kumazawa, Y., Nakanishi, K., Yasueda, H., Motoki, M., 1996. Purification and characterization of transglutaminase from walleye pollack liver. Fish Science 62, 959–964.

Kumazawa, Y., Sano, K., Seguro, K., Yasueda, H., Nio, N., Motoki, M., 1997. Purification and characterization of transglutaminase from Japanese oyster (*Crassostrea gigas*). Journal of Agricultural and Food Chemistry 45, 604–610.

Kuo, J.M., Pan, B.S., 1991. Effect of lipoxygenase on formation of cooked shrimp flavour compound-5,8,11-tetradecatrien-2-one. Agricultural and Biological Chemistry 55, 827–848.

Lakshmanan, P.T., Antony, P.D., Gopakumar, K., 1996. Nucleotide degradation and quality changes in mullet (*Liza corsula*) and pearlspot (*Etroplus suratensis*) in ice and at ambient temperatures. Food Control 7 (6), 277–283.

Lanier, T.C., 1994. Functional food protein ingredients from fish. In: Sikorski, Z.E., Pan, B.S., Shahidi, F. (Eds.), Seafood Proteins. Chapman & Hall, London, pp. 127–159.

Lee, Y.Z., Simpson, B.K., Haard, N.F., 1982. Supplementation of squid fermentation with proteolytic enzymes. Journal of Food Biochemistry 6, 127–134.

Lee, H.G., Lanier, T.C., Hamann, D.D., Knopp, J.A., 1997. Transglutaminase effects on low temperature gelation of fish protein sob. Journal of Food Science 62, 20–24.

Lerner, A.B., 1953. Metabolism of phenylalanine and tyrosine. Advances in Enzymology 14, 73–128.

Liao, P.H., Jones, L., Lau, A.K., Walkemeyer, S., Egan, B., Holbek, N., 1997. Composting of fish wastes in a full-scale in-vessel system. Bioresource Technology 59, 163–168. http://dx.doi.org/10.1016/S0960-8524(96)00153-8.

Liaset, B.D., Lied, E., Espe, M., 2000. Enzymatic hydrolysis of by-products from the fish-filleting industry; chemical characterisation and nutritional evaluation. Journal of the Science of Food and Agriculture 80, 581–589. http://dx.doi.org/10.1002/(SICI)1097-0010(200004)80:5<581::AID-JSFA578>3.0.CO;2-I.

Lopez, M.D., Carreno, F.L.G., 2000. Applications of fish and shellfish enzymes in food and feed products. In: Haard, N.F., Simpson, B.K. (Eds.), Seafood Enzymes-Utilization and Influence on Postharvest Seafood Quality. Marcel Dekker, Inc., New York, pp. 571–618.

Manu-Tawiah, W., Haard, N.F., 1987. Recovery of carotenoprotein from the exoskeleton of snow crab, *Chionectes opilio*. Canadian Institute of Food Science and Technology Journal 20, 31–33.

Martone, C.B., Borla, O.P., Sanchez, J.J., 2005. Fishery by-product as a nutrient source for bacteria and archaea growth media. Bioresource Technology 96, 383–387. http://dx.doi.org/10.1016/j.biortech.2004.04.008.

Maruyama, N., Nozawa, H., Kumura, I., Satake, M., Seki, N., 1995. Transglutaminase induced polymerisation of a mixture of different fish myosins. Fish Science 61, 495–500.

Melendo, J.A., Beltran Jose, A., Pedro, R., 1997. Tenderisation of squid (*Loligo vulgaris* and *Illex coindetit*) with bromelain and a bovine spleen lysosomal-enriched extract. Food Research International 30, 335–341.

Melendo, J.A., Beltran, J.A., Roncalcs, P., 1998. Characterization of a crude lysosomal extract from bovine spleen for its use in the processing of muscle foods. Food Biotechnology 12, 239–262.

Motoki, M., Seguro, K., 1994. Trends in Japanese soy protein research. Inform 5, 308–313.

Motoki, M., Seguro, K., 1998. Transglutaminase and its uses for food processing. Trends in Food Science & Technology 9, 204–210.

Muzaddadi, A.U., 2015. Minimization of fermentation period of *Shidal* from Barbs (*Puntius* spp.). Fishery Technology 52, 34–41.

Myhrman, R., Bruner-Lorand, J., 1970. Lobster muscle transpeptidase. Methods in Enzymology 19, 765–770.

Myrnes, B., Johansen, A., 1994. Recovery of lysozyme from scallop waste. Preparative Biochemistry 24, 69–80.

Nambudiri, D.D., 1987. Enzyme Reactions as Freshness of Fish and Shellfish (Ph.D. thesis). Cochin University of Science and Technology, Cochin, India.

Neilsen, P.M., 1995. Reactions and potential industrial applications of transglutaminase. Review of literature and patents. Food Biotechnology 9, 119–156.

Nozawa, H., Mamagoshi, S., Seki, N., 1997. Partial purification and characterization of six transglutaminases from ordinary muscles of various fishes and marine invertebrates. Comparative Biochemistry and Physiology Part B: Biochemistry and Molecular Biology 118B, 313–317.

Otwell, W.S., Marshall, M.R., 1986. Studies on the Use of Sulfites to Control Shrimp Melanosis (Blackspot): Screen Alternatives to Sulfiting Agents to Control Shrimp Melanosis. Florida seagrant technical paper no. 46., pp. 1–10.

Owens, J.D., Mendoza, L.S., 1985. Enzymatically hydrolysed and bacterially fermented fishery products. Journal of Food Technology 20, 273–283.

Pan, B.S., Kuo, J.M., 1994. Flavor of shellfish and kamaboko flavorings. In: Shahidi, F., Botta, J.R. (Eds.), Seafoods: Chemistry, Processing Technology and Quality. Elsevier, New York, USA, pp. 85–110.

Pan, B.S., Kuo, J.M., 2000. Polyphenoloxidase. In: Haard, N.F., Simpson, B.K. (Eds.), Seafood Enzymes-Utilization and Influence on Postharvest Seafood Quality. Marcel Dekker, Inc., New York, pp. 317–336.

Pan, B.S., Tsai, J.R., Chen, L.M., Wu, C.M., 1997. Lipoxygenase and sulfur-containing amino acid in seafood flavor formation. In: Shahidi, F., Cadwallader, K.R. (Eds.), Flavor and Lipid Chemistry of Seafoods, Washington DCAmer. Chem. Soc. Symp. Series, vol. 674, pp. 64–75.

Pandey, P.C., Pandey, V., 1991. Urease purification from the seeds of *Cajanus cajan* and its application in a biosensor construction. Applied Biochemistry and Biotechnology 31, 247–252.

Park, J.K., Morita, K., Fukumoto, I., Yamasaki, Y., Nakagawa, T., Kawamukai, M., Matsuda, H., 1997. Purification and characterization of the chitinase (ChiA) from *Enterobacter* sp. G-1. Bioscience, Biotechnology and Biochemistry 61, 684–689.

Raa, J., Gildberg, A., 1982. Fish silage. A review. Critical Reviews in Food Science and Nutrition 16, 383–419.

Raa, J., Hjelmeland, K., Gilberg, A., 1985. Processing offish and squid by controlled proteolysis. In: Ravindran, K. (Ed.), Harvest and Post-harvest Technology of Fish. Society of Fishery Technologists, India, p. 593.

Raa, J., 1986. Modern biotechnology: impact on aquaculture and the fish processing industry. In: Paper Presented at the 5th World Productivity Congress, Jakarta, Indonesia, 13–16 April 1986.

Raksakulthai, N., Lee, Y.Z., Haard, N.R., 1986. Effect of enzyme supplements on the production of fish sauce from male capelin (*Matlotus villosus*). Canadian Institute of Food Technology Journal 19, 28–33.

Ramesh, C., Lewis, N.F., 1980. Effect of lysozyme and sodium EDTA on shrimp microflora. European Journal of Applied Microbiology and Biotechnology 10, 253–258.

Rebah, F.B., Miled, N., 2013. Fish processing wastes for microbial enzyme production: a review. Biotech 3 (4), 255–265.

Reed, G., 1993. Introduction. In: Nagodawithana, T., Reed, G. (Eds.), Enzymes in Food Processing, third ed. Academic Press, San Diego, pp. 1–5.

Rehbein, H., 1988. Relevance of trimethylamine oxide demethylase activity and haemoglobin content to formaldehyde production and texture deterioration in frozen stored minced fish muscle. Journal of the Science of Food and Agriculture 43, 261–276.

Saisithi, P., 1994. Traditional fermented fish: fish sauce production. In: Martin, A.M. (Ed.), Fisheries Processing: Biotechnological Applications. Chapman & Hall, London, pp. 111–131.

Saito, T., Arai, K., Matsuyoshi, M., 1959. A new method for estimating the freshness of fish. Bulletin of the Japanese Society of Scientific Fisheries 24, 749–750.

Seki, N., Uno, H., Lee, N.H., Kimura, I., Toyoda, K., Fujita, T., Arai, K., 1990. Transglutaminase activity in Alaska pollack muscle and surimi, and its reaction with myosin B. Nippon Suisan Gakkaishi 56, 125–132.

Shen, K.T., Chen, M.H., Chan, H.Y., Jeng, J.H., Wang, Y.J., 2009. Inhibitory effects of chitooligosaccharides on tumor growth and metastasis. Food and Chemical Toxicology 47, 1864–1871.

Shirai, K., Guerrero, I., Huerta, S., Saucedo, G., Castillo, A., Gonzalez, R.O., Hall, G.M., 2001. Effect of initial glucose concentration and inoculation level of lactic acid bacteria in shrimp waste ensilation. Enzyme and Microbial Technology 28, 446–452.

Sikorski, Z.E., Gildberg, A., Ruiter, A., 1995. Tish products. In: Ruiter, A. (Ed.), Fish and Fishery Products – Composition, Nutritive Properties and Stability. CAB International, Wallingford, UK, pp. 315–346.

Simpson, B.K., Haard, N.F., 1984a. Trypsin from Greenland cod as a food-processing aid. Journal of Applied Biochemistry 6, 135–143.

Simpson, B.K., Haard, N.F., 1984b. Purification and characterization of trypsin from Greenland cod *Gadus ogac*). 1. Kinetic and thermodynamic characteristics. Canadian Journal of Biochemistry and Cell Biology 62, 894–900.

Simpson, B.K., Haard, N.F., 1985. The use of proteolytic enzymes to extract carotenoproteins from shrimp wastes. Journal of Applied Biochemistry 7, 212–222.

Simpson, B.K., Haard, N.F., 1987. Cold-adapted enzymes from fish. In: Knorr, D. (Ed.), Food Biotechnology. Marcel Dekker, New York, pp. 495–527.

Simpson, B.K., 2000. Digestive proteinases from marine animals. In: Haard, N.F., Simpson, B.K. (Eds.), Seafood Enzymes-Utilization and Influence on Postharvest Seafood Quality. Marcel Dekker, Inc., New York, pp. 191–213.

Simpson, B.K., Gagne, N., Simpson, M.V., 1994. Bioprocessing of chitin and chitosan. In: Martin, A.M. (Ed.), Fisheries Processing: Biotechnological Applications. Chapman and Hall, London, pp. 155–173.

Stefánsson, G., Steingrímsdóttir, U., 1990. Application of enzymes for fish processing in Iceland—present and future aspects. In: Voigt, M.N., Botta, J.R. (Eds.), Advances in Fisheries Technology and Biotechnology for Increased Profitability. Technomic Publishing, Lancaster, PA, pp. 237–250.

Stefansson, G., 1988. Enzymes in the fishing industry. Food Technology 42 (3), 64–66.

Sugihara, T., Yashima, C., Tamura, H., Kawasaki, M., Shimizu, S., 1973. Process for preparation of ikura (salmon egg). US Patent No. 3759718.

Tavares, J.F.P., Baptista, J.A.B., Marcone, M.F., 1997. Milk–coagulating enzymes of tuna fish waste as a rennet substitute. International Journal of Food Sciences and Nutrition 48, 169–176.

Tokunaga, F., Yamada, M., Miyata, T., Ding, Y.L., Hiranaga, M., Muta, T., Iwanaga, S., 1993. Limulus hemocyte transglutaminase. Its purification and characterization, and identification of the intracellular substrates. Journal of Biological Chemistry 268, 252–261.

Tomé, E., Levy, B.A., Bello, R.A., 1995. Proteolytic activity control in fish silage [Spanish]. Archivos Latinoamericanos de Nutrición 45, 317–321.

Toppe, J., Albrektsen, S., Hope, B., Aksnes, A., 2007. Chemical composition, mineral content and amino acid and lipid profiles in bones from various fish species. Comparative Biochemistry and Physiology Part B 146, 395–401. http://dx.doi.org/10.1016/j.cbpb.2006.11.020.

Tsai, G.I., Wu, Z.Y., Su, W.H., 2000. Antibacterial activity of a chitooligosaccharide mixture prepared by cellulose digestion of shrimp chitosan and its application to milk preservation. Journal of Food Protection 63, 747–752.

Tsukamasa, Y., Sato, K., Shimizu, Y., Imai, C., Sugiyama, M., Minegishi, Y., Kawabata, M., 1993. ε-(γ—Glu-tamyl) lysine crosslink formation in sardine myofibril sol during setting at 25°C. Journal of Food Sci-ence 58, 785–787.

Uematsu, K., Kitano, M., Morita, M., Iijima, N., 1992. Presence and ontogeny of intestinal and pancreatic phospholipase A2-like proteins in the red sea bream, *Pagrus major*. An immunocytochemical study. Fish Physiology and Biochemistry 9, 427–438.

Venugopal, V., 1994. Production of fish protein hydrolysates by microorganisms. In: Martin, A.M. (Ed.), Fisheries Processing, Biotechnological Applications. Chapman and Hall, London, pp. 223–243.

Venugopal, V., Shahidi, F., 1995. Value-added products from underutilized fish species. Critical Reviews in Food Science and Nutrition 35, 431–453.

Venugopal, V., Shahidi, F., 1998. Traditional methods to process underutilized fish species for human con-sumption. Food Reviews International 14, 35–97.

Venugopal, V., Lakshmanan, R., Doke, S.N., Bongirwar, D.R., 2000. Enzymes in fish processing, biosensors and quality control: a review. Food Biotechnology 14 (1–2), 21–77. http://dx.doi.org/10.1080/08905430009549980.

Vilhelmsson, O., 1997. The state of enzyme biotechnology in fish processing industry. Trends in Food Sci-ence & Technology 8, 266–270.

Voskresensky, N.A., 1965. Salting of herring. In: Borgstrom, G. (Ed.), Fish as Food, vol. 3. Academic Press, London, pp. 107–115.

Wang, S.L., Chio, S.H., 1998. Deproteinization of shrimp and crab shell with protease of *Pseudomonas aeru-ginosa* K 187. Enzyme and Microbial Technology 22, 629–633.

Wang, S.L., Liang, T.W., Yen, Y.H., 2011. Bioconversion of chitin-containing wastes for the production of enzymes and bioactive materials. Carbohydrate Polymers 84, 732–742.

Wen, C.M., Tseng, C.S., Cheng, C.Y., Li, Y.K., 2002. Purification, characterization and cloning of a chitinase from *Bacillus* sp. NCTU2. Biotechnology and Applied Biochemistry 35, 213–219.

Whitaker, J.R., 1994. Classification and nomenclature of enzymes. In: Principles of Enzymology for the Food Sciences, second ed. Marcel Dekker, New York, pp. 367–385.

Xu, W., Yu, G., Xue, C., Xue, Y., Ren, Y., 2008. Biochemical changes associated with fast fermentation of squid processing by-products for low salt fish sauce. Food Chemistry 107, 1597–1604. http://dx.doi.org/10.1016/j.foodchem.2007.10.030.

Ya, T., Simpson, B.K., Ramaswamy, H., Yaylayan, V., Smith, J.P., Hudon, C., 1991. Carotenoproteins from lobster waste as a potential feed supplement for cultured salmonids. Food Biotechnology 5, 87–93.

Yamahoto, M., Mackey, J., 1981. An enzymic method for reducing curd formation in canned salmon. Journal of Food Science 46, 656–657.

Yamamoto, M., Saleh, F., Ohtsuka, A., Hayashi, K., 2005. New fermentation technique to process fish waste. Animal Science Journal 76, 245–248. http://dx.doi.org/10.1111/j.1740-0929.2005.00262.x.

Yano, Y., Oikawa, H., Satomi, M., 2008. Reduction of lipids in fish meal prepared from fish waste by a yeast *Yarrowia lipolytica*. International Journal of Food Microbiology 121, 302–307. http://dx.doi.org/10.1016/j.ijfoodmicro.2007.11.012.

Yasueda, H., Kumazawa, H., Motoki, M., 1994. Purification and characterization of a tissue type transgluta-minase from red sea bream (*Pagrus major*). Bioscience, Biotechnology and Biochemistry 58, 2041–2045.

Yasumaga, K., Abe, Y., Yamazawa, M., Arai, K.I., 1996. Heat-induced change in myosin heavy chains in salt-ground meat with a food additive containing transglutaminase. Nippon Suisan Gakkaishi 62, 659–668.

Zhu, Y., Rinzema, A., Trampe, J., Bol, J., 1995. Microbial transglutaminase - a review of its production and application in food processing. Applied Microbiology and Biotechnology 44, 277–282.

FURTHER READING

Bárzana, E., García-Garibay, M., 1994. Production of fish protein concentrates. In: Martin, A.M. (Ed.), Fish-eries Processing: Biotechnological Applications. Chapman & Hall, London, pp. 206–222.

CHAPTER 10

Enzymes for Nutritional Enrichment of Agro-Residues as Livestock Feed

J.O. Ugwuanyi
University of Nigeria, Nsukka, Nigeria

INTRODUCTION

Over the past two decades or more, improvements in global agricultural productivity have led to improvement in quality of nutrition on the average global scale. But global averages do not count for much when nations and regions are taken separately. In some regions and several countries the growth in production has been eclipsed by growth in human population resulting, in some cases, in near negative change in nutrition status of the population. Protein–energy malnutrition remains an important problem in many such countries and regions, made worse by declining purchasing power. The pressure on food production for human consumption means that supply of quality feed for animal production is often of secondary importance. A consequence is that many developing countries have deficits in animal feed supply with far-reaching implications for animal production. In such countries the use of food-grade grains, legumes, pulses, and tubers for animal nutrition is avoided whenever possible. Yet, traditional forages, being nutritionally deficient, cannot sustain intensive animal production.

Use of refuse and waste from food processing and agro-food industries and agricultural value chain including spent brewer's grain, rice bran and husk, corn bran, cassava process waste, fibrous residues, biofuel coproducts, etc., for animal nutrition can be an important consideration for reducing human–animal competition for protein and energy (Swain and Barbuddhe, 2008; Abalaka and Daniyan, 2010; Cooper and Weber, 2012). Unfortunately, the feeding value of such wastes is often too low to secure economic animal production and processes for improving the nutritional value of such wastes need to be considered (El-Boushy and van der Poel, 2001). Disposal pressure and costs mean that whenever wastes and refuse can be used for animal nutrition as feed ingredient there is value added already in the disposal cost.

Agro-food wastes originate from a variety of primary agricultural production steps (straw, culls, leaves, trimmings, and press cakes), intensive animal farming, fishery, and aquaculture, and from processing plants. These wastes and refuse have significant energy and protein value that may be valorized in animal nutrition following only minimal reprocessing depending on purpose. Clear advantages will also accrue from reduction in

Agro-Industrial Wastes as Feedstock for Enzyme Production
ISBN 978-0-12-802392-1
http://dx.doi.org/10.1016/B978-0-12-802392-1.00010-1

233

disposal costs and reduction in human–animal competition for scarce food-grade grains, legumes, and tubers in those countries where this is acute. The greatest value will accrue from the use of these by-products with minimal modifications, such as by simple addition of enzymes to enhance digestion of structural polysaccharides and proteins, and degradation of antinutrients. Unfortunately, the nutritional value and digestibility of various wastes with respect to specific animals are hardly known. This is an area that merits active investigation if the different agro-food and industrial wastes are to have a sustainable place in animal nutrition.

Major constraints to the use of agro-food waste (and this includes waste from biofuel industries) in animal nutrition relate to both nutritional and technical challenges (El-Boushy and van der Poel, 2001), which include:

- Variability in nutrient level, quality, and need for and cost of supplementation.
- Disproportionate content of antinutritional and/or toxic factors and components.
- Presence of pathogenic microorganisms and need for sterilization.
- Bulkiness, wetness, and/or texture and need for reprocessing, such as by drying and pelleting.

Inefficient utilization of feed by various domestic and food animals due to either structural inaccessibility of the substrate or enzymatic incompetence or insufficiency on the part of animals sometimes result in less than 75% utilization of available feed nutrients by animals, particularly monogastric animals (poultry and pigs). This is a major challenge to the global economic production of food animals. The addition of exogenous enzymes to feed has evolved as a way of getting around these challenges to feed utilization. This is particularly important since the highest cost component in animal production relates to the cost of feed. Specifically, use of exogenous enzymes in feed is indicated when it is important to degrade antinutrients in feed. They are also used to increase the availability of otherwise digestible polymers, such as starch when they are enclosed in structural polymers or bound up in nondigestible complexes, and to digest fibrous polymers that are not readily utilized by target animals or assist juvenile and incompetent animals that produce limited amounts of digestive enzyme. The problems that necessitate the use of exogenous enzymes are usually aggravated when by-products and coproducts are used in feed because of disproportionate partitioning of such polymers and antinutrients in these products.

Consequently, the feed enzyme market is dominated by four main classes of enzymes: fiber-degrading enzymes (cellulase complexes, xylanases, pectinases), proteases, amylolytic enzymes, and phytic enzymes. These enzymes that are mostly of microbial origin have been used alone or as complex multienzyme preparations to improve nutritional value and utilization of different feeds. They lead to improvement in feed digestibility by the targeted degradation of structural or nonstarch polysaccharides (NSPs), leading to release of monosaccharides, by exposing starch/energy carbohydrates and proteins for degradation, or by hydrolyzing antinutritional factors that hinder digestion of substrates or absorption of otherwise available nutrients. Other enzymes improve nutrient quality by degrading complexes such as phytate for which animals have limited competence. Appropriate use of exogenous enzymes can have significant positive effects on animal nutrition, and a variety of enzymes are available commercially for use as feed supplements. A critical ingredient of successful supplementation is to match enzyme and substrate in the formulation, for the digestive track environment of target animal and the storage environment of the feed (Adeola and Cowieson, 2004). Use of enzyme supplements can help to achieve nutritional uniformity across feed variety and normalize quality differentials between different feeds (Bedford and Schuze, 1998; Bedford, 2000; Choct, 2006; Cowieson and Adeola, 2005; Awati, 2014). This is an important consideration for enzyme use since the most value of a feed enzyme is derived when it is used to achieve improvement of suboptimal feeds.

FEED SUBSTRATES FOR ANIMAL NUTRITION

A variety of carbohydrate and protein sources are used in formulation of animal feed. These vary tremendously from region to region depending on the food security situation of the society, the available crop types, and status of agro-related industries. The latter determines type and abundance of by-products and coproducts. In much of the developed countries the principal energy sources for animal nutrition are grain carbohydrates, including corn, wheat, barley, oat, rye, sorghum, triticale, etc. In those countries, protein sources may include soybeans and a variety of other grain legumes. The energy and protein efficiency of these crops can be high, and depending on the animals (ruminants or monogastric) and their competency (as a function of age), these feed sources may be used with limited or no need for enzyme supplementation. When certain whole grains and legumes or oil seeds are used in animal nutrition, the content of antinutrients, when they exist, may be compensated by the high energy and protein value of the feed, making the need for enzyme supplementation noncritical. In such cases the nutritional sufficiency of the feed may make the real value (any improvement that may result) of enzyme supplementation appear equivocal. In many cases, by-products of the processing of these grains and legumes for confectionery, food, or industry (including bioenergy) are used in animal nutrition. In such cases there may be disproportionate partitioning of antinutrients, structural and nonstarch polysaccharides in the by-products to make the use of enzyme supplements important (Cooper and Weber, 2012). In such cases the feed quality improvements that result from the use of enzyme supplements can be very obvious and the economic gain of the supplementation significant.

In the developing countries with sometimes acute food security challenges, the energy sources available for animal production that are shared with humans may be a narrower range of crop types, mostly roots and tubers. Intensive ruminant production can be limited in such countries with grass forage being the principal sources of energy (and even protein). Poultry and pigs are more frequently fed using a limited range of grains, such as sorghum and millet, but mostly by-products of agro-food processing, such as spent brewer's grains as well as root and tuber crops (Adesehinwa et al., 2008). Protein sources are also more limited. These economies would appear to have more compelling need for enzyme supplementation of feed but have less access to enzymes and the technology for economic production of enzymes. The use of enzymes in aquaculture is almost entirely limited to the more-developed economies, which are also the countries with developed fisheries (Davis et al., 1998; George and Otubusin, 2007; Davis and Gouveia, 2008).

Energy/Carbohydrate Sources

In economies that employ grains, including maize, wheat, barley oats, rye, etc., as principal sources of carbohydrate for animal feeding, the level of nonstarch polysaccharides in feed relative to the total feed carbohydrate can be low. It is for this reason among others that the value of enzyme addition in feeds seems to be equivocal (Cozannet et al., 2012). However, for the other countries where food security challenges persist and grains are not readily available for use in animal feed, the by-products and wastes/residues from agro-food and beverage industry are significant energy sources in feed (Owen and Jayasuriya, 1989; Adeshinwa et al., 2008). These include distillers' soluble and brewer's spent grains, including spent malted barley and more recently in several developing sub-Saharan African countries, sorghum and spent malted sorghum. Unfortunately, the majority of NSPs and antinutrients partition into these by-products. The available digestible energy in these by-products can be quite low, suggesting important roles for nutritional enzymes supplementation in order to get maximum value from them. NSP content of root and tuber crops is low, and in countries with abundant root and tuber crops they can play important roles in animal feeding without major consideration for enzyme supplements. However, the availability of the roots and tubers for use in animal nutrition may also be low. As such, wastes arising from the processing of these crops for human nutrition can become an important part of the animal feed regime (Adeshinwa et al., 2008). Cassava process wastes that are high in NSPs are mostly used for animal nutrition. In the absence of enzyme supplements, these root and tuber wastes would be more useful in ruminant than in monogastric animal nutrition, yet it is routinely used to feed swine. The presence of high content of NSP in animal feed can pose significant challenges due to the effect on nitrogen loss from intestine as well as increase in digesta viscosity (Knuckles et al., 1989; El-Bousy and van der Poel, 2001; Hussain et al., 2008).

In young animals, such as broilers of less than 3 weeks of age (and even young or newly weaned piglets), starch digestion may be poor even with maize-based energy feeds and much more in feed containing other grains. In such instances, the use of amylase and other carbohydrase supplements in the feed may be indicated as assistance to the limited endogenous enzymes (Bedford and Schuze, 1998; Cowieson et al., 2005; Centeno et al., 2006; Ao et al., 2010a,b; Asmare, 2014).

Protein Substrates

Various raw materials are employed as protein sources in animal feed. In most of the developed and even developing countries soybean (meal) is the major protein source in animal diets (Ghazi et al., 1996a,b, 1997a,b, 2002, 2003; Marsman et al., 1997). In addition to soybean a range of other oil seed meals and legumes are also used (Khan et al., 2006). Depending on availability, these may include palm kernel cake and meals (Albuquerque et al., 2012), groundnut meal and other grain legumes or residue from processing of oil seed. These protein sources can be associated with antinutrient components, such as lectins and trypsin inhibitors that interfere with protein degradation and/or amino acid mobilization and absorption from the intestine (Simbaya et al., 1996; Ghazi et al., 2003). In such instances, it is recommended to employ appropriate proteases to enhance the digestion of proteins and mobilization of amino acids from the feed. In juvenile animals with a rapid growth rate, insufficiency in production of protease may interfere with the ability of the animal to metabolize complex proteins leading to nutritional imbalance even in the presence of otherwise adequate protein in the diet (Ao et al., 2010a,b; Ao, 2011). For such animals, the use of appropriate exogenous proteases may enhance digestion of proteins (Café et al., 2002; Cowieson et al., 2006a,b). The use of proteases to ensure protein mobilization may also be indicated when animals are fed high-molecular-weight protein, such as in a variety of agro-process waste and by-products (meat process wastes).

In many situations, the limited availability of a protein source for a particular animal may be related to association of protein with complex polysaccharides and antinutrients or inhibitors. The use of enzymes other than protease may enhance protein availability by degrading the structural polysaccharides that associate with and protect protein from degradation by available protease (Kornegay, 2001; Drew et al., 2005; Garcia-Ruiz et al., 2006; Farhangi and Carter, 2007; Han and BeMiller, 2008; Ngxumeshe and Gous, 2009; Du Plessis and van Resenburg, 2014).

Phytate/Phosphorus

Phosphorus is a critical mineral in the physiology and metabolism of all animals. It is a major requirement for the mineralization of bones, rapid growth, development of immunity, and for development and maturation of reproductive functions. A majority of the grain-based diets used in human nutrition and feeding of food animals contain enough phosphorus to sustain the growth of animals. Unfortunately, a majority of this phosphorus (60–80%) is present in a storage form bound up in structural phytic acid complex or phytates, which are mostly nondegradable by animals. Phytic acid, myo-inositol 1, 2, 3, 4, 5, 6 hexakis dihydrogen phosphate is an essential component of all seeds, accumulated rapidly by seeds as part of the ripening process (Asada et al., 1969). The location of phytate varies from seed to seed (Pallauf and Rimbach, 1997). In small grains, phytate lies mainly in the bran (aleurone layer, testa, and pericarp), and in the case of maize, it is found mainly in the germ. As a consequence of location, phytate partitions disproportionately in grain process by-products and wastes arising from grain processing. Therefore when these are used in animal nutrition, such feeds contain even greater amount of phytate than those compounded from whole grains. In the less developed economies where these by-products play important roles in animal nutrition, it is to be expected that the problem of antinutrient related to phytate will be even more pronounced. Weight gains and feed conversion efficiency in animals fed these by-products can be severely impaired. In most legumes, phytate is located in the cotyledons while in soybean it is more generally distributed, associated with protein bodies throughout the seed. The content and distribution of phytate in a variety of plant seeds have been documented (Eeckhout and de Paepe, 1994; Ravindran et al., 1994, 1995). Kornegay (2001) and Selle et al. (2010) reviewed the content of phytate and phytase enzyme in a variety of cereals and grain legumes. The contribution of the individual seeds to the phytate content of feed depends on the level of incorporation of the whole grain or its components.

The antinutrient effects of phytates are not limited to the nonavailability of bound phosphorus. Phytate is a very strong chelating agent with unique capacity to bind up divalent and trivalent cations and hence render it insoluble and unavailable for metabolic use. Zinc, copper, calcium, cobalt, manganese, iron, and magnesium, all important minerals, are tightly bound by phytate in insoluble complexes in the digestive tract. Phytate is also able to form enzyme-resistant complexes with proteins and interfere with amino acid metabolism in the acidic environment of the stomach and small intestine by forming complexes with amino groups, such as lysyl, histidyl, and arginine (De Rham and Jost, 1979; Ravindran, 1995; Storebakken et al., 1998; Ravindran et al., 2000, 2001; Akinmusire and Adeola, 2009; Choct et al., 2010). It has been reported that phytate may inhibit the action of proteolytic enzymes, pepsin and trypsin, in the gastrointestinal tract of animals (Knuckles et al., 1989). Under neutral conditions, the carboxyl groups of some amino acids may bind to phytate through a divalent or trivalent mineral. The result of such complex, phytate–protein or phytate–mineral–protein, may reduce the utilization of protein (Kornegay, 2001). Although much less important than protein–phytate complex, reversible starch–phytate complexes are also readily formed and may interfere with the ability of animals to effectively digest starch (Thompson et al., 1987).

Some animals produce phytate-degrading enzyme or phytase (Nys et al., 1996), but this is not always enough to ensure adequate mobilization of phosphorus from phytate complex. In other animals the enzyme is located downstream of the region of the digestive tract where phosphorus is well absorbed (Kornegay, 2001). As such, its value in phosphorus metabolism is limited. Microorganisms in the large intestine also produce some phytase that supplements animal phytase but the level of activity is insufficient to ensure adequate mobilization of phosphorus (Ravindran et al., 1999a,b, 2006). The use of exogenous phytase in the feed can lead to degradation of phytic acid and liberation of phosphorus for absorption by the animals (Robinson et al., 2002; Selle and Ravindran, 2007; Selle et al., 2007). This can lead to significant reduction in the cost of provision of phosphorus in feed and also in the pollution potential of animal waste resulting from the excretion of phosphates in feces (Inborr and Ogle, 1988; Debnath, 2003; Debnath et al., 2005; Baruah et al., 2004; Wu et al., 2004; Nagaraj et al., 2007; Cowieson, 2010).

FORAGE, SILAGE, AND RELATED FEED SUBSTRATES

The mainstay of ruminant production is high-fiber substrates, including a variety of grasses, legumes, and forage. These are fed as either fresh or dried forage or as prede-graded and preserved feed (silage). A variety of grass forage has been used traditionally for production of ruminants. Although these have historically sustained animal production, the need for intensive production has led to various levels of supplementation and replacement with higher-energy grains. Ruminants are equipped with the capacity to degrade forage because of their association with rumen bacteria that degrade fiber in the rumen. Notwithstanding the ability of ruminants to digest forages and grasses, the need for efficient mobilization of energy from high-fiber diet, particularly for intensively reared animals, has led to use of a number of fibrolytic enzyme supplements in forage and silage-based feed (Beauchemin et al., 2003; Adesogan, 2005; Adesogan et al., 2007; Gaafar et al., 2010; Atrian and Shahryar, 2012). The efficiency and results obtained have been extremely varied and even equivocal in instances. In the process of silage making, a number of fibrolytic enzymes have been applied and these include cellulase, xylanase, mannanase, and β-glucanase (McAllister et al., 1999; Meng et al., 2005; Hussain et al., 2008; McCarthy et al., 2013). These enzymes are primarily used to enhance the degradation of fiber to improve the silage-making process. However, it is possible that some of it may survive the silage process and carry through into animal digestive tract and contribute to nutrient mobilization in ruminants. When applied directly to silage, a number of fibrolytic enzymes have been shown to improve nutrient mobilization from silage (McAllister et al., 1999; Adesogan et al., 2007; Atrian and Shahryar, 2012).

Agricultural and Food Industry Wastes, Refuse, and Residues

A wide range of agricultural and food industry wastes and refuse accumulate all over the world and are mostly disposed of in landfills and to some extent in water bodies where they become major pollutants. These agro-food industry refuse and residues include fruit and vegetable process waste as well as spent grains from grain process industries including breweries, distilleries as well as those from sugar cane, roots and tubers, pulses and oil seed, and legumes. Refuse from developing biofuel industries are also beginning to add to the global stock of high-energy agro-food and bioindustry refuse that can find application in animal feeding. To a large extent, most of these residues have low available energy because of their limited digestibility and high content of antinutrients. It is for this reason that they may be more valuable as feed ingredients for ruminants than for monogastric animals. Even for ruminants with capacity to degrade them, they may be used only in limited quantities because of low-energy, low-protein, and limited mineral nutrient content (Owen and Jayasuriya, 1989). In sugar-processing industries, large quantities of high-energy beet and cane refuse including pulp and trimmings are generated and can be used with minimal modification in animal feeding (IFC, 2007). Fiber from date process has high-energy content that can be harnessed with minimal enzymatic supplementation for animal feeding (Shafiei et al., 2010). The other agricultural refuse that remains attractive for use in animal nutrition relates mostly to fruit, root, and tuber process wastes. These refuse contain enough energy resources to be important in the nutrition of animals with little or no modification when used in limited quantities. Most are bulky, however, restricting their use to within the vicinity of their generation, which can be a major drawback to their widespread use in animal nutrition. In many rural agricultural communities, they are used in cottage animal production without any modification. Most are highly perishable and may be dried or ensiled to extend storage life if they are generated in large quantities. The few that have high content of fiber may be modified by use of exogenous enzymes to enhance digestibility either directly as part of the feeding regime or in the process of silage making. Some of the major agro-food wastes with potential for use in animal nutrition are discussed in the following section.

FRUIT WASTES

Apples

Over 20% of global production of this fruit is not marketed because the apples are damaged. When this is added to the pomace and pomace sludge arising from the processing of another 20% or more for apple juice, then there arises a huge stock of potentially usable apple waste. Most of these wastes are disposed of to land, landfills, or other waste treatment options at considerable costs. At the moment the most use for apple process wastes with potential for value addition would appear to be as feedstock for fermentative production of high-volume, low-value products such as organic acid and enzymes, in addition to use as feedstock for extraction of biochemicals (Dhillon et al., 2013). However, huge potential exists for the use of this abundant energy-rich waste for animal feeding. Apple pomace contains about 8% protein on a dry weight basis and very significant metabolizable energy. Pomace may be used as generated or following ensiling with forage for feeding ruminants. It may also be used in low percentages to replace grains for feeding nonruminants including pigs and poultry. The use of this waste stream in animal nutrition currently is restricted by problems associated with its low digestibility, high-sugar content and low-vitamin and -mineral content. Supplementation of pomace with enzyme complex–containing amylase, cellulose complex, and protease is recommended for efficient utilization by nonruminants. Protein enrichment of apple waste may also be achieved by growth of feed-grade yeasts such as *Candida* and *Pichia* spp. prior to feeding to animals, and the resulting enriched waste can be used in high concentrations to replace grains in the feed of both ruminants and monogastric animals (Dhillon et al., 2013).

Plantains and Bananas

Global production of these important fruits is increasing annually, and their roles in food security are increasing also. However, it is estimated that up to 10% of the annual produce of plantains and bananas is lost in the fields, and this number may be higher in remote and difficult-to-access locations. While plantains and banana are valuable foods for human nutrition, the wastes parts can be quite useful in animal nutrition. In addition to the waste produce, peels and leaves of these crops have been used directly or following ensiling for feeding ruminants. In addition, damaged and small-sized bananas as well as harvested pseudostems may also be fed to ruminants and pigs to replace part of the conventional feed and reduce cost of animal production. Banana peels contain up to 40% starch and together with the leaves are generally considered readily digested by ruminants and monogastric animals requiring little or no enzyme supplementation. Banana and plantain wastes have been successfully used in high concentration in the feed of pigs, poultry, and ruminants.

Citrus Wastes

The residue left after the extraction of citrus juice is pulp, which can account for up to 60% of the weight of fresh fruit. This is made up of peel, internal tissue, and seeds. Citrus pulp has a high content of water and soluble sugar, which makes it readily digestible, highly perishable, and bulky unless it is sun dried to extend the storage life. Citrus pulp is very valuable for feeding ruminants but much less so for use in monogastric animals on account of the high-fiber content. In ruminants, it may be used to replace cereal or forage in the diet and is well tolerated by most ruminants. Use of citrus waste in animal nutrition benefits in particular from the fact that the citrus harvest season coincides with the dry season when grass forage is scarce. Citrus waste can be used alone to replace significant amount of feed material and can also be ensiled with dry forage. In this form it has a pleasant aroma and is well received by cattle. Citrus seed meal may also be used as other seed meals for feeding ruminants but not monogastric animals in which it may be toxic.

Pineapple

Pineapple pomace arising from the processing of pineapple to juice may account for up to 50% of the weight of harvest and can be used to feed ruminants. Pineapple wastes are very palatable and readily digestible having high content of soluble sugars, fiber, and pectin. Pineapple pomace may be fed to ruminants fresh, ensiled, or dried and in high concentrations without adverse effects. Pineapple waste has been incorporated in up to

50% level in the feed of ruminants without any adverse effects (Makinde et al., 2011; Wadhwa and Bakshi, 2013). Like citrus pomace, it is more valuable for use in ruminant than in monogastric animals, but it is more restricted in distribution than citrus and so has attracted less attention as animal feed.

Other Fruit and Vegetable Wastes and Residues

The processing of other fruits and vegetables results in large quantities of refuse and residues that may be used for feeding of ruminants either directly or following drying or ensiling (Zepf and Jin, 2013). These include mango peels and seed kernels, sugar beet pulp, carrot pulp, melon peels and pulp, cabbage leaves, cauliflower leaves, tomato pomace, baby corn, grape marc, etc.

Grain and Other Crop Residues

Production of the major crops of the world results in the generation of very large quantities of residues. To a large extent, the term *residues* has routinely been used to refer to by-products or residues arising from cereal, sugarcane, oilseed, pulses, and some root and tuber crop production. These are high-fiber wastes and have mostly low-nutritive value. Utilization of crop residues as feed has been the subject of intense discussions and R&D activities. Initially, this was mostly driven by the developed economies but has since advanced also in the developing economies of the tropics where protein–energy malnutrition is more acute. Several works that have been done on the use of residues in animal nutrition (Owen and Jayasuriya, 1989). However, viable commercial application of residues for animal nutrition, severally touted as the possible solution to human–animal competition for calories, remains to be seen. Most of the residues still end up being left in the field or burnt to dispose or generate power (particularly in the case of sugar cane bagasse). In a very limited way they have also been applied to ruminant feeding in some cases following some treatment, such as ensiling with other crops and wastes. The use of enzyme supplements (mainly fibrolytic activities) to enhance the digestion of residues has been reported and may have prospects as the better and more efficient microbial cellulase complexes are produced.

Biofuel Wastes and By-Products as Animal Feed

Global growth in the production and use of liquid biofuel as bioethanol and biodiesel has also resulted in the massive production of by-products of these processes. These two biofuels are derived from agricultural commodities, such as grain and sugar (cane and beet) for bioethanol and oilseeds for biodiesel. The production processes result in the generation of vast amounts of refuse that are then used as feed (Cooper and Weber, 2012; Kalscheur et al., 2012). Globally, bioethanol is currently made from grains, cane and beet sugar and less significantly from cassava, potato, and other carbohydrate sources. Commonly used grains include maize, wheat, barley, sorghum, rye, triticale, and oats.

Coproducts known as distiller's grain and solubles of varying compositions (depending on the source grain and process) result from the production of bioethanol, and these are used directly or following drying as components of animal feed (Galyean et al., 2012). The production of ethanol from beet sugar results in accumulation of beet pulp that is also used in animal nutrition. In addition to beet pulp, molasses arising from sugar production can also be used in animal feeding. Production of biodiesel from oilseeds, including rapeseed, canola, soybeans, palm oil, castor, Jatropha, and neem result in the accumulation of oilseed meals as for production of edible oil (Albuquerque et al., 2012; Zahari et al., 2012). In the case of edible oilseeds, the by-products are used in animal nutrition as protein sources as in the case of food-grade edible oil. The coproducts of biofuel production have found use in feeding of various animals including ruminants (Kalscheur et al., 2012; Erickson et al., 2012; Galyean et al., 2012), swine (Shurson et al., 2012), and poultry as well as in aquaculture. For the less-conventional or inedible oilseeds as well as algal biomass resulting from production of algal lipids, various studies are underway to use the resulting meals in animal nutrition, and these may in future contribute to global supply of animal feed. However, at the moment commercial production of biofuel is limited to only a few countries and the coproducts are used in the producing countries or may be exported.

ENZYMES USED IN ANIMAL FEED SUPPLEMENTATION

Carbohydrases

Amylases

Animals and poultry receive diets rich in grain starch and NSPs or fibrous structural polysaccharides. In some countries, particularly in the tropics, starch and carbohydrate from roots and tubers are important sources of energy for animal feed. Depending on the nature of the grain, digesta viscosity arising from high content of NSPs and structural polysaccharides and the presence of antinutrient complexes in feed may lead to reduction in feed conversion efficiency. Broilers that receive these forms of feed have benefited from supplementation of the feed with exogenous amylases (and other carbohydrases) (Gracia et al., 2003; Choct, 2006; Abudabos, 2010; Ao, 2011). The precise bases for the benefits of using exogenous amylase supplement are quite contentious because starch as the principal substrate for this enzyme is routinely digestible by most food animals and poultry. It is believed that the usefulness of exogenous amylase supplement in animal feed may accrue from enzymatic reduction of intestinal (digesta) viscosity leading to improvement in feed digestibility and intake as well. Amylases may also play roles in ameliorating the impact of antinutritive component of grains (Lazaro et al., 2004). Young monogastric animals including poultry and pigs secrete limited amount of amylase and other enzymes. Therefore, addition of exogenous amylase to feed may complement the limited amylase secreted by young animals (Inborr, 1990; Gracia et al., 2003; Selle et al., 2007).

In many cases, exogenous carbohydrases have helped to improve utilization of feed protein and amino acids presumably through degradation of carbohydrate–protein complexes that expose otherwise inaccessible proteins for protease action (Vahjen et al., 2007). Reduction in digesta viscosity is important in feed based on grains with high content of NSP, such as wheat and barley (Wu et al., 2004; Du Plessis and van Resenburg, 2014) and much less so in corn- or sorghum-based feeds (except in so far as structural or NSP in these grains may encapsulate other nutrients) (Mahagna et al., 1995; Gracia et al., 2003; Cowieson, 2010) or those in which the carbohydrate components are derived from tuber or root crops. The benefit of using exogenous amylase in broilers is most clearly demonstrated in the young birds where the capacity to produce the appropriate quantity of amylase and other digestive enzymes is limited (Uni et al., 1995; Café et al., 2002; Jiang et al., 2008).

Successful application of amylase in the improvement of feed efficiency in ruminants including beef steer and dairy cattle have been reported, and these are associated with feed containing high concentrations of grain (Hristov et al., 2008; Klingerman et al., 2009; Gencoglu et al., 2010; DiLorenzo et al., 2011; Weiss et al., 2011; McCarthy et al., 2013; Noziere et al., 2014). Success has also been reported in the amylase supplementation of feed containing high concentration of grain by-product (McCarthy et al., 2013). Exogenous amylase used in ruminant diet are of value, however, only if the amylase is resistant to ruminal degradation, as ruminal degradation of exogenous enzymes can be a problem with use of such feed additives. Few studies exist on the amylase supplementation of ruminant feed, and the exact mechanisms by which it achieves feed improvement remain unexplored. This is understandable as the majority of the works on enzyme supplementation of ruminant diets have centered on the use of fibrolytic enzymes.

Information on the use of exogenous amylase as single preparation in feed is scarce and this enzyme has often been used as part of a complex that may include some or all of xylanase, protease, cellulase, and phytase. This has tended to make interpretation of the exact role and value of exogenous amylase contentious. However, Ritz et al. (1995) and Gracia et al. (2003) reported that the use of exogenous amylase alone in poultry feed resulted in significant improvement in weight gain, growth rate, feed conversion and also a reduction in mortality compared to control that received no enzyme. Unfortunately, considerable discrepancies in response abound in the literature (Centeno et al., 2006; Singh et al., 2012) and these have been attributed to enzyme source, grain type, feed form, enzyme activity, and dose level as well as susceptibility of exogenous amylase to degradation by exogenous or endogenous protease (Yuan et al., 2008) and the quality of the basal feed (Ngxumeshe and Gous, 2009; Asmare, 2014). Quality of the basal feed has been a major consideration in the evaluation of the contribution of exogenous enzymes in general, and there seems to be a consensus that exogenous enzymes are of most value in improving deficient feeds. In nutritionally adequate and balanced feeds, the value of external enzyme addition becomes routinely equivocal.

Cellulase and Xylanase Complexes (Fibrolytic Enzymes)

Enzyme supplementation of feed has variously been acclaimed to improve feed digest-ibility in both ruminants and monogastric animals. This is most pronounced in relation to digestion of NSPs, fibers, and other structural polysaccharides. In monogastric ani-mals, the improvements are related mostly to the release of inaccessible nutrients rather than the availability or release of energy compounds directly by the action of the enzyme. In ruminants, the use of fibrolytic enzymes of carbohydrase complexes includ-ing cellulase and xylanase activities may complement the action of ruminant microbes to facilitate the digestion of fibers and structural polysaccharides. These have been reported to improve ruminal digestibility, weight gain, and milk production in rumi-nants (Yang et al., 1999, 2000). However, the major use of cellulase- and xylanase-rich supplements in animal feeds relates to the need to hydrolyze complex carbohydrates in nonruminants that are not physiologically equipped to degrade such polysaccharides. The compounds are present as part of the structural integrity of the substrate, as cell wall component, and often shield otherwise digestible polymers, such as starch, pro-teins, or even lipids from their corresponding enzymes (Adeola and Cowieson, 2004; Vahjen et al., 2007). Cellulose is a major component of all fiber-rich feed materials including forage and important agricultural process refuse, such as cane sugar bagasse. These are metabolized by rumen microorganisms and can supply the energy needs of ruminants but are not metabolized by nonruminants. The inclusion of cellulase and xylanase complex in the diet of both ruminants and monogastric animals has been reported to aid the mobilization of energy resources from these complex polysaccha-rides. In monogastric animals, this role may be related to exposing nutrients that are enveloped and shielded by the polymers besides the clearly established roles in reducing digesta viscosity when used with highly viscous feed and ingredients (Adeola and Bed-ford, 2004; Kerr and Shurson, 2013). On the other hand, such enzymes may contribute directly to the degradation of polymers in the rumen of ruminants (Kiarie et al., 2007). In these animals, digesta viscosity may be less pronounced and important than in nonruminants.

Several studies have been published in which improved protein metabolism was linked with use of carbohydrase and fibrolytic supplement in feed (Tahir et al., 2008; Yin et al., 2010). Some of the studies established that degradation of pectic substances in cell wall of feed components improved protein digestion, apparently by exposing protein to the action of proteases.

Proteases

Globally, the most predominant source of protein (over 60%) for animal feed use is soy-bean. The extent varies from region to region. In the United States and rest of North America, the contribution of soybean to animal protein need may be over 90% but is much less in several other countries. The availability and high quality of soybean meal has made it particularly acceptable in animal nutrition. In several other countries where soybean is less available or important, different legumes and oil seeds are sources of protein in animal nutrition. In oil palm–producing regions, for instance, a lot of animal protein comes from palm kernel cake, and the same may be the case in regions that produce other legumes and oil seeds, such as groundnut. Other prominent sources of protein in animal nutrition include sunflower meal, canola meal, groundnut meal, cottonseed meal, sesame meal, and palm kernel meal. These protein sources are not uniformly digested by food animals. In some instances, protease supplementation of feed containing these protein sources have helped to improve digestion and protein metabolism. The presence of protease inhibitor and antinutrients in these proteins has also necessitated the use of protease supplements in feeds containing them.

Although the use of protease supplement in animal nutrition is less established than that of other enzymes, including carbohydrases and phytases, it is by no means new. Many studies have explored and reported on the beneficial use of stand-alone proteases in animal nutrition (Sun, 2007) and several commercial heat-stable protease preparations are currently available to farmers. However, in most cases proteases have been used as part of enzyme complexes that often include xylanases, cellulase complex and other activities (Cowieson and Adeola, 2005), which has made the interpretation of value contribution of protease supplementation complicated in some cases (Swift et al., 1996; Schang et al., 1997; Marsman et al., 1997; Zanella et al., 1999; Ghazi et al., 2003; Ao et al., 2010a,b). In these complexes the role of protease in enhancement of the nutrient value of the feed has been mostly contentious (Asmare, 2014). The reasons for the variety and inconsistencies in the response to protease addition are varied. These may include differences in the pH optima (acid vs alkaline protease) for the activity of the supplemental protease and the nature of the feed, as well as the repressive effect of exogenous protease on intestinal protease secretion (Sun, 2007).

In most studies on the protease supplementation of feed, the principal protein component was an oil seed whose protein is mostly digestible (Raza et al., 2009). Improvements in protein metabolism following the addition of exogenous protease have been explained in instances to be due to degradation of antinutrients, such as lectins and trypsin inhibitors (Ghazi et al., 2002). To some extent improvements also result from the digestion of protein (Khan et al., 2006; Raza et al., 2009; Oxenboll et al., 2011; Romero et al., 2013), which has benefit in improved nitrogen utilization, better production environment, and poultry health (Nagaraj et al., 2007). Exogenous protease may also improve protein (and starch) digestion by breaking starch–protein cross-links (Han and BeMiller, 2008).

Enzymes are produced by food animals to sustain growth on most regular diets; however, specific activities necessary to break down some compounds in feed are not found or are present in low concentration in the digestive tract. In the absence of exogenous enzyme supplementation, such animals may suffer growth impediment. This can be particularly important in monogastric animals, such as poultry and pigs. Yet the use of exogenous protease in monogastric animal nutrition appears to be the least studied of enzyme supplements. Protease use in animal nutrition is mostly limited to the use of enzyme complexes that include this activity to achieve degradation of antinutritional components, including carbohydrate–protein complexes in grain and legume meals (Swift et al., 1996; Schang et al., 1997; Marsman et al., 1997; Zanella et al., 1999; Ghazi et al., 2003; Ao et al., 2010a,b). Few studies have used stand-alone proteases of bacterial, fungal, or plant origin in feed supplementation and reported nutritional enhancement related to improvement in feed digestibility (Ghazi et al., 1996a,b; Singh et al., 2011) and also to the degradation of antinutrient components of soybean meal (Rooke et al., 1998; Ghazi et al., 2002). Considering the digestibility of most protein sources used in feed, it seems that the principal purpose of adding proteases in feed containing protein meals for monogastric animals is to destroy or inactivate the antinutritional factors, such as residual trypsin inhibitors, lectins, and antigenic proteins (Mahagna et al., 1995). The use of protease supplements to aid the digestion of structural proteins present in animal wastes that may find value in animal nutrition has not been well studied, thus this presents a growth front in the use of waste materials for animal nutrition.

Phytases

Phytases occur widely in nature in tissues of animals and plants and in microorganisms, including bacteria, yeasts, and fungi. Phytases are able to break down phytate in cereals and legumes to liberate inorganic phosphate for animal nutrition. There are two main classes of phytase: 3-phytase (EC 3.1.3.8), which are microbial phytases that hydrolyze the phosphate group at the C3 position; and 6-phytase (EC 3.1.3.26) of plant origin, which acts first at the C6 position. In the context of animal nutrition, phytases may originate from the intestine of food animal, from endogenous plant phytase present in some feed ingredients, from intestinal bacteria, or be added as exogenous enzyme in the feed (Nys et al., 1996). Exogenous phytases are the most important in animal nutrition essentially because monogastric animals, pigs, and poultry that have the most need of phytase produce negligible enzyme activity whose value in dietary phosphorus digestibility remains speculative. Phytases are also produced by intestinal microorganisms, but the value of such enzymes in phytate degradation and availability of dietary phosphorus in the intestine is questionable given the limited absorption of phosphorous from the large intestine where these microorganisms reside and produce the enzyme. Most cereals and legumes contain negligible amounts of phytase and although they have been demonstrated to be capable digesting phytate, the disparity between their pH optima of 5 and above and the pH of stomach may render them incapable of contributing significantly to phytate digestion and phosphate metabolism in animals particularly in pigs. It has been demonstrated that in vivo, and on a unit/unit basis, microbial phytases are more potent in the degradation of phytate in animal intestine than cereal phytase (Eeckhout and de Paepe, 1994). Digestion of phytate that is of value in phosphate metabolism takes place in the stomach with a very low pH. Therefore, any enzyme without activity in the stomach is unlikely to contribute significantly to the availability of phytate phosphorus in monogastric animals. In addition, much of the plant enzymes are likely to get inactivated in the process of feed processing and therefore not likely to be carried through to the animal intestine.

The phytases that have found commercial application in the feed industry are of microbial origin (Jackson et al., 1996; Baruah et al., 2004), produced using *Aspergillus* sp. Traditional commercial phytases are produced using *Aspergillus niger* or *Aspergillus oryzae*, which employed either homologous gene or by the expression of genes derived from *Aspergillus ficuum*. The commercial success of *Aspergillus*-derived phytase has been driven by the thermal stability of the enzyme, which enables it to survive feed pelleting. In addition the enzyme is able to function at the pH of the animal digestive tract, especially in the stomach of pigs and crop and proventriculus of broilers (before the small intestine where proteolytic inactivation may become important).

Combined Enzymes and Enzyme Complexes

Optimization of the value of exogenous enzymes in animal nutrition has been of primary interest in studies on the use of enzymes in animal feed. It is for this reason that different workers have emphasized the use of enzyme cocktails that have emerged as the preferred forms of exogenous enzymes in animal nutrition (Janssens et al., 2000). Most commercial feed enzymes include activities of protease, phytase, and complex carbohydrases (Miller et al., 2008a,b). The use of enzyme cocktails including some or all of these activities has been reported to help improve weight gain, feed efficiency, apparent ileal digestibility of starch and protein, and apparent total tract digestibility of NSPs in birds fed a wheat-, wheat screening-, soybean meal- (SBM), canola meal-, and peas-based diet (Meng et al., 2005). Similarly, improvement in the digestibility of calcium, phosphorous, and amino acids was reported in birds fed enzyme cocktails including activities of xylanase, amylase, protease, and phytase (Simbaya et al., 1996; Zanella et al., 1999; Cowieson et al., 2006a,b). The use of enzyme cocktails is not restricted to poultry as they have also been used successfully in other monogastric animals as well as in ruminants. In any event, the value of an enzyme cocktail depends on there being activity that targets and improves the availability (or digestibility) of the most efficiency-limiting (or low-availability) feed ingredient. For instance, in the absence of exogenous phytase in low-phosphorus diet, an enzyme cocktail that includes carbohydrase and protease would not improve phosphorous availability (Troche et al., 2007).

USE OF EXOGENOUS ENZYMES IN ANIMAL NUTRITION

Enzymes in Swine Nutrition

The global production of pork as major protein source is increasing. This increase is even higher in the developing economies that have borne the brunt of protein malnutrition.

Unfortunately, the growth is constrained by the availability and cost of feed in such economies. In much of the third world, maturation time for swine is longer and weight gain slower than that in the more developed economies due to limited availability of quality feed. The cost of feed and input for pig production has caused slower development of swine industry and recourse to a variety of agro-food wastes and coproducts from biofuel and bioindustries as additives in swine feed to reduce the cost of feed. The by-products and coproducts contain disproportionate amounts of antinutrients and toxins relative to the source grains or legumes, necessitating reprocessing or use of additives to help improve nutrient availability.

In addition to various forms of processing, the use of exogenous enzymes of microbial origin has gained acceptability as a means of modifying various swine feeds. Today varieties of commercial enzyme preparations are available and are used to improve the production of pigs using feeds based on different agricultural products and especially by-products. These feed enzymes are composed of carbohydrases, proteases, and phytases. Phytases are recognized to improve feed efficiency by improving availability of phosphorous, while various carbohydrases help to improve energy intake. The results from the use of proteases in enzyme supplements are much less consistent.

Some of the more prominent considerations in the use of enzyme additives in feed include: (1) Nature of the substrate and the feed. For an enzyme to be useful in a feed, it is recognized that its substrate or the availability of its substrate must be limiting in the feed. It is also important that the enzyme be able to act on the substrate in vivo in the gut segment of the animal where the enzyme product will be absorbed or in vitro prior to consumption; and (2) age of the animal. It is important to consider the age of the animal being fed with enzyme-supplemented feed as the state of development of animal gut is important in the ability of the animal to derive the benefits of enzyme application (Adeshinwa et al., 2008).

Enzymes in Ruminant Nutrition

Ruminants have a digestive system that is more complicated than that of monogastric animals. This has in part accounted for the slower development or success of enzyme supplementation in the diet of these animals. It has also tended to complicate the interpretation of data arising from the supplementation of ruminant diets with exogenous enzymes. Most research and the commercial application of enzymes in the nutrition of ruminants have focused mainly on the value of fibrolytic enzymes even though amylases and proteases have also been reported to improve feed efficiency in ruminants. The use of exogenous enzyme in ruminants has grown because of the limited digestion of feed organic matters in rumen and also from the various attempts to use low-value, high-volume fibrous plant materials for ruminant feeding and the belief that fibrolytic enzymes may assist in the breakdown of such materials (Annison, 1997; Yang et al., 2000).

Enzymes may be used as silage additives to improve silage quality or incorporated into feed to achieve partial digestion of feed material including various forages prior to or following ingestion by the animal. The enzymes that have found use in ruminant nutrition include cellulase complex, xylanases, amylases, and pectinases (Eun and Beauchemin, 2007; Miller et al., 2008a,b). These have mostly been used as multienzyme complexes in animal nutrition. When cereal grains are used as the major source of carbohydrates in ruminant nutrition, amylases have been used to improve digestion, while xylanases and enzymes of the cellulose complex have also been used successfully to reduce digesta viscosity and improve feed conversion (particularly in wheat- and barley-based grain feeds).

Notwithstanding the complexity of the ruminant digestive system, there has been significant growth in the market for ruminant feed enzymes due to reported improvement across a range of variables including milk production (Yang et al., 2000; Beauchemin et al., 2003; Adesogan, 2005; Atrian and Shahryar, 2012). Other researchers however, have been unable to demonstrate clear improvement in animal performance including milk production as a result of exogenous enzyme use in feed (ZoBell et al., 2000; Miller et al., 2008a,b). Although results of the use of ruminant enzymes have been variable, it has been postulated that greatest improvement in feed efficiency occurs when enzymes are used in situations of compromised fiber digestion and when energy supply is the most potent limiting factor in animal production (Beauchemin et al., 2003). The degree of variability reported in ruminant response to exogenous enzyme addition is a major impediment to the growth of the ruminant feed enzyme market, and this is likely to remain the case until a clearer understanding of the bases for and control of the variations is achieved.

Enzymes in Poultry Nutrition

The use of exogenous enzymes in poultry nutrition is by far the most developed and established commercial application of enzymes in animal feed supplementation. It has evolved over the past 30 years or more. Today, the market for poultry enzymes is large, spanning nearly all the countries of the world. And as nations reduce the use of prophylactic antibiotics in poultry and farm animal nutrition, the role of enzyme supplements in poultry is likely to increase even further. Use of enzymes in poultry feed like other nutritional enzymes is driven by the fact of feed containing ingredients that the poultry are unable to digest or that impede digestive enzymes (Khattak et al., 2006; Café et al., 2002; Abudabos, 2010; Awati, 2014).

Large portions of commercial poultry feed contain barley, oats, peas, or wheat (feed ingredients capable of causing increase in digesta viscosity) and spent grains as energy sources as well as a vegetable protein source, commonly soybean meal. Today a range of other agricultural wastes, by-products, and coproducts are used in feed formulation for poultry particularly as the price of grain rises in response to feed–food competition. These can all contribute to the need for more enzyme supplementation as these unconventional feed ingredients are likely to have disproportionate amounts of components that are either not readily digested by the birds or are capable of interfering with the digestive process (El-Boushy and van der Poel, 2001; Choct et al., 2010; Coope and Weber, 2012).

A range of enzymes have found commercial application in poultry intended to aid reduction in digesta viscosity and also to reduce the effect of antinutrients in the feed. Use of enzyme additives in poultry can be particularly important in young chicks with limited capacity to produce endogenous enzymes. Enzymes that have been commercially applied in poultry nutrition include enzymes of the cellulase complex and hemicellulases as well as amylases and phytases (Ghazi et al., 1997a,b; Cowieson and Adeola, 2005; Cowieson et al., 2006a,b; Cowieson and Ravindra, 2008; Centeno et al., 2006; Cowieson, 2010; Du Plessis and van Resenburg, 2014). To lesser extents, lipases and proteases have also been employed in poultry but mostly as part of constituents of feed enzyme complex.

Enzymes in Fish Nutrition

Protein remains the principal component of the feed of cultured fish and the major cost component in the production of aquaculture. Both the quality and quantity of protein in the diet of fish are critical to the success of aquaculture. By far the major source of protein for commercial aquaculture is fish meal, a limited resource that accounts for up to 50% of the total protein needs of farmed fish. As a consequence, both the price of fish feed and the growth of farmed fish industry depend largely on the availability and price of fish meal. This is not sustainable for a human population and fish market that keep growing. In the tropical developing countries, such as many countries of sub-Saharan Africa that depend almost completely on fish meal import, the development of aquaculture is tightly bound to the availability and price of meal. The consequence is the slower-than-expected growth of aquaculture in regions that have to depend largely on fish for protein supply (Davies and Gouveia, 2008). The need to replace fish meal partially or completely with cheaper and more readily available protein sources in commercial fish feed has become compelling. Fish nutritionists have been working to improve the nutritional and protein quality of a number of alternative protein sources, particularly plant protein, for fish farming. Unfortunately, the imbalance in the amino acid and lipid profile of plant feed ingredients makes their ready use for replacement of fish meal difficult. In addition to this imbalance is the content of antinutrient components that these plants produce. The presence of phytate in the grains and oil seed meals, for instance, is a strong disincentive for its use in aquaculture.

A focus of studies on alternative protein sources for fish farming has been the use of a variety of exogenous enzyme supplements to improve digestibility and utilization of feed ingredients. This effort is a direct outgrowth of the successful use of exogenous enzymes in farm animal nutrition. It has been established that the use of exogenous enzymes in formulation of fish feed can produce the same effects as in other animals, resulting in improved performance of fish (Farhangi and Carter, 2007; Soltan, 2009). As a consequence, the use of a variety of exogenous enzymes in the formulation of fish feeds is growing significantly. Exogenous enzyme complexes containing lipase, protease, phytase, and amylase have been demonstrated to improve feed conversion efficiency and growth of a number of cultured fish, including tilapia, salmon, *Clarias* spp, and catfish, among others (Jackson et al., 1996; Refstie et al., 1999; Giri et al., 2003; Drew et al., 2005; Debnath et al., 2005; Lin et al., 2007; Yildrin and Turan, 2010). Use of microbial phytase has been demonstrated to improve the utilization of phosphorus in cultured fish (Robinson et al., 2002; Debnath, 2003). Exogenous enzymes supplement fish endogenous enzyme, increasing digestibility and utilization of lipid, protein, starch, and phytate phosphorus in the diet. Nutritional phytase supplement can effectively replace dietary phosphate supplement in the feed of cultured fish. Besides improvement in phosphorus availability, microbial phytase supplements have been demonstrated to improve protein conversion in fish (Storebakken et al., 1998; Debnath, 2003) through the degradation of phytate–protein complexes that reduced the availability of protein. In addition to the effect on protein nutrition, phytase use in fish feed improves mineral availability and can reduce or completely eliminate the need for mineral supplementation in fish feed. Plant protease, papain, has been reported as an effective growth promoter by helping to improve the utilization of plant protein sources (ground nut and soybean meal) and contribute to phosphorus metabolism by the degradation of phytate in the feed (Singh et al., 2011). Although claims have been made that use of exogenous enzymes may help to improve secretion of endogenous enzymes, the target of use is actually to aid the direct degradation and utilization of feed components (Hardy, 2000).

The use of plant-based feed components as replacement for limited and expensive fish meal is critical for the development of the cultured-fish industry. Such feed must be formulated in a manner to achieve growth and production that is comparable to the use of fish meal. Plant components have limitations related to digestibility that limit their use as effective replacement for fish meal. At the moment considerable work still needs to be done to be able to achieve feed conversion efficiency that will enable the significant or wholesale replacement of fish meal in fish feeds. The use of various exogenous enzymes including phytase, protease, and carbohydrase in the formulation of fish feeds will play important roles in the development of the fish feeds of the future in which fish meal will not play a dominant role.

CONCLUSIONS

Enzymes are used in formulation of animal feed with the aim of increasing feed efficiency to be able to achieve increased animal weight (gain) and improved products including milk and egg for unit feed and given time. The first application of enzymes in animal feed formulation was the use of β-glucanase in barley-based feed to reduce feed viscosity in chicken gut. Subsequently, a variety of enzymes have found use in the formulation of feed for a number of food animals, including cattle, goats, fish, pigs, and poultry, and the market for feed enzyme is growing rapidly. Of particular significance in the use of enzymes in feed formulation is the potential to enhance the value of a variety of agro-food wastes and by-products in animal nutrition. This could enhance global food security by reducing animal–human competition for food, which can be particularly challenging in the less-developed economies with limited commercial agriculture and food insecurity. As interest in the use of feed enzymes grows, it is hoped that procedures will evolve for rapid screening of ingredient samples, which will allow the formulation of feed to match the bottleneck substrate with the enzyme and also allow the enzyme dose to be adjusted depending on the content of target substrate in the ingredient or complete diet.

ACKNOWLEDGMENTS

The author wishes to acknowledge support from EU-ACP EDULINK-Food Security and Biotechnology in Africa (FSBA) project (Grant: FED/2013/320-152). Opinions expressed in this article are those of the author.

REFERENCES

Abalaka, M.E., Daniyan, S.Y., 2010. Assessment of the performance of chicks fed with cereal wastes enriched with single cell protein – *Candida tropicalis*. Australian Journal of Technology 13, 261–264.

Abudabos, A., 2010. Enzyme supplementation of corn-soybean meal diets improves performance in broiler chicken. International Poultry Science 9, 292–297.

Adeola, O., Bedford, M.R., 2004. Exogenous dietary xylanase ameliorates viscosity-induced anti-nutritional effects in wheat-based diets for White Pekin ducks (*Anas platyrinchos domesticus*). British Journal of Nutrition 92, 87–94.

Adeola, O., Cowieson, A.J., 2004. Opportunities and challenges in using exogenous enzyme to improve non ruminant animal production. Journal of Animal Science 89, 3189–3218.

Adesehinwa, A.O.K., Dairo, F.A.S., Olagbegi, B.S., 2008. Response of growing pigs to cassava peel based diets supplemented with Avizyme® 1300: growth, serum and haematological indices. Bulgarian Journal of Agricultural Science 14, 491–499.

Adesogan, A.T., Kim, S.-C., Arriola, K.G., Dean, D.B., Staples, C.R., 2007. Strategic addition of dietary fibrolytic enzymes for improved performance of lactating dairy cows. In: Proceedings of the 18th Florida Ruminant Nutrition Symposium. University of Florida, Gainesville, Florida, pp. 92–110.

Adesogan, A.T., 2005. Improving forage quality and animal performance with fibrolytic enzymes. Florida Ruminant Nutrition Symposium 91–109.

Akinmusire, A.S., Adeola, O., 2009. True digestibility of phosphorus in canola and soybean meals for growing pigs: influence of microbial phytase. Journal of Animal Science 87, 977–983.

Albuquerque, N.I., Guimaraes, D.A.A., Dias, H.L.T., Teixeira, P.C., Moreira, J.A., 2012. Utilization of palm kernel cakes (*Elaeis guineensis* and *Orbignya phalerata*) co-products of the biofuel industry, in collared peccary (*Pecari tajacu*) feed. In: Makkar, H.P.S. (Ed.), Biofuel Co-products as Livestock Feed; Opportunities and Challenges. FAO, Rome, pp. 263–274.

Annison, G., 1997. The use of exogenous enzymes in ruminant diets. In: Recent Advances in Animal Nutrition in Australia. University of New England, Armidale, NSW2351, Australia, pp. 8–16.

Ao, T., Cantor, A.H., Pescatore, A.J., Pierce, J.L., Dawson, K.A., 2010a. Effects of citric acid, alpha- galactosidase and protease on *in vitro* nutrient release from soybean meal and trypsin inhibitor content in raw whole soybeans. Animal Feed Science and Technology. http://dx.doi.org/10.1016/j.anifeedsci.2010.08014.

Ao, T., Pierce, J.L., Hoskins, B., Paul, M., Pescatore, A.J., Cantor, A.H., Ford, M.J., King, W.D., 2010b. Allzyme SSF® increased AMEn of the corn-soy diet and improved performance of boilers. Poultry Science 89 (Suppl. 1), 863. Abstract #1111.

Ao, T., 2011. Using exogenous enzymes to increase the nutritional value of soybean meal. In: El-Shemy, H. (Ed.), Poultry Diet, Soybean and Nutrition Available from: http://www.intechopen.com/books/soybean-and-nutrition/using-exogenous-enzymes-to-increase-thenutritional-value-of-soybean-meal-in-poultry-diet.

Asada, K., Tanaka, K., Kasai, Z., 1969. Formation of phytic acid in cereal grains. Annals of the New York Academy of Science 165, 801–814.

Asmare, B., 2014. Effect of common feed enzymes on nutrient utilization of monogastric animals. International Journal of Biotechnology and Molecular Biology Research 5, 27–34.

Atrian, P., Shahryar, H.A., 2012. Effect of fibrolytic enzyme treated alfalfa on performance in olstein beef cattle. European Journal of Experimental Biology 2, 270–273.

Awati, A., 2014. Combined enzyme and probiotic solution unlocks full feed potential. Poultry international 2014, 32–35.

Baruah, K., Sahu, N.P., Pal, A.K., Debnath, D., 2004. Dietary phytase: an ideal approach for a cost effective and low polluting aqua-feed. NAGA, World Fish Center Quarterly 27, 15–19.

Beauchemin, K.A., Colombatto, D., Morgavi, D.P., Yang, W.Z., 2003. Use of exogenous fibrolytic enzymes to improve feed utilization by ruminants. Journal of Animal Science 81 (E. Suppl. 2), E37–E47.

Bedford, M.R., Schuze, H., 1998. Exogenous enzymes for pigs and poultry. Nutrition Research Reviews 11, 91–114.

Bedford, M.R., 2000. Exogenous enzymes in monogastric nutrition-their current value and future benefits. Animal Feed Science and. Technology 86, 1–13.

Café, M.B., Borges, C.A., Fritts, C.A., Waldroup, P.W., 2002. Avizyme improves performance of broilers fed corn-soybean meal-based diets. Journal of Applied Poultry Research 11, 29–33.

Centeno, M.S.J., Ponte, P.I.P., Ribeiro, T., Prates, J.A.M., Ferreira, L.M.A., Soares, M.C., Gilbert, H.J., Fontes, C.M.G.A., 2006. Galactanases and mannanases improve the nutritive value of maize and soybean meal based diets for broiler chicks. Journal of Poultry Science 43, 344–350.

Choct, M., Dersjant-Li, Y., McLeish, J., Peisker, M., 2010. Soy oligosaccharides and soluble non starch polysaccharides: a review of digestion, nutritive and anti-nutritive effects in pigs and poultry. Asian–Australian Journal of Animal Science 23, 1386–1398.

Choct, M., 2006. Enzymes for the feed industry: past, present and future. Worlds Poultry Science Journal 62, 5–15.

Cooper, G., Weber, J.A., 2012. An outlook on world biofuel production and its implication for the animal feed industry. In: Makkar, H.P.S. (Ed.), Biofuel Co-products as Livestock Feed. FAO, pp. 1–12.

Cowieson, A.J., Adeola, O., 2005. Carbohydrase, protease and phytase have an additive beneficial effect in nutritionally marginal diets for broiler chicks. Poultry Science 84, 1860–1867.

Cowieson, A.J., Ravindra, V., 2008. Sensitivity of broiler starters to three doses of an enzyme cocktail in maize-based diets. British Poultry Journal 49, 340–346.

Cowieson, A.J., Hruby, M., Faurschou Isaksen, M., 2005. The effect of conditioning temperature and exogenous xylanase addition on the viscosity of wheat-based diets and the performance of broiler chickens. British Poultry Science 46, 717–724.

Cowieson, A.J., Singh, D.N., Adeola, O., 2006a. Prediction of ingredient quality and the effect of a combination of xylanase, amylase, protease and phytase in the diets of broiler chicks. 1. Growth performance and digestible nutrient intake. British Poultry Science 47, 477–489.

Cowieson, A.J., Singh, D.N., Adeola, O., 2006b. Prediction of ingredient quality and the effect of a combination of xylanase, amylase, protease and phytase in the diets of broiler chicks. 2. Energy and nutrient utilisation. British Poultry Science 47, 490–500.

Cowieson, A.J., 2010. Strategic selection of exogenous enzymes for maize/soy-based poultry diets. Journal of Poultry Science 47, 1–7.

Cozannet, P., Preynat, A., Noblet, J., 2012. Digestible energy values of feed ingredients with or without addition of enzymes complex in growing pigs. Journal of Animal Science 90, 209–211.

Davies, S.J., Gouveia, A., 2008. Enhancing the nutritional value of pea seed meal (*Pisum sativum*) by thermal treatment of specific isogenic selection with comparison to soybean meal for African catfish *Clarias gariepinus*. Aquaculture 283, 116–122.

Davis, A.D., Johnson, W.L., Arnold, C.R., 1998. The use of enzyme supplements in shrimp diets. In: IV International Symposium on Aquatic Nutrition, La Paz, Mexico.

De Rham, O., Jost, T., 1979. Phytate protein interactions in soybean extracts and low-phytate soy protein products. Journal of Food Science 44, 596.

Debnath, D., Pal, A.K., Sahu, N.P., Jain, K.K., Yengkokpam, S., Mukherjee, S.C., 2005. Effect of dietary microbial phytase supplementation on growth and nutrient digestibility of *Pangasius pangasius* (Hamilton) fingerlings. Aquatic Research 36, 180–187.

Debnath, D., 2003. Effect of Dietary Microbial Phytase Supplementation on Growth Performance and Body Composition of *Pangasius pangasius* Fingerling (M.F.Sc. thesis). Central Institute of Fisheries Education Versona, Mumbai, India.

Dhillon, G.S., Kaur, S., Brar, S.K., 2013. Perspective of apple processing waste as low-cost substrates for bioproduction of high value products: a review. Renewable and Sustainable Energy Reviews 27, 789–805.

DiLorenzo, N., Smith, D.R., Quinn, M.J., May, M.L., Ponce, C.H., Steinberg, W., Engstrom, M.A., Galyean, M.L., 2011. Effects of grain processing and supplementation with exogenous amylase on nutrient digestibility in feedlot diets. Livestock Science 137, 178–184.

Drew, M.D., Raez, R., Gauthier, R., Thiessen, D.L., 2005. Effect of adding protease to coextruded flax-pea and canola-pea products on nutrient digestibility and growth performance of rainbow trout (*Oncorhynchus mykiss*). Animal Feed Science and Technology 119, 117–128.

Du Plessis, R.E., van Resenburg, C.J., 2014. Carbohydrase and protease supplementation increased performance of broilers fed maize-soybean-based diets with restricted metabolizable energy content. South African Journal of Animal Science 44, 262–270.

Eeckhout, W., de Paepe, M., 1994. Total phosphorus, phytate-phosphorus and phytase activity in plant feedstuffs. Animal Feed Science and Technology 47, 19–29.

El-Boushy, van der Poel, A.F.B., 2001. Formulating feed from waste and by-products. World Poultry 17, 34–36.

Erickson, G.E., Klopfeinstein, T.J., Watson, A.K., 2012. Utilization of feed coproducts from wet or dry milling for beef cattle. In: Makkar, H.P.S. (Ed.), Biofuel Co-products as Livestock Feed; Opportunities and Challenges. FAO, Rome, pp. 77–100.

Eun, J.-S., Beauchemin, K.A., 2007. Enhancing in vitro degradation of alfalfa and corn silage using feed enzymes. Journal of Dairy Science 90, 2839–2851.

Farhangi, M., Carter, C.G., 2007. Effect of enzyme supplementation to dehulled lupin based diets on growth, feed efficiency, nutrient digestibility and carcass composition of rainbow trout *Oncorhynchus mykiss* (Walbaum). Aquatic Research 38, 1274–1282.

Gaafar, H.M.A., Abdel-Raouf, E.M., El-Reidy, K.F.A., 2010. Effect of fibrolytic enzyme supplementation and fibre content of total mixed ration on productive performance of lactating buffaloes. Slovak Journal of Animal Science 43, 147–153.

Galyean, M.L., Cole, N.A., Brown, M.S., MacDonald, J.C., Ponce, C.H., Schutz, J.S., 2012. Utilization of wet distillers grain in high-energy beef cattle diets based on processed grain. In: Makkar, H.P.S. (Ed.), Biofuel Co-products as Livestock Feed; Opportunities and Challenges. FAO, Rome, pp. 61–76.

Garcia-Ruiz, A.I., Garcia-Palomares, J., Garcia-Rebollar, P., Chamorro, S., Carabano, R., de Blas, C., 2006. Effect of protein source and enzyme supplementation on ileal protein digestibility and fattening performance in rabbits. Spanish Journal of Agricultural Research 4, 297–303.

Gencoglu, H., Shaver, R.D., Steinberg, W., Ensink, J., Ferraretto, L.F., Bertics, S.J., Lopes, J.C., Akins, M.S., 2010. Effect of feeding a reduced-starch diet with or without amylase addition on lactation performance in dairy cows. Journal of Dairy Science 93, 723–732.

George, F.A.O., Otubusin, S.O., 2007. Fish feed development for sustainable aquaculture fish production in Africa. In: Paper Presented at Farm Management Association of Nigeria Conference, Ayetoro, Nigeria.

Ghazi, S., Rooke, J.A., Galbraith, H., Bedford, M.R., 1996a. Effect on nitrogen digestibility in growing chicks and broilers of treating soybean meal with different proteolytic enzymes. British Poultry Science 37, S53–S54.

Ghazi, S., Rooke, J.A., Galbraith, H., Bedford, M.R., 1996b. The potential for improving soya-bean meal in diets for chickens: treatment with different proteolytic enzymes. British Poultry Science 37, S54–S55.

Ghazi, S., Rooke, J.A., Galbraith, H., Bedford, M.R., 1997a. Effect of adding protease and alpha-galactosidase enzyme to soybean meal on nitrogen retention and true metabolisable energy in broilers. British Poultry Science 38, S28.

Ghazi, S., Rooke, J.A., Galbraith, H., Bedford, M.R., 1997b. Effect of feeding growing chicks semi-purified diets containing soybean meal and different amounts of protease and alpha-galactosidase enzymes. British Poultry Science 38, S29.

Ghazi, S., Rooke, J.A., Galbraith, H., Bedford, M.R., 2002. The potential for the improvement of the nutritive value of soybean meal by different proteases in broiler chicks and broiler cockerels. British Poultry Science 43, 70–77.

Ghazi, S., Rocke, J.A., Galbraith, H., 2003. Improvement of the native value of soybean meal by protease and alpha-galactosidase treatment in broiler cockerels and broiler chicks. British Poultry Science 44, 410–418.

Giri, S.S., Sahoo, S.K., Sahu, A.K., Meher, P.K., 2003. Effect of dietary protein levels on growth, survival, feed utilization and body composition of hybrid Clarias catfish (*Clarias batrachus* × *Clarias gariepinus*). Animal Feed Science and Technology 104, 169–178.

Gracia, M.I., Aranibar, M.J., Lazaro, R., Medel, P., Mateos, G.G., 2003. α-Amylase supplementation of broiler diets based on corn. Poultry Science 82, 436–442.

Han, J.A., BeMiller, J.M., 2008. Effects of protein on crosslinking of normal maize, waxy maize, and potato starches. Carbohydrate Polymers 73, 532–540.

Hardy, R.W., 2000. New developments in aquatic feed ingredients and potential of enzyme supplements. In: Cruz-Suárez, L.E., Ricque-Marie, D., Tapia-Salazar, M., Olvera-Novoa, M.A., Civera-Cerecedo, R. (Eds.), Avances en Nutrición Acuícola V. Memorias del V Simposium Internacional de Nutrición Acuícola. 19–22 Noviembre, 2000. Mérida, Yucatán, Mexico.

Hristov, A.N., Basel, C.E., Melgar, A., Foley, A.E., Ropp, J.K., Hunt, C.W., et al., 2008. Effect of exogenous polysaccharide degrading enzyme preparations on ruminal fermentation and digestibility of nutrients in dairy cows. Animal Feed Science and Technology 145, 182–193.

Hussain, A., Nisa, M., Sarwar, M., Sharif, M., Javaid, A., 2008. Effect of exogenous fibrolytic enzymes on ruminant performance. Pakistani Journal of Agricultural Science 45, 297–306.

IFC, 2007. Environmental, Health and Safety Guidelines for Sugar Manufacturing. International Finance Corporation.

Inborr, J., Ogle, R.B., 1988. Effect of enzyme treatment of piglet feeds on performance and post weaning diarrhoea. Swedish Journal of Agricultural Research 18, 129–133.

Inborr, J., 1990. Enzymes: catalysts for pig performance. Feed Management 41, 22–30.

Jackson, L.S., Li, M.H., Robinson, E.H., 1996. Use of microbial phytase in channel catfish *Ictaluru punatatus* diets to improve utilization of phytate phosphorus. Journal of World Aquaculture Society 27, 309–313.

Janssens, G.P.J., Hesta, M., Debal, V., De Wilde, R.O.M., 2000. The effect of feed enzymes on nutrient and energy retention in young racing pigeons. Annals of Zootech 49, 151–156.

Jiang, Z., Zhou, Y., Lu, F., Han, Z., Wang, T., 2008. Effects of different level of supplementary alpha-amylase on digestive enzyme activities and pancreatic amylase mRNA expression of young broilers. Asian-Australian Journal of Animal Science 21, 97–102.

Kalscheur, K.F., Garcia, A.D., Schingoethe, D.J., Royon, F.D., Hippen, A.R., 2012. Feeding biofuel coproducts to dairy cattle. In: Makkar, H.P.S. (Ed.), Biofuel Co-products as Livestock Feed; Opportunities and Challenges. FAO, Rome, pp. 115–154.

Kerr, B.J., Shurson, G.C., 2013. Strategies to improve fibre utilization in swine. Journal of Animal Science and Biotechnology 4 (11), 1–12.

Khan, S.H., Sardar, R., Siddique, B., 2006. Influence of enzymes on performance of broiler fed sunflower-corn based diets. Pakistan Veterinary Journal 26, 109–114.

Khattak, F.M., Pasha, T.N., Hayat, Z., Mahmud, A., 2006. Enzymes in poultry nutrition. Journal of Animal and Plant Science 16, 1–7.

Kiarie, E., Nyachoti, C.M., Slominski, B.A., Blank, G., 2007. Growth performance, gastrointestinal microbial activity and nutrient digestibility in early weaned swine fed diets containing flaxseed and carbohydrase enzyme. Journal of Animal Science 85, 2982–2993.

Klingerman, C.M., Hu, W., McDonell, E.E., DerBedrosian, M.C., Kung Jr., L., 2009. An evaluation of exogenous enzymes with amylolytic activity for dairy cows. Journal of Dairy Science 92, 1050–1059.

Knuckles, B.E., Kuzmicky, D.D., Gumbmann, M.R., Betschart, A.A., 1989. Effect of myoinositol phosphate esters on *in vitro* and *in vivo* digestion of protein. Journal of Food Science 54, 1348–1350.

Kornegay, E.T., 2001. Digestion of phosphorus and other nutrients: the role of phytases and factors influencing their activity. In: Bedford, M.R., Partridge, G.G. (Eds.), Enzymes in Farm Animal Nutrition. CAB International, Wallingford, UK, pp. 237–298.

Lazaro, R., Latorre, M.A., Medel, P., Gracia, M., Mateos, G.G., 2004. Feeding regimen and enzyme supplementation to rye based diets for broilers. Poultry Science 83, 152–160.

Lin, S., Mai, K., Tan, B., 2007. Effect of exogenous enzyme supplementation in diets on growth and feed utilization in tilapia, *Oreochromis niloticus* × *O. aureus*. Aquatic Research 38, 1645–1653.

Mahagna, M., Nir, I., Larbier, M., Nitsan, Z., 1995. Effect of age and exogenous amylase and protease on development of the digestive tract, pancreatic enzyme activities and digestibility of nutrients in young meat-type chicks. Reproduction Nutrition Development 35, 201–212.

Makinde, O.A., Odeyinka, S.M., Ayandiran, S.K., 2011. Simple and Quick Method for Recycling Pineapple Waste into Animal Feed, vol. 23. Livestock Research for Rural Development. Article #188. Retrieved February 16, 2015, from: http://www.lrrd.org/lrrd23/9/maki23188.htm.

Marsman, G.J.P., Gruppen, H., Van der Poel, A.F.B., Kwakkel, R.P., Verstegen, M.W.A., Voragen, A.G.J., 1997. The effect of thermal processing and enzyme treatments of soybean meal on growth performance, ileal nutrient digestibilities, and chime characteristics in broiler chicks. Poultry Science 76, 864–872.

McAllister, T.A., Oosting, S.J., Popp, J.D., Mir, Z., Yanke, L.J., Hristov, A.N., Treacher, R.J., Cheng, K.J., 1999. Effect of exogenous enzymes on digestibility of barley silage and growth performance of feedlot cattle. Canadian Journal of Animal Science 79, 353–360.

McCarthy, M.M., Engstrom, M.A., Azem, E., Gressley, T.F., 2013. The effect of an exogenous amylase on performance and total-tract digestibility in lactating dairy cows fed a high by-product diet. Journal of Dairy Science 96, 3075–3084.

Meng, X., Slominski, B.A., Nyachoti, C.M., Campbell, L.D., Guenter, W., 2005. Degradation of cell wall polysaccharides by combinations of carbohydrase enzymes and their effect on nutrient utilization and broiler chicken performance. Poultry Science 84, 37–47.

Miller, D.R., Elliott, R., Norton, B.W., 2008a. Effects of an exogenous enzyme, Roxazyme® G2, on intake, digestion and utilisation of sorghum and barley grain-based diets by beef steers. Animal Feed Science and Technology 145, 159–181.

Miller, D.R., Granzin, B.C., Elliott, R., Norton, B.W., 2008b. Effects of an exogenous enzyme, Roxazyme® G2 Liquid, on milk production in pasture fed dairy cows. Animal Feed Science and Technology 145, 194–208.

Nagaraj, M., Hess, J.B., Bilgili, S.F., 2007. Evaluation of a feed-grade enzyme in broiler diets to reduce pododermatitis. Journal of Applied Poultry Research 16, 52–61.

Ngxumeshe, A.M., Gous, R.M., 2009. Effect of varying level of thermostable xylanase, amylase and protease (TXAP) composite enzyme supplement on body growth of broiler chickens. South African Journal of Animal Science 39 (Suppl. 1), 312–315.

Noziere, P., Steinberg, W., Silberberg, M., Morgavi, D.P., 2014. Amylase addition increases starch ruminal digestion in first-lactation cows fed high and low starch diets. Journal of Dairy Science 97, 1–10.

Nys, Y., Frapin, D., Pointillart, P., 1996. Occurrence of phytase in plants, animals and microorganisms. In: Coelho, M.B., Kornegay, E.T. (Eds.), Phytase in Animal Nutrition and Waste Management. BASF Corporation, Mount Olive, New Jersey, pp. 213–240.

Owen, E., Jayasuriya, M.C.N., 1989. Use of crop residues as animal feeds in developing countries. Research and Development in Agriculture 6, 129–138.

Oxenboll, K.M., Pontoppidan, K., Fru-Nji, F., 2011. Use of a protease enzyme in poultry feed offers promising environmental benefits. International Journal of Poultry Science 10, 842–848.

Pallauf, J., Rimbach, G., 1997. Nutritional significance of phytic acid and phytase. Archive of Animal Nutrition 50, 301–319.

Ravindran, V., Ravindran, G., Sivalogan, S., 1994. Total and phytate phosphorus contents of various foods and feedstuffs of plant origin. Food Chemistry 50, 133–136.

Ravindran, V., Bryden, W.L., Kornegay, E.T., 1995. Phytates: occurrence, bioavailability and implications in poultry nutrition. Poultry and Avian Biology Reviews 6, 125–143.

Ravindran, V., Cabahug, S., Ravindran, G., Bryden, W.L., 1999a. Influence of microbial phytase on apparent ileal amino acid digestibility in feedstuffs for broilers. Poultry Science 78, 699–706.

Ravindran, V., Hew, L.I., Ravindran, G., Bryden, W.L., 1999b. A comparison of ileal digesta and excreta analysis for the determination of amino acid digestibility in food ingredients for poultry. British Poultry Science 40, 266–274.

Ravindran, V., Cabahug, S., Ravindran, G., Selle, P.H., Bryden, W.L., 2000. Response of broiler chickens to microbial phytase supplementation as influenced by dietary phytic acid and non-phytate phosphorus levels. II. Effects on apparent metabolisable energy, nutrient digestibility and nutrient retention. British Poultry Science 41, 193–200.

Ravindran, V., Selle, P.H., Ravindran, G., Morel, P.C.H., Kies, A.K., Bryden, W.L., 2001. Microbial phytase improves performance, apparent metabolizable energy and ileal amino acid digestibility of broilers fed a lysine-deficient diet. Poultry Science 80, 338–344.

Ravindran, V., Morel, P.C.H., Partridge, G.G., Hruby, M., Sands, J.S., 2006. Influence of an *E. coli*-derived phytase on nutrient utilization in broiler starters fed diets containing varying concentrations of phytic acid. Poultry Science 85, 82–89.

Ravindran, V., 1995. Phytases in poultry nutrition. An overview. In: Proceedings, Australian Poultry Science Symposium, vol. 7, pp. 135–139.

Raza, S., Ashraf, M., Pasha, T.N., Latif, F., 2009. Effect of enzyme supplementation of broiler diets containing varying level of sunflower meal and crude fibre. Pakistan Journal of Botany 45, 2543–2550.

Refstie, S., Svihus, B., Shearer, K.D., Storebakken, T., 1999. Nutrient digestibility in Atlantic salmon and broiler chickens related to viscosity and non-starch polysaccharide content in different soybean products. Animal Feed Science and Technology 79, 331–345.

Ritz, C.W., Hulet, R.M., Self, B.B., Denbow, D.M., 1995. Growth and intestinal morphology of male turkeys as influenced by dietary supplementation of amylase and xylanase. Poultry Science 74, 1329–1334.

Robinson, E.H., Li, M.H., Manning, B.B., 2002. Comparison of microbial phytase and dicalcium phosphate for growth and bone mineralization of pond-raised channel catfish *Ictalurus punctatus*. Journal of Applied Aquaculture 12, 81–88.

Romero, L.F., Parsons, C.M., Utterback, P.L., Plumstead, P.W., Ravindran, V., 2013. Comparative effects of dietary carbohydrases without or with protease on the ileal digestibility of energy and amino acids and AMEn in young broilers. Animal Feed Science and Technology 181, 35–44.

Rooke, J.A., Slessor, M., Fraser, H., Thomson, J.R., 1998. Growth performance and gut function of piglets weaned at four weeks of age and fed protease-treated soya-bean meal. Animal Feed Sciene and Technology 70, 175–190.

Schang, M.J., Azcona, J.O., Arias, J.E., 1997. Effects of a soya enzyme supplement on performance of broilers fed corn/soy or corn/soy/full-fat soy diets. Poultry Science 76 (Suppl. 1), 132 (abstract).

Selle, P.H., Ravindran, V., 2007. Microbial phytase in poultry nutrition. Animal Feed Science and Technology 135, 1–41.

Selle, P.H., Gill, R.J., Scott, T.A., 2007. Effects of pre-pelleted wheat and phytase supplementation on broiler growth performance and nutrient utilization. In: Proceedings, Australian Poultry Science Symposium, 19, pp. 182–185.

Selle, P.H., Ravindran, V., Cowieson, A.J., Bedford, M.R., 2010. Phytate and phytase. In: Bedford, M.R., Partridge, G.G. (Eds.), Enzymes in Farm Animal Nutrition, second ed. CAB International, Oxfordshire, pp. 160–205.

Shafiei, M., Karimi, K., Taherzadeh, J.S., 2010. Palm date fibres: analysis and enzymatic hydrolysis. International Journal of Molecular Sciences 11, 4285–4296.

Shurson, G.C., Zijlstra, R.T., Kerr, B.J., Stein, H.H., 2012. Feeding biofuel co-products to pigs. In: Makkar, H.P.S. (Ed.), Biofuel Co-products as Livestock Feed; Opportunities and Challenges. FAO, Rome, pp. 175–208.

Simbaya, J., Slominski, B.A., Guenter, W., Morgan, A., Campbell, L.D., 1996. The effects of protease and carbohydrase supplementation on the nutritive value of canola meal for poultry: in vitro and in vivo studies. Animal Feed Science and Technology 61, 219–234.

Singh, P., Maqsood, S., Samoon, M.H., Phulia, V., Danish, M., Chalal, R.S., 2011. Exogenous supplementation of papain as growth promoter on diet of fingerlings of *Coprinus carpio*. International Aquaculture Research 3, 1–9.

Singh, A., Masey O'Neill, H.V., Ghosh, T.K., Bedford, M.R., Haldar, S., 2012. Effects of xylanase supplementation on performance, total volatile fatty acids and selected bacterial population in caeca, metabolic indices and peptide concentrations in serum of broiler chickens fed energy restricted maize-soybean based diets. Animal Feed Science and Technology 177, 194–203.

Soltan, M.A., 2009. Effect of dietary fish meal replacement by poultry by-product meal with different grain source and enzyme supplementation on performance, feed recovery, body composition and nutrient balance of *Nile tilapia*. Pakistan Journal of Nutrition 8, 395–407.

Storebakken, T., Shearer, K.D., Roem, A.J., 1998. Availability of protein, phosphorus and other elements in fishmeal, soy-protein concentrate and phytase-treated soy-protein concentrate based diets to Atlantic salmon, *Salmo salar*. Aquaculture 161, 365–379.

Sun, X., 2007. Effect of Corn Quality and Enzyme Supplementation on Broiler Performance, Gastrointestinal Enzyme Activity, Nutrient Retention, Intestinal Mucin and Jejunal Gene Expression (Ph.D. dissertation). Virginia Polytechnic Institute and State University. 131pp.

Swain, B.K., Barbuddhe, S.B., 2008. Use of Agro-industrial By-products to Economise Feed Cost in Poultry Production. ICAR Technical Bulletin No. 13.

Swift, M.L., van Keyserlingk, M.A.G., Leslie, A., Teltge, D., 1996. The effect of Allzyme Vegpro supplementation and expander processing on the nutrient digestibility and growth of broilers. In: 12th Annual Symposium on Biotechnology in the Feed Industry Lexington, Kentucky. Supplement 1, Enclosure Code UL 2.1.

Tahir, M., Saleh, F., Ohtsuka, A., Hayashi, K., et al., 2008. An effective combination of carbohydrases that enables reduction of dietary protein in broilers: importance of hemicellulose. Poultry Science 87, 713–718.

Thompson, L.U., Button, C.L., Jenkins, D.J.A., 1987. Phytic acid and calcium affect the *in vitro* rate of navy bean starch digestion and blood glucose response in humans. American Journal of Clinical Nutrition 46, 467–473.

Troche, C., Sun, X., McElroy, A.P., Remus, J., Novak, C.L., 2007. Supplementation of Avizyme 1502 to corn-soybean meal-wheat diets fed to Turkey tom poults: the first fifty-six days of age. Poultry Science 86, 496–502.

Uni, Z., Noy, Y., Sklan, D., 1995. Posthatch changes in morphology and function of the small intestines in heavy and light strain chicks. Poultry Science 74, 1622–1629.

Vahjen, W., Osswald, T., Schafer, K., Simon, O., 2007. Comparison of a xylanase and a complex of non-starch polysaccharide degrading enzymes with regard to performance and bacterial metabolism in weaned piglets. Archives of Animal Nutrition 61, 90–102.

Wadhwa, M., Bakshi, M.P.S., 2013. Utilization of Fruit and Vegetable Wastes as Livestock Feed and as Substrates for Generation of Other Value Added Products. RAP Publications, FAO, Rome. 67pp.

Weiss, W.P., Steinberg, W., Engstrom, M.A., 2011. Milk production and nutrient digestibility by dairy cows when fed exogenous amylase with coarsely ground dry corn. Journal of Dairy Science 94, 2492–2499.

Wu, Y.B., Ravindran, V., Morel, P.C.H., Hendriks, W.H., Pierce, J., 2004. Evaluation of a microbial phytase produced by solid-state fermentation in broiler diets. 1. Influence on performance, toe ash contents, and phosphorus equivalency estimates. Journal of Applied Poultry Research 13, 373–383.

Yang, W.Z., Beauchemin, K.A., Rode, L.M., 1999. Effects of an enzyme feed additive on extent of digestion and milk production of lactating dairy cows. Journal of Dairy Science 82, 391–403.

Yang, W.Z., Beauchemin, J.A., Rode, L.M., 2000. A comparison of methods of adding fibrolytic enzymes to lactating cow diets. Journal of Dairy Science 83, 2512–2520.

Yildrin, Y.B., Turan, F., 2010. Effects of exogenous enzyme supplementation in diets on growth and feed utilization in African catfish, *Claria gariepinus*. Journal of Animal and Veterinary Advance 9, 327–331.

Yin, Y.L., Zhang, Z., Huang, J., Yin, Y., 2010. Digestion rate of dietary starch affects systemic circulation of amino acids in weaned swine. British Journal of Nutrition 103, 1404–1412.

Yuan, J., Yao, J., Yang, F., Yang, X., Wan, X., Han, J., Wang, Y., Chen, X., Liu, Y., Zhou, Z., Zhou, N., Feng, X., 2008. Effect of supplementing different level of commercial enzyme complex on performance, nutrient availability, enzyme activity and gut morphology of broilers. Asian-Australian Journal of Animal Science 21, 692–700.

Zahari, M.W., Alimon, A.R., Wong, H.K., 2012. Utilization of oil palm co-products as feed for livestock in Malaysia. In: Makkar, H.P.S. (Ed.), Biofuel Co-products as Livestock Feed; Opportunities and Challenges. FAO, Rome, pp. 243–262.

Zanella, I., Sakomura, N.K., Silversides, F.G., Fiqueirdo, A., Pack, M., 1999. Effect of enzyme supplementation of broiler diets based on corn and soybeans. Poultry Science 78, 561–568.

Zepf, F., Jin, B., 2013. Bioconversion of grape marc into protein rich animal feed by microbial fungi. Chemical Engineering and Process Techniques 1, 1011.

ZoBell, D.R., Wiedmeier, R.D., Olson, K.C., Treacher, R., 2000. The effect of an exogenous enzyme treatment on production and carcass characteristics of growing and finishing steers. Animal Feed Science and Technology 87, 279–285.

FURTHER READING

Douglas, M.W., Parsons, C.M., Bedford, M.R., 2000. Effect of various soybean meal sources and Avizyme on chick growth performance and ileal digestible energy. Journal of Applied Poultry Research 9, 74–80.

Eeckhout, W., de Paepe, M., 1991. The quantitative effects of an industrial microbial phytase and wheat phytase on the apparent phosphorus absorbability of a mixed feed by piglets. Medical Faculty Landbouwwetenschappen Rijksuniversiteit Gent 56, 1643–1647.

Ezejiofor, T.I.N., Enebaku, U.E., Ogueke, C., 2014. Waste to wealth- value recovery from agro-food processing wastes using biotechnology: a review. British Biotechnology Journal 4, 418–481.

Hristov, A.N., McAllister, T.A., Cheng, K.J., 2000. Intraruminal supplementation with increasing levels of exogenous polysaccharide-degrading enzymes: effects on nutrient digestion in cattle fed barley grain diets. Journal of Animal Science 78, 477–487.

Krause, M., Beauchemine, K.A., Rode, L.M., Farr, B.I., Norgaard, P., 1998. Fibrolytic enzyme treatment of barley grain and source of forage in high-grain diets fed to growing cattle. Journal of Animal Science 76, 2912–2920.

Onderci, M., Sahin, N., Sahin, K., Cikim, G., Aydin, A., Ozercam, I., Aydin, S., 2006. Efficacy of supplementation of α-amylase producing bacterial culture on the performance, nutrient use and gut morphology of broiler chickens fed a corn based diet. Poultry Science 85, 505–510.

Owusu-Asiedu, A., Baidoo, S.K., Nyachoti, C.M., 2002. Effect of heat processing on nutrient digestibility in pea and supplementing amylase and xylanase to raw, extruded or micronized pea based diets on performance of early-weaned pigs. Canadian Journal of Animal Science 82, 367–374.

Passos Dos, A.A., Kim, S.W., 2014. Use of enzymes in pig diets. In: VI Congresso Latino Americano de Nutricao Animal- SALA SUINOS; September 2014.

Yoruk, M.A., Gul, M., Hayirli, A., Karaoglu, M., 2006. Multienzyme supplementation to peak producing hens fed corn-soybean meal based diets. International Journal of Poultry Science 5, 374–380.

CHAPTER 11

Potential Applications of Enzymes in Brewery and Winery

M.R. Spier[1], A. Nogueira[2], A. Alberti[2], T.A. Gomes[1], G.S. Dhillon[3]
[1]Federal University of Paraná, Curitiba, Brazil; [2]State University of Ponta Grossa, Ponta Grossa, Brazil; [3]University of Alberta, Edmonton, AB, Canada

INTRODUCTION

Worldwide beer production was estimated to be 197.3 billion liters in 2013 (Barth-Haas Group, 2013). China remains the world's largest producer, with 50 billion liters in 2013, followed by the United States (22.4 billion). Brazil is in the third position, producing 13.55 billion liters of beer, followed by Germany and Russia.

In 2013, the world's largest beer company, Anheuser-Busch InBev, produced 39.9 billion liters (20.2%), followed by SABMiller (18.7 billion liters, or 9.5%), Heineken, Carlsberg, and China Resources Snow Breweries (CRB). Together, the five breweries account for over 50% of the world's beer production (Barth-Haas Group, 2013; Research and Markets, 2013). Anheuser-Busch InBev, SABMiller, Heineken, and Carlsberg occupy a total of 46% market share (Research and Markets, 2013).

According to International Organization of Vine and Wine (OIV) (2014a), the world's wine consumption in 2013 was 238.7 million hL. The United States became the biggest market in the world in terms of volume. In 2013 bottled wines and sparkling wines made up the vast majority of the world wine market. France is the world's largest wine consumer and biggest wine-producing country, with 46.2 million hL of wine production in the 2014 harvest. Italy produced 44.4 million hL of wine (OIV, 2014b) and is the second largest producer. The United States and China are the main drivers of consumption growth globally. According to Morgan Stanley Research (2013), the world's biggest top 10 wine producing countries in descending order are: France, Italy, Spain, United States, China, Chile, Argentina, Australia, South Africa, and Germany.

Enzymes are biological macromolecules produced by living organisms and act as catalysts to bring specific biochemical reactions (Gurung et al., 2013). Historically, they are involved in a variety of applications, such as beer, yogurt, bread, cheese, and vinegar, conferring particular characteristics in each product (Whitehurst and Van Oort, 2010). In wine making, the use of pectinases was introduced in the 1970s to improve white must clarification (Bruchmann and Fauveau, 2010).

Agro-Industrial Wastes as Feedstock for Enzyme Production
ISBN 978-0-12-802392-1
http://dx.doi.org/10.1016/B978-0-12-802392-1.00011-3

Today, enzymes are bringing improvements in processing steps, nutritional value and functionality, and sensorial properties (such as texture and flavor) (Van Oort, 2010; Gurung et al., 2013). The enzymes play a key role in the brewery and winery process. They accelerate the release of readily digestible sugars and other nutrients for fermentation by the yeasts. Different enzymes may improve mashing, lautering, and filtration, to increase fermentability rates, to produce low-calorie beer, to improve beer and wine stabilization, for corrective actions, and to enhance maturation.

This chapter discusses different enzymes involved in beer and wine processing.

ENZYMES IN BREWING

Duke Wilhelm IV enacted the Purity Law of 1516 in Bavaria, which stated that only barley, hops, yeast, and water are allowed in brewing (German Beer Institute, 2008). This is the oldest law governing food and beverage, named "*Reinheitsgebot*," and was recognized by UNESCO as a piece of the world's "intangible heritage" (BBC News, 2013).

Nowadays there are great beer varieties that can be produced with total substitution of malted barley or partially by mixing malted barley, unmalted barley, and/or adjunct sources, or even other malted grains, such as malted wheat (eg, "Weiss Beer" production) or other forms of sugar, in varying quantities to complement the sugars extracted from the grains. Besides, ale or lager yeasts contribute to obtain different types of beer that vary also according to the recipe or formulation. Concerning alcoholic content, beer contains up to 12% (v/v) alcohol concentration, but it may present higher variation, eg, from 0.05% to 0.5% (v/v) in alcohol free beers to 12.5% alcohol (v/v) in strong beers (Baxter and Hughes, 2001).

Brewers have learned not only how the endogenous enzymes contribute to issues such as fermentability, filterability, foam, clarity, flavor, etc. but also how to take advantage of exogenous enzymes (Bamforth, 2009). The enzymatic groups involved in the first stage in the brewing process are those endogenous enzymes biosynthesized during the malting process.

Malting Process: A Natural Brewing Enzymes Factory

Malting is one of the most important processes in brewing with the aim to stimulate endogenous enzyme production present naturally in the barley grains. The common endogenous enzymes in malt are α-amylase, β-amylase, α-glucosidase, limit dextrinase, carboxypeptidase, α-glucosidase, endo-e exopeptidase, endo-β-glucanase, lipoxygenase, xylo-acetylesterase, feruloyl esterase, xylanase, α-L-arabinofuranosidase, endo-β-1,3-glucanase, and endo β-1,4-glucanase (Linko et al., 1998; Li et al., 2005; Bamforth, 2009; Van Oort, 2010). Various other endogenous enzymes are responsible for reactions occurring during malting, such as acid and alkaline phosphatases, peroxidases, catalases, polyphenoloxidases, lipoxygenases, phospholipases, phytases, and others (Bamforth, 2009; Van Oort, 2010).

Brewer malt or barley malted for brewer's purposes is the product obtained by the transformation of the grain by forced and controlled germination under special conditions of humidity and temperature of barley *Hordeum* sp. (*Hordeum distichum*, the most indicated species). The malting process consists of three main steps:

1. *Sleeping:* the grains are immersed in a water bath with external aeration. Water and air reaches the embryo through the micropyle, stimulating it to start germination and also serving to distribute the enzymes for endosperm modification (Van Oort, 2010). The biosynthesis is initiated with the increase of α- and β-amylases, carboxypeptidase, and β-glucanase activities internally in germ seed. These hydrolytic enzymes are secreted by the germ, crossing though the scutellum membrane to reach the starch endosperm. Initially, β-glucanases hydrolyzes β-glucans (membrane involving starch granules) and the starch modification is initiated by α-amylases. In parallel, proteases act on the protein content in the grain. The aleurone layer also releases enzymes during germination. Sugars, amino acids, and other simple molecules in reverse nourish the seed to start the grain germination. The process may take 40–68 h and varies according to the aeration system, temperature, and grain quality. Finally, the radicle (little root of the grain) appears substantially as a signal of the onset of germination (Briggs et al., 1981; Briggs, 1998; Van Oort, 2010).

2. *Germination:* The grains from the sleeping step presenting with suitable hydration are distributed in perforated trays or in large mats with a perforated bottom at controlled relative humidity of 100%, temperature of 14–16°C for five days. The hydrated embryo of barley grains need to respire, so the grains are turned regularly allowing air exchange and to control temperature. The grain produces growth hormones named gibberellins, but also needs internal nutrient reserves, such as lipids and sugar reserves for the respiration process. The glutinous and mucilaginous matter is reduced, the taste becomes sweet, and the texture is so loose that it easily bruises or crumbles to powder between the fingers; this change is considered to be sufficient when the radicle has come nearly to the end of seed and is just ready to burst out. The radicle achieves 3/4 to 4/5 of the grain length, as more than that will reduce sugar content of the grain (Briggs et al., 1981; Briggs, 1998; Van Oort, 2010).

3. *Kilning:* This step consists of a gradual drying of the barley grains, aiming to stabilize it for storage until industry application. Air drying, hot and dry malt passes through the bed in an upward or downward flow. It is possible to obtain different malt types depending on temperature and time of drying. For example, mild temperature produces light malt (Pilsen malt) and the enzyme content remains largely intact. In contrast, high temperature is applied for obtaining dark malt, which presents poor enzyme activity. The temperature during kilning is raised from ~15°C to 85°C in a controlled manner (temperature, airflow, relative humidity, time) (Briggs et al., 1981; Briggs, 1998; Van Oort, 2010).

The temperature range from 55°C to 62°C preserves enzymatic content active in malt. The heating temperature of around 62°C is suitable for the activity of endogenous malt enzymes (Van Oort, 2010). Higher temperatures generate different types of malt, although with a deficient enzymatic content. The temperature is adjusted according to the kind of malt we wish to make. If not controlled, the exposure at high temperatures causes scorching of the grains (Webster et al., 1856; Van Oort, 2010; Preedy, 2011). Malt has an aperture to permit to escape of the heated air and vapor. The malt is spread out on the floor about 4 in. in thickness so that it can dry uniformly. The heated air ascends through the holes in the floor, passes up through the malt, and makes its way out through the roof, carrying the moisture along with it. Initially, the temperature should not be higher than 90°C and should rise gradually to reach 140°C (Webster et al., 1856; Sydney, 1873; Curtis, 2013).

Beer Production (Brewhouse Processing)

The brewhouse processing consists of several steps starting from raw material, such as malted barley and water. The process consists of malt milling, mashing, lautering or primary filtration, boiling with hops addition, whirlpool/cooling, yeast inoculation, fermentation, maturation (or secondary fermentation), cooling, addition of additives, secondary filtration, finishing with CO_2 level adjustments, packaging, and distribution.

The malt grains are milled to expose endosperm, ensure good hydration and good access of the malt enzymes to their substrates. Milled malt and water, and adjuncts if any, are mixed for the mashing step. The mixture is heated until 50°C for proteolysis, 62°C for gelatinization/liquefaction, 72°C for saccharification, and 78°C for mashing-off and for malting enzymes inactivation (Van Oort, 2010). Each temperature range is the optimum temperature of each enzymatic group. The endogenous malt-derived proteolytic enzymes (proteases, carboxypeptidases, amino peptidases, and dipeptidases) with endo-β-1,4-glucanases and pentosanases are most active and have optimum temperature ~50°C in the first step of mashing. These enzymes could help to degrade barley endosperm proteins (such as hordein and glutelin), which are considered as undesirable compounds/contaminants and are normally precipitated out in the barley spent grains (Dhillon et al., 2016).

The second step consists of three processes: gelatinization, liquefaction, and saccharification. Then the mash temperature is raised to 70°C and enzymatic inactivation is observed (Briggs et al., 1981; Briggs, 1998; Esslinger, 2009; Van Oort, 2010; Briess, 2013). The starch of barley gelatinizes at 61–65°C, so it is available to be hydrolyzed by the enzymes α- and β-amylases present in the malt. Alpha-amylases randomly hydrolyze starch to dextrins while β-amylases attack the starch and dextrins from the nonreducing ends, stripping off pairs of glucose molecules (maltose), but it is inactivated after longer times (more than 30 min) at 65°C saccharification period.

Glucoamylases and pullulanases participate in the last step of starch degradation, specifically in saccharification. These enzymes are exogenous; they are not produced during malting by endogenous barley enzymatic complex. So they are applied during the mashing step. Glucoamylases, also known as amyloglucosidases (AMGs) break all the linkages between the glucose molecules including branch point α-1,6 bonds, although the branch points are broken more slowly. Pullulanase is also important because it permits AMG to rapidly convert the debranched molecules to glucose. Pullulanases specifically and rapidly attack the branch chain link α-1,6 bonds (Briess, 2013).

The four main enzymatic reactions and respective enzymes involved during the mashing step in beer production are presented in Table 11.1. Endoproteases and exopeptidases act in the breakdown of malt proteins. Endoproteases randomly break the large protein molecules into polypeptide chains. The exopeptidases (such as carboxypeptidases and aminopeptidases) attack the polypeptides from a particular end, stripping off small units to produce amino acids. For example, carboxypeptidases attack the proteins from the carboxyl end and aminopeptidases attack the proteins from the amino end (Van Oort, 2010). Deficient proteolysis may cause negative effect on colloidal stability of the beer and excess of proteolysis may impact negatively on foam stability of the beer by reducing the level of foam.

Enzymatic Processes in Yeast Fermentation in Beer

Beer fermentation is traditionally performed by *Saccharomyces* yeasts strains, such as *Saccharomyces cerevisiae, Saccharomyces uvarum*, and *Saccharomyces carlsbergensis*. Depending on the beer type (ale or lager), distinct yeast strains are used. There are physiological and

Table 11.1 Performance of Endogenous Malting Enzymes During Enzymatic Reactions in Mashing

Enzymatic Group	Enzymes	Function in the Four Main Enzymatic Reactions During Mashing
Proteolytic	Endoproteases and exopeptidases	Hydrolysis of proteins into peptides and free amino acids; although most proteolysis occurs during malting. Reduction of haze problems caused by proteins, and foam stability of beer.
	Endopeptidase (carboxypeptidases and aminopeptidases)	Amino acids and peptides contribute to color and flavor of beer.
Amylolytic	α-, β-amylases and glucoamylases (amyloglucosidases)	Breakdown of gelatinized starch into fermentable carbohydrates (glucose, fructose, sucrose, maltose, and maltotriose).
Glucanolytic	Endo β-1,4-glucanases, endo barley β-glucanases	Degradation of β-glucan and arabinoxylan chains.
Pentosanolytic	Pentosanases	Degradation of pentosans into arabinose and xylose.

Table 11.2 Characteristics of Ale and Lager Brewing Yeasts

Type	Ale	Lager
Strains	*Saccharomyces cerevisiae, Saccharomyces bayanus*	*Saccharomyces pastorianus, Saccharomyces uvarum* (carlsbergensis)
Fermentation rate	🍺🍺🍺	🍺🍺🍺
Efficient use of maltotriose	🍺	🍺🍺🍺
Melibiose cleavage by α-galactosidases	No	🍺🍺🍺
Sulfur dioxide production	🍺	🍺🍺🍺

🍺 = means low; 🍺🍺🍺 = high or efficient.

genetic differences between ale and lager strains. Both ale and lager yeast strains present high-fermentation rates, but only lager strains, such as *S. uvarum* (*carlsbergensis*) possess the gene MEL (Table 11.2). This gene gives them the capacity to cleavage the oligosaccharide melibiose into glucose and galactose by the biosynthesis of α-galactosidases (melibiases) (Stewart and Russel, 1998). At static fermentation in a vessel containing 16°C all-malt wort, lager strains used maltotriose more efficiently than ale strains, whereas maltose utilization efficiency was not dependent on the type of brewing strain (Zheng et al., 1994). Maltose and maltotriose are transported into the cell across cell membrane, and then the two sugars are hydrolyzed to glucose units by the α-glucosidase (maltase) system (Van Oort, 2010). Future manipulated yeast strains in brewing must show capacity for synthesis of the extracellular amylases, β-glucanases, and β-glucosidase (Stewart et al., 2013) because wort components, such as dextrins, β-glucan, and soluble proteins are not metabolized by strains of brewer's yeast.

Enzyme Action for Enhancing Maturation

Diacetyl is a common off-flavor in beer, and the reduction of diacetyl concentration is mainly achieved by controlling the fermentation temperature and the length of the beer maturation process (Teranishi et al., 1999). Diacetyl is a by-product of amino acid metabolism and is formed by the oxidative decarboxylation of α-acetolactate during the exponential growth phase of the yeast during fermentation (Van Oort, 2010). So, diacetyl is taken up by yeast and reduced to acetoin by the enzyme diacetyl reductase during maturation. Acetoin has no effect on beer flavor (Teranishi et al., 1999). In concentrations above $0.1\,mg\,L^{-1}$, diacetyl is considered a flavor defect and results in a butterscotch flavor in beer (Van Oort, 2010). In order to solve this problem, the microbial enzyme acetolactate decarboxylases, as external enzymes supplementation may be interesting to convert acetolactate directly to acetoin. It is found in various microorganisms, but not in yeast (Teranishi et al., 1999). Van Oort (2010) reported acetolactate decarboxylase from

Bacillus species. Another acetolactate decarboxylase was also obtained from *Brevibacterium acetylicum* (Oshiro et al., 1989).

Enzyme Action in Low-caloric, Low-carbohydrate, and Gluten-free Beer Production

One of the most significant recent developments in the brewing industry is the use of microbial enzymes for the production of low-caloric or "light" beers (Nagodawithana and Reed, 1993). Light beer (low-carbohydrate content) can be produced by the addition of some enzymes during wort production or during fermentation. The main components responsible for caloric value of beer are alcohol and the unfermentable carbohydrate content (represented by dextrin limit). In addition, an increase in fermentability can be obtained by adding an enzyme, AMG (amyloglucosidase or glucoamylase). Then, low/zero caloric beer can be produced by adding fungal α-amylase along with AMG, to ensure complete breakdown of the α-1,4 linkages. These enzymes can be added during mashing process or directly to the fermenter (Van Oort, 2010).

Different approaches for the production of gluten free beer were discussed by Hager et al. (2014). These approaches are precipitation of hordeins reported by (Siebert and Lynn, 1998; Siebert, 1999; Van Landschoot, 2011), enzymatic treatments by using prolyl endopeptidases (Pasternack et al., 2008; Van Landschoot, 2011; Guerdrum and Bamforth, 2012), or by using alternative cereals or pseudocereal materials or even gluten-free barley malt (Van Landschoot, 2011).

Two enzymatic methods have been tried to reduce gluten from raw material of beer. One is through genetically modified yeast expressing a specific enzyme to degrade gluten in malt grain and adjuncts containing gluten; the other is by adding exogenous enzymes to modify gluten fraction. In addition, transglutaminase can be added exogenously to modify the gluten fraction (Esslinger, 2009). Guerdrum and Bamforth (2012) and Acton (2012) studied the production of gluten-free beer by using an enzyme, endoproteinase, which lowered the concentration of prolamin to levels below reliable limit of detection. Also Hartmann et al. (2006) demonstrated that proteases from germinated wheat, rye, and barley rapidly cleave celiac toxic peptides into nontoxic fragments with less than nine amino acids.

Endoproteinases and exopeptidases (mainly carboxypeptidases) are the enzymes responsible for reducing storage proteins, such as hordeins (prolamins) and hordenins (glutelins) to soluble proteins (von Bothmer et al., 2003), which may reduce the gluten complex of barley for gluten-free beer production. In addition, cysteine proteinases can play a major role in the degradation of storage proteins during mashing and malting. Table 11.3 shows exogenous enzymes applied in brewing, which are not found in malt grains or in brewery yeasts.

Dextrins, arabinoxylan, and β-glucan are the highest molecular weight substances found in beer that increase its viscosity. Arabinoxylan had the largest effect on filterability

Table 11.3 Exogenous Enzyme Applications in Brewing
Brewery Process Step

Application	Enzyme and Sources	Action
Aid in the production of sugars for yeast fermentation	α-amylases: *Aspergillus* spp.	Barley starch hydrolysis during malting process, or in high adjunct mashes, or containing raw barley grain (unmalted)
	Bacillus spp. *Microbacterium imperiale*	Heat-stable exogenous α-amylases (from *Bacillus* spp.) during mashing can reduce the wort starch problem to the lautering step and sparging
Mashing or fermenter	Glucoamylases: *Aspergillus niger*	Participate in saccharification (the last step of starch degradation) for releasing glucose units
Filtration aids, prevention of haze	β-glucanase: *Aspergillus* spp. (*A. niger*, *Aspergillus oryzae*) *Trichoderma reesei*	Hydrolyze β-glucans in beer mashes, but this enzyme in unmalt barley is used as an adjunct, is not sufficient to hydrolyze it. Hence, addition of exogenous β-glucanases is required
Improve extract yields, reduce run-off times, improve lautering velocity	*Bacillus subtilis*, *Penicillium emersonii*, *Lactic acid bacteria* (*Lactobacillus plantarum*, *Pediococcus pentosaceus*)	Also applied in fermenter or during maturation to assist in degradation of residual glucans, which may cause the filters to block, slow run-off times, low extract yields, and/or development of haze in the final product
Mashing	Glucanases, hemicellulases, and xylanases: Fungal	Cell wall breakdown. Applied mainly if poorly modified malt or unmalted adjuncts, such as barley, wheat, or oat are used. In this case these enzymes have corrective action
	Aspergillus spp. *A. niger*	Xylanases eliminate pentosans, especially when use wheat as adjunct
Mashing	Proteinase	Production of free amino nitrogen, especially in high adjunct mashes
Fermenter	α-acetolactate decarboxylase: *Bacillus* sp. *Brevibacterium acetylicum*	Enhance maturation, by converting α-acetolactate into acetoin, before it can be converted into diacetyl

Continued

Table 11.3 Exogenous Enzyme Applications in Brewing—cont'd
Brewery Process Step

Application	Enzyme and Sources	Action
Filtration and beer haze Stored beer	Papain (mixture of cysteine proteinases)	Elimination of haze-forming components (proteins, or polypeptides and polyphenols), improving beer stabilization
Stored beer	Prolyl endopeptidase	Elimination of haze-forming polypeptides; potential value in producing beer for celiacs
Packaged beer	Glucose oxidase/catalase	Elimination of oxygen

Briggs, D.E., et al., 1981. Malting and Brewing Science: Malt and Sweet Wort, second ed. Springer Science & Business Media, 914 p.; Oshiro, T., et al., 1989. Purification and characterization of a-acetolactate decarboxylase from *Brevibacterium acetylicum*. Agricultural and Biological Chemistry 53, 1913–1918; Briggs, D.E., 1998. Malts and Malting. Blackie Academic & Professional, London, 796 p.; Teranishi, R., et al., 1999. Flavor Chemistry: Thirty Years of Progress. Kluwer Academic, New York, 439 p.; Linko, M., et al., 1998. Recent advances in the malting and brewing industry. Journal of Biotechnology 65, 85–98; Galante, Y.M., et al., 1998. Application of *Trichoderma* enzymes in food and feed industries. In: Harman, G.F., Kubicek, C.P. (Eds.), Trichoderma & Gliocladium—Enzymes, Biological Control and Commercial Applications, vol. 2. Taylor & Francis, London, pp. 327–342; Bhat, M.K., 2000. Cellulases and related enzymes in biotechnology. Biotechnology Advances 18, 355–383; Laitila, A., et al., 2006. *Lactobacillus plantarum* and *Pediococcus pentosaceus* starter cultures as a tool for microflora management in malting and for enhancement of malt processability. Journal Agricultural Food Chemistry 54, 3840–3851; Li, Y., et al., 2005. Studies on water-extractable arabinoxylans during malting and brewing. Food Chemistry 93, 33–38; Lewis, M.J., Bamforth, C.W., 2011. Essays in Brewing Science, first ed. Springer, United States, 179 p.; Dhillon, G.S., Kaur, S., Brar, S.K., 2012. Flocculation and haze removal from crude fermented beer using in-house produced laccase via koji fermentation with *Trametes versicolor* using brewery spent grain. Journal of Agriculture and Food Chemistry 60 (32), 7895–7904; Van Oort, M., 2010. Enzymes in food technology – introduction. Chapter 1. In: Whitehurst, R.J., Van Oort, M. (Eds.), Enzymes in Food Technology, second ed. Wiley-Blackwell, Chichester, UK, pp. 1–17.

because it forms highly viscous solutions and is still present in malt at rather high levels (Fincher and Stone, 1986; Viëtor et al., 1991). Arabinoxylans also participates in haze formation (Coote and Kirsop, 1976). No significant interactions between the carbohydrates with respect to viscosity or beer filterability have been observed (Sadosky et al., 2002). Although nonstarch polysaccharides (arabinoxylans and (1–3),(1–4)-β-D-glucans) and their products affect brewing filtration process and physical stability of beer, they can improve the sensory properties of the beer (Coote and Kirsop, 1976; Bamforth, 1982; Viëtor and Voragen, 1993). Addition of xylanolytic enzymes, such as endo-(1,4)-β-xylanase, has been shown to improve brewhouse yield by lowering the wort viscosity (Ducroo and Frelon, 1989).

Trends of Enzyme Use in Brewing

The potential of using enzymes for obtaining different beer products by improving quality or offering different attributes may encourage researchers in the development of other new beer products. Some of the beer products developed by using enzymes include light beer, low-carbohydrate beer, and gluten-free beer. There is a demand for innovative beer products seeking new tastes, flavors, functional appeal, or appealing to a specific audience.

ENZYMES IN WINERY

Wine making

Wine can be defined as a natural product resulting from a biotechnological process, which begins during ripening of the grapes and continues during harvesting, with effects from the alcoholic fermentation, clarification, and bottling (Romano et al., 2003). They can be classified according to the type of grape (*Vitis labrusca* and *Vitis vinifera*), the final color (red, rose, and white) and the sugar content (dry, demi-sec, and sweet). Furthermore, there is a category of "special wines," such as sparkling wines, fortified wines, and flavored wines (Torresi et al., 2011).

The main unitary operations that involve the wine making of red wines, after the choice of grape cultivar, consists in the harvesting, separation of berries and stems, crushing of berries, fermentation and maceration, pressing, conclusion of fermentation, raking (lees), malolactic fermentation (if required), raking (lees), maturation in oak barrels, clarification, filtration, and bottling. White wine processing is similar to red wine processing. The basic differences are the selection of the grape cultivar and the maceration is not performed (Pérez-Serradilla and Castro, 2008). Sparkling wines are obtained by a secondary fermentation of a base wine. Depending on the production technology, they can be classified into sparkling wines produced by a traditional refermentation in the bottle and sparkling wines produced by a secondary fermentation in hermetically sealed tanks (Torresi et al., 2011).

Biotechnological techniques have been of fundamental importance in oenology; among these are technological innovations, enzymatic treatments by commercial preparation in free or immobilized form, selected yeasts, improvement of microbial starters, and enzyme immobilization. The use of exogenous enzyme preparations helps to overcome the problem of the insufficient activity of endogenous enzymatic activity in the grapes. But the use of the enzymes in the wine industry remains limited for several reasons that can be summarized as follows: traditionalism of winemakers, influence on enzymatic activities related to physicochemical characteristics of musts and wines (Romano et al., 2003; Palmeri and Spagna, 2007).

Enzymes in Winery

Enzymes in wine making are considered as a processing aid and their use is regulated in various countries. Enzyme preparations are required to comply with the specifications from the Joint FAO/WHO Expert Committee on Food Additives and from the Food Chemical Codex for food-grade enzymes. The European regulations for wine making are made by the Common Market Organization and International Organization of Vine and Wines (OIV). The aims of the OIV are to validate oenological treatments, to introduce some changes to wine making practices, and establish applicable restrictions (Curvelo-Garcia, 2005; Reynolds, 2010). In general, the other wine–producing countries outside Europe have been following the OIV recommendations. However, in the United States wine regulations are made by the Alcohol and Tobacco Tax and Trade Bureau, which has listed more enzymes (protease, amylase, catalase, and glucose-oxidase) for use in wine making (Reynolds, 2010).

The use of enzymatic preparations in the unit operations in the wine processing, such as maceration, clarification, stabilization, filtration, and extraction of aromatic precursors of grapes present in the grape must and wine should be in accordance with specifications of the International Oenological Codex published by the OIV. Thus, the enzyme preparations approved by resolution of the OIV were evaluated for safety and efficacy. These enzymes have an ideal application concentration (influenced by the variety of grape, processing, must pH, temperature (10–30°C), and contact time). However, the concentration of enzyme preparations in the wine must be low. During the clarification and filtration, a percentage of the enzyme concentration is reduced. Therefore, the residual enzyme concentration is low and does not cause allergy problems for consumers.

Enzymes in Grapes

Most enzymes present in the grape berry are involved in the maturation process. Invertase is considered one of the key enzymes during ripening. The function of invertase is to hydrolyze saccharose, synthesized during prematuration of the grapes, to glucose and fructose. The sugar accumulation is evidenced by the grape ripening, about 2.5 mmol of sugars/hour/berry (Robinson and Davies, 2000; Ribéreau-Gayon et al., 2006). Tyrosinase is responsible to catalyze redox reaction of the white grape berry, promoting oxidization of phenols into quinones (undesirable browning) (Van Rensburg and Pretorius, 2000).

Pectinases are naturally found in grapes and catalyze reactions of polysaccharide degradation of the cell wall. They are useful in the maceration process, promoting the release of phenolic compounds, anthocyanin, and flavor precursors, and contribute to the quality of the final product (Pinelo et al., 2006). The concentration of pectinases depends on the grape varieties (Pardo et al., 1999; Pinelo et al., 2006). Other endogenous grape enzymes are glycosidases that are responsible for the hydrolysis of aroma precursors (Van Rensburg and Pretorius, 2000). These enzymes are important in wine making but have low concentration and/or activity in grapes. Pectinases and glycosidases are commercially available.

Enzymes Released by Yeasts

Several nonconventional yeasts are used in oenology, however, *Saccharomyces* sp. are the main genus used in wine fermentation. This yeast releases various enzymes during the wine fermentation and cell lysis. Invertase (β-fructofuranosidase) is one of the most important enzymes and it catalyzes the hydrolysis of saccharose present in grape must into glucose and fructose molecules (Ribereau-Gayon et al., 2006). Beta-1,3 glucanases are released by yeast during wine fermentation and are related to the cell wall hydrolysis and release of mannoproteins into medium during autolysis of the yeast. However, to increase hydrolysis and accelerating aging in lees, exogenous β-glucanase can be employed to enhance wine structure and stability (Spagna et al., 2002).

Pectinases catalyze the degradation of pectic substances and can also be found in yeasts (Pardo et al., 1999; Merín et al., 2014).

The pectinases from yeasts are preferable over fungal pectinases, due to their specificity and producing a single type of enzyme, the polygalacturonase.

The β-glucosidase from *S. cerevisiae* has been studied to improve the release of aromatic compounds in wine. The yeasts *Debaryomyces pseudopolymorphus*, *Hanseniaspora uvarum*, *Candida oleophila*, *Debaryomyces polymorphus*, and *Brettanomyces* spp. showed high β-glucosidase activity (Villena et al., 2007).

Commercial Enzymes Applied in Winery

Various commercial enzymes are used to increase the extraction of phenolic compounds, color, and aroma compounds and thus to enhance the wine stability, mouthfeel, and complexity (Table 11.4) (Ortega-Regules et al., 2008; Petropulos et al., 2014). Many of these commercial enzymes are present in grapes or in the microorganisms involved in wine making, however, a significant effect was not observed due to low activity (Van Rensburg and Pretorius, 2000).

Enzymes for wine making are available in liquid and powder forms. Liquid enzymes should be stored at temperatures ranging from 0°C to 10°C and a loss of 3–5% of their activity per year has been observed. The activity loss increases at ambient temperatures to 20–30%. Powder forms have more stability and do not require storage at low temperatures and only 1–2% loss of activity has been observed. However, powdered formulations should be manipulated with caution due to potential risk of allergy. The liquid formulations are stabilized either with potassium chloride, glycerol, or sodium chloride and might contain preservatives, such as potassium sorbate or sodium benzoate.

Pectinases are the main enzymes used in wine making and are added to the must during maceration step or débourbage of the must or wine (clarification). When enzymes are used at maceration, this is usually a mixture of pectinases (polygalacturonase, pectin lyase, pectin methylesterase), cellulases, and hemicellulases, which causes a degradation of the cell wall and therefore improves the extraction of the must during pressing. In addition, macerating enzymes increases the extraction of phenolic compounds, especially anthocyanins that contribute to color intensity of the beverage (Resolution OIV Oeno 13/2004). The better extraction of tannins and proanthocyanidins increases formation of the anthocyanin-derived pigments and enhances copigmentation (Ducasse et al., 2010; Romero-Cascales et al., 2012).

Table 11.4 Enzymes Used in Wine Making

Enzymes	Maceration	Color stability	Clarification	Pressing Whites	Aromatic Whites*	Clarify White Must	Yeast lysis	Red Color (extraction)	Red Color (stability)	Filtration	Kill GPB
Pectinase	1,4		1,4	1		1		1	1	1,4	
Cellulase	1	1		1				1	1		
Hemicellulase	1	1		1				1	1		
Glucanase/β-Glucanase			1			1	3			1	
Glycosidase	7			1							
Polygalacturonase						1				1	
β-Glucosidase**					1						
Rhamnosidase					6						
Apiosidase					6						
Arabinofurosidase					1						
Laccase		2							2		
Lysozyme											5

Note: * use to release bound terpenes; ** inhibited by sugar, use only after alcoholic fermentation; GPB (gram-positive bacteria, specifically lactic acid bacteria). The subscript number in wine cup indicates the following references: [1] Van Rensburg and Pretorius (2000), [2] Minussi et al. (2007), [3] Torresi et al. (2014), [4] Canal-Llaubères (1993), [5] Guzzo et al. (2011), [6] Mateo and Jiménez (2000), and [7] Romero-Cascales et al. (2012).

Similar to pectinases, glucanases also facilitate the clarification and filtration process. The glucanases act on glucans (polymers of glucose) that cause turbidity and filtration problems. The bentonite and other filtering agents are not effective for glucans, mandating the use of commercial β-glucanases in wine making (Humbert-Goffard et al., 2004) according to Résolutions OIV Oeno 14/2004; OIV 3/85 and OIV Oeno 18/2004. Average doses are 3–5 g per $100 kg^{-1}$ added in crushed grape for must extraction and 1–3 g per $100 L^{-1}$ in wine clarification.

An important quality parameter of wines is the aromatic composition. The main volatile compounds of the wine come from the grapes that can be released during maceration. Part of these flavor compounds are in the form of odorless, nonvolatile glycosides. Glycosidases can act on these compounds releasing terpenes, C_{13}-norisoprenoids, benzene derivatives, and long-chain aliphatic alcohols, which can improve the aroma quality of the grape musts and wines. A concentration of 6 g $100 L^{-1}$ is sufficient to obtain good results (Miguel et al., 1998) and is allowed by the Resolution OIV Oeno 16/2004.

The use of maceration and clarification enzymes can adversely affect the content of the terpenic and norisoprenoid compounds (Armada et al., 2010; Palmeri and Spagna, 2007), probably by precipitation at clarification step. Commercial β-glucosidase from *Aspergillus niger* should be used in the final stages of processing, since they are partly inhibited by glucose. Despite the beneficial effect on the wine, the activity of this enzyme could affect the anthocyanin molecule that is a problem to color stability of red wines (Le Traon-Masson and Pellerin, 1998).

To avoid stuck and sluggish fermentation, the addition of urea as a nitrogen source is a common practice. However, ethanol reacts with the added urea and forms ethyl carbamate, a potential carcinogen. To avoid this reaction, ureases can be employed to hydrolyze urea to ammonia and carbon dioxide. Preferentially, the addition should be made to the wine after sedimentation of the lees; dosage should be done based on the urea content in the wine, and urease will be eliminated in the subsequent wine filtration (Resolution OIV 2/95) (Curvelo-Garcia, 2005).

Enzymes can also be used to avoid bacterial contamination and wine spoilage, allowing a reduction in the concentration of sulfur dioxide. Lysozyme acts against gram-positive bacteria, eg, lactic acid bacteria (*Oenococcus*, *Pediococcus*, and *Lactobacillus*). Moreover, if partial or delayed malolactic fermentation is desired, this enzyme can be used (Guzzo et al., 2011). However, the lysozyme concentration in wines cannot exceed $500 mg L^{-1}$ (Resolution OIV 6/97).

Despite the positive effect of the use of enzymes, the presence of the side activities of crude enzymatic preparations like polyphenoloxidases, esterases, and antocianases can be detrimental to wine quality (Palmeri and Spagna, 2007). Moreover, the effectiveness of the enzyme treatments are related to some parameters, such as the cultivar, the ripening stage, and the wine making techniques (Ortega-Regules et al., 2008; Petropulos et al., 2014).

Trends of Enzyme Use in Winery

Enzymes application is a widespread oenological practice during pressing, clarification, and filtration, to enhance the extraction of phenolic compounds and aromas (Van Rensburg and Pretorius, 2000; Armada et al., 2010; Ducasse et al., 2010). Side activities of the commercial enzymatic preparations pose problems in wine making. Excessive activity of methyl esterase, β-glucosidase, and cinnamyl esterase may cause anthocyanin degradation, aromatic phenols, and methanol formation that are detrimental to wine quality.

Research is currently underway on the development, isolation, and purification of new enzymes. A β-glucosidase isolated from *Issatchenkia terrícola* showed a good tolerance to high-glucose concentrations when applied to white Muscat wine fermentation (González-Pombo et al., 2011). Callejón et al. (2014) identified an enzyme that is able to degrade biogenic amines in wine. The biogenic amines cause human health problems when consumed. Furthermore, there is a search for fermentative yeasts with significant production of enzymes of interest. Manzanares et al. (2003) report on the increase on monoterpenol content, especially linalool, when a genetically modified industrial strain that produced α-L-rhamnosidase and β-D-glucosidase was used.

Therefore, the production of purified enzymes and identification of new products to address wine problems is challenging to researchers.

LIST OF ABBREVIATIONS

°P Plato
hL Hectoliter, a metric unit of volume equal to 100 L
mhL Million hectoliter
OIV International Organization of Vine and Wines
v/v Volume of alcohol by volume of beverage

ACKNOWLEDGMENTS

The authors would like to thank the collaboration and support by CNPq, Capes, and Fundação Araucária.

REFERENCES

Acton, Q.A., 2012. Enzymes and Coenzymes – Advances in Research and Application: 2012 Edition. Scholarly Editions™, Atlanta. 2602 p.
Armada, L., et al., 2010. Influence of several enzymatic treatments on aromatic composition of white wines. LWT – Food Science and Technology 43 (10), 1517–1525.
Bamforth, C.W., 1982. Brewers Digest 57 (6), 22–27.
Bamforth, C.W., 2009. Current perspectives on the role of enzymes in brewing. Journal of Cereal Science 50, 353–357.
Barth-Haas Group, 2013. Barth-Report 2012/2013, United Nations and World Bank Statistics Beer Production. Market Leaders and Their Challengers in the Top 40 Countries in 2012. Joh. Barth & Sohn GmbH & Co., Nurnberg. 16 p.

Baxter, E.D., Hughes, P.S., 2001. Beer: Quality, Safety and Nutritional Aspects. Nutritional Aspects of Beer. (Chapter 5). The Royal Society of Chemistry, UK. 138 p.

BBC News, December 02, 2013. German Beer Brewers in 'World Heritage' Appeal. Business.

von Bothmer, R., et al., 2003. Diversity in Barley (*Hordeum vulgare*). Elsevier Science B.V., Amsterdam. 300 p.

Briess, 2013. Mashing for Optimal Yield. Ellis Rich. Division Manager Brew East. Briess Malt and Ingredients Company.

Briggs, D.E., et al., 1981. Malting and Brewing Science: Malt and Sweet Wort, second ed. Springer Science & Business Media. 914 p.

Briggs, D.E., 1998. Malts and Malting. Blackie Academic & Professional, London. 796 p.

Bruchmann, A., Fauveau, C., 2010. Enzymes in potable alcohol and wine production. In: Whitehurst, R.J., van Oort, M. (Eds.), Enzymes in Food Technology, second ed. Blackwell Publishing, UK, pp. 195–210.

Callejón, S., et al., 2014. Identification of a novel enzymatic activity from lactic acid bacteria able to degrade biogenic amines in wine. Applied Microbiology and Biotechnology 98, 185–198.

Canal-Llaubères, R.M., 1993. Enzymes in winemaking. In: Fleet, G.H. (Ed.), Wine Microbiology and Biotechnology, vol. 886. Hardwood Academic Publishers, Washington, DC, pp. 477–506.

Coote, N., Kirsop, B.H., 1976. A haze consisting largely of pentosan. Journal of the Institute of Brewing 82, 34.

Curtis, T., 2013. A London Encyclopaedia, or Universal Dictionary of Science, Art, Literature and Practical Mechanics, vol. 13. Forgotten Books, London, pp. 484–485 (Original work published 1829).

Curvelo-Garcia, A.S., 2005. Práticas enológicas internacionalmente reconhecidas. Ciência e Técnica Vitivinícola 20 (2), 105–130.

Dhillon, G.S., Kaur, S., Brar, S.K., 2012. Flocculation and haze removal from crude fermented beer using in-house produced laccase via koji fermentation with *Trametes versicolor* using brewery spent grain. Journal of Agriculture and Food Chemistry 60 (32), 7895–7904.

Dhillon, G.S., Kaur, S., Oberoi, H.S., Spier, M.R., Brar, S.K., 2016. Agricultural-based protein byproducts: characterization and applications. In: Protein By-products: Transformation From Environmental Burden into Value-added Products. Elsevier Publisher.

Ducasse, M.A., et al., 2010. Effect of macerating enzyme treatment on the polyphenol and polysaccharide composition of red wines. Food Chemistry 118 (2), 369–376.

Ducroo, P., Frelon, P.G., 1989. European brewery convention. In: Proceedings of the 22nd Congress, Zurich, pp. 445–452.

Esslinger, H.M., 2009. Handbook of Brewing: Processes, Technology, Markets. John Wiley & Sons, Weinheim. 676 p.

Fincher, G.B., Stone, B.A., 1986. In: Pomeranz, Y. (Ed.), Advances in Cereal Science and Technology, vol. 8. Am Assoc Cereal Chem, St. Paul, MN, pp. 207–295.

Galante, Y.M., et al., 1998. Application of *Trichoderma* enzymes in food and feed industries. In: Harman, G.F., Kubicek, C.P. (Eds.), Trichoderma & Gliocladium—Enzymes, Biological Control and Commercial Applications, vol. 2. Taylor & Francis, London, pp. 327–342.

German Beer Institute, 2008. German Beer Primer for Beginners. Federal Republic of Germany.

González-Pombo, P., et al., 2011. A novel extracellular β-glucosidase from *Issatchenkia terricola*: isolation, immobilization and application for aroma enhancement of white Muscat wine. Process Biochemistry 46 (1), 385–389.

Guerdrum, L.J., Bamforth, C.W., 2012. Prolamin levels through brewing and the impact of prolyl endoproteinase. Journal of the American Society of Brewing Chemists 70 (1), 35–38.

Gurung, N., et al., 2013. A broader view: microbial enzymes and their relevance in industries, medicine, and beyond. BioMed Research International 2013:329121 18 p.

Guzzo, F., et al., 2011. The inhibitory effects of wine phenolics on lysozyme activity against lactic acid bacteria. International Journal of Food Microbiology 148, 184–190.

Hager, A.S., Taylor, J.P., Waters, D.M., Arendt, E.K., 2014. Gluten free beer – a review. Trends in Food Science and Technology 36 (1), 44–54.

Hartmann, G., et al., 2006. Rapid degradation of gliadin peptides toxic for coeliac disease patients by proteases from germinating cereals. Journal of Cereal Science 44, 368–371.

Humbert-Goffard, A., et al., 2004. An assay for glucanase activity in wine. Enzyme and Microbial Technology 34, 537–543.

Laitila, A., et al., 2006. *Lactobacillus plantarum* and *Pediococcus pentosaceus* starter cultures as a tool for microflora management in malting and for enhancement of malt processability. Journal Agricultural Food Chemistry 54, 3840–3851.

Le Traon-Masson, M.P., Pellerin, P., 1998. Purification and characterization of two β-D-glucosidases from an *Aspergillus niger* enzyme preparation: affinity and specificity toward glucosylated compounds characteristic of the processing of fruits. Enzyme and Microbial Technology 22 (5), 374–382.

Li, Y., et al., 2005. Studies on water-extractable arabinoxylans during malting and brewing. Food Chemistry 93, 33–38.

Linko, M., et al., 1998. Recent advances in the malting and brewing industry. Journal of Biotechnology 65, 85–98.

Manzanares, P., et al., 2003. Construction of a genetically modified wine yeast strain expressing the *Aspergillus aculeatus* RHAA gene, encoding an α-l-rhamnosidase of enological interest. Applied and Environmental Microbiology 69, 7558–7562.

Mateo, J.J., Jiménez, M., 2000. Monoterpenes in grape juice and wines. Journal of Chromatography A 881, 557–567.

Merín, M.G., et al., 2014. Pectinolytic yeasts from viticultural and enological environments: novel finding of *Filobasidium capsuligenum* producing pectinases. International Journal of Basic Microbiology 54, 835–842.

Miguel, M., et al., 1998. Enzimas pectolíticas com actividades β-glicosidásicas na vinificação de vinhos brancos. Enologia 31-32, 9–15.

Minussi, R.C., et al., 2007. Phenols removal in musts: strategy for wine stabilization by laccase. Journal of Molecular Catalysis B: Enzymatic 45, 102–107.

Nagodawithana, T., Reed, G., 1993. Enzymes in Food Processing, third ed. Academic Press Inc., California. 480 p.

OIV, 13 May, 2014a. International Organization of Vine and Wine. The Wine Market: Developments and Trends. Available in: http://www.oiv.int.

OIV, 2014b. OIV Confirms France as World's Top Wine Producing Country. 27 out 2014. Available in: http://www.harpers.co.uk/news.

Ortega-Regules, A., et al., 2008. Changes in skin cell wall composition during the maturation of four premium wine grape varieties. Journal of the Science of Food and Agriculture 88 (3), 420–428.

Oshiro, T., et al., 1989. Purification and characterization of a-acetolactate decarboxylase from *Brevibacterium acetylicum*. Agricultural and Biological Chemistry 53, 1913–1918.

Palmeri, R., Spagna, G., 2007. β-Glucosidase in cellular and acellular form for winemaking application. Enzyme and Microbial Technology 40 (3), 382–389.

Pardo, F., et al., 1999. Effect of diverse enzyme preparations on the extraction and evolution of phenolic compounds in red wines. Food Chemistry 67, 135–142.

Pasternack, R., Marx, S., Jordan, D., 2008. Prolamin Reduced Beverages and Methods for the Preparation Thereof. Patent 2008 0003327 Al.

Pérez-Serradilla, J.A., Castro, M.D.L.D., 2008. Role of lees in wine production: a review. Food Chemistry 111, 447–456.

Petropulos, V.I., et al., 2014. Study of the influence of maceration time and oenological practices on the aroma profile of Vranec wines. Food Chemistry 165, 506–514.

Pinelo, M., et al., 2006. Upgrading of grape skins: significance of plant cell-wall structural components and extraction techniques for phenol release. Trends in Food Science & Technology 17, 579–590.

Preedy, V.R., 2011. Beer in Health and Disease Prevention. Academic Press. 1248 p.

Research and Markets, 2013. Global and Chinese Beer Industry Report, 2012–2014. 115 p.

Reynolds, A.G., 2010. Managing Wine Quality: Oenology and Wine Quality, vol. 2. Woodhead Publishing Limited, UK. 651 p.

Ribéreau-Gayon, P., et al., 2006. second ed. Handbook of Enology – the Microbiology of Wine and Vinifications, vol. 1. John Wiley & Sons, Ltd.

Robinson, S.P., Davies, C., 2000. Australian Journal of Grape and Wine Research 6, 175.

Romano, P., et al., 2003. Function of yeast species and strains in wine flavour. International Journal of Food Microbiology 86 (1–2), 169–180.

Romero-Cascales, I., et al., 2012. The effect of a commercial pectolytic enzyme on grape skin cell wall degradation and colour evolution during the maceration process. Food Chemistry 130 (3), 626–631.

Sadosky, P., et al., 2002. Effect of arabinoxylans, β-glucans, and dextrins on the viscosity and membrane filterability of a beer model solution. Journal of the American Society of Brewing Chemists 60 (4), 153–162.

Siebert, K.J., Lynn, P.Y., 1998. Mechanisms of beer colloidal stabilization. Journal of the American Society of Brewing Chemists 55 (2), 73–78.

Siebert, K.J., 1999. Effects of protein–polyphenol interactions on beverage haze, stabilization, and analysis. Journal of Agricultural and Food Chemistry 47 (2), 353–362.

Spagna, G., et al., 2002. Properties of endogenous β-glucosidase of a *Saccharomyces cerevisiae* strain isolated from Sicilian musts and wines. Enzyme and Microbial Technology 31, 1030–1035.

Stewart, G.G., et al., 2013. 125th Anniversary Review: developments in brewing and distilling yeast strains. Journal of the Institute of Brewing 119, 202–220.

Stewart, G.G., Russell, I., 1998. An Introduction to Brewing Science & Technology: series III: Brewer's Yeast. The Institute of Brewing, London. 108 p.

Sydney, 16 January 1873. The Manufacture of Malt. Empire (Sydney, NSW: 1850–1875), p. 4. Available in: http://nla.gov.au/nla.news-article63226247.

Teranishi, R., et al., 1999. Flavor Chemistry: Thirty Years of Progress. Kluwer Academic, New York. 439 p.

Torresi, S., et al., 2011. Biotechnologies in sparkling wine production. Interesting approaches for quality improvement: a review. Food Chemistry 129, 1232–1241.

Torresi, S., et al., 2014. Effects of a β-glucanase enzymatic preparation on yeast lysis during aging of traditional sparkling wines. Food Research International 55, 83–92.

Van Landschoot, A., 2011. Gluten-free barley malt beers. Cerevisia 36 (3), 93–97.

Van Oort, M., 2010. Enzymes in food technology – introduction. Chapter 1. In: Whitehurst, R.J., Van Oort, M. (Eds.), Enzymes in Food Technology, second ed. Wiley-Blackwell, Chichester, UK, pp. 1–17.

Van Rensburg, P., Pretorius, I.S., 2000. Enzymes in winemaking: harnessing natural catalysts for efficient biotransformations—a review. South African Journal for Enology and Viticulture 21, 52–73.

Viëtor, R.J., Voragen, A.G.J., 1993. Composition of non-starch polysaccharides in wort and spent grain from brewing trials with malt from good malting quality barley and a feed barley. Journal of the Institute of Brewing 99, 243–248.

Viëtor, R.J., et al., 1991. Journal of Cereal Science 14, 73–83.

Villena, M.A., et al., 2007. β-Glucosidase activity in wine yeasts: application in enology. Enzyme and Microbial Technology 40, 420–427.

Webster, T., et al., 1856. The American Family Encyclopedia of Useful Knowledge, or Book of 7223 Receipts and Facts: A Whole Library of Subjects Useful to Every Individual. Derby & Jackson. 1238 p.

Whitehurst, R.J., Van Oort, M. (Eds.), 2010. Enzymes in Food Technology, second ed. Wiley-Blackwell, Chichester, UK. 371 p.

Zheng, X., et al., 1994. Factors influencing maltose utilization during wort fermentations. Journal of the American Society of Brewing Chemists 52, 41–47.

FURTHER READING

Bamforth, C.W., 2011. Beer: A Quality Perspective. Academic Press, New York. 304 p.

Munar, J.M., Sebree, B., 1997. Gushing—a malster's view. Journal of the American Society of Brewing Chemists 55, 119–122.

Oliver, G., 2011. The Oxford Companion to Beer. Oxford University Press, New York. 960 p.

Yadav, P.R., 2006. Plant Product Biotechnology. Discovery Publishing House, New Delhi. 314 p.

CHAPTER 12

Recent Applications of Enzymes in Personal Care Products

K. Sunar, U. Kumar, S.K. Deshmukh
The Energy and Resources Institute, New Delhi, India

INTRODUCTION

Enzymes, unique protein molecules that catalyze most of the reactions in living organisms, are rightly termed as catalytic machinery of the living system. Current applications of enzymes are focused on many different markets including pulp and paper, leather, detergents, textiles, pharmaceuticals, chemicals, food and beverages, biofuels, animal feeds, and personal care, among others (Adrio and Demain, 2014). At the same time, the end use market for industrial enzymes is extremely widespread with numerous industrial commercial applications (Adrio and Demain, 2005). Over 500 industrial products are being made using enzymes (Johannes and Zhao, 2006; Kumar and Singh, 2013). It was in 1833 that scientists discovered that a thermolabile substance was able to convert starch into sugar and called it "diastase," which is now known as "amylase." Later, in 1926 the protein nature of enzymes were finally confirmed when Summer (1926) successfully crystallized urease enzymes from Jack beans. Then an era of utilizing cell-free enzymes started with renin, an aspartic protease in cheese making. In the last 20 years, the global beauty market has grown by 4.5% a year on average (CAGR), with annual growth rates ranging from around 3–5.5%, also known as cosmetics and toiletries or personal care products (PCPs) (Barbalova, 2011). The world's cosmetic industry is worth tens of billions of US dollars, and the industry is constantly seeking new products with ingredients that have specific actions for which enzymes have been the most preferred choice for enhancement of PCPs. The first commercial enzyme was prepared by Rohm in Germany in 1914. The trypsin enzyme was isolated from animals; it degraded proteins and was used in detergents. The larger-scale commercialization of enzymes started with microbial protease derived from *Bacillus*, which was used in washing powders. In 1959, Novozyme of Denmark made it a huge business when they started manufacturing detergents with these microbial enzymes. In addition to the cheese industry, enzymes had been used in various fields, such as food industries and manufacturing of fruit juice, since 1930, however, a major breakthrough started around the 1960s when enzymes were industrially manufactured and were used in the starch industry. The traditional acid

Agro-Industrial Wastes as Feedstock for Enzyme Production
ISBN 978-0-12-802392-1
http://dx.doi.org/10.1016/B978-0-12-802392-1.00012-5

hydrolysis of starch was completely replaced by alpha amylases and glucoamylases that could entirely convert starch into glucose. The starch industry became the second largest industry to use enzymes after the detergent industry. Over the years, biotechnology has shown that it is now possible to utilize and harness the use of these enzymes in diverse fields. Enzymes have recently been started to be used by cosmetic scientists in developing PCPs for wide acceptability as they have been found to have good consumer appeal and improved performance. However, they have always been poorly evaluated for their functionality in cosmetic science. Proteolytic enzymes like bromelain, papain, etc. have been used in PCPs for skin peeling and smoothing for many years, however, the general problem associated with such use is the irritation caused by some enzymes on the skin surfaces due to their proteolytic activities. The area where the topical applications of enzymes are widely explored and have shown significant benefits is in skin protection, with enzymes having excellent stability. The enzymes used for skin protection have profound abilities to capture free radicals caused by environmental pollution, microorganisms, sunlight, radiations etc. The recent trend of application of enzymes in PCPs shows ample variability in terms of enzymes used from different types of classes for their specific roles and function. Studies of enzyme formulations suitable for topical use have also shown that such dosage forms are relatively easy to handle. However, the choice of base, surface active agent, etc., is important to provide for a stable formulation, and proper vehicle selection is also critical for the proper activity. Another futuristic approach to cosmetics and skin care product development is to increase the efficacy of existing ingredients that might improve skin functioning. Many new topical ingredients—from mushrooms to salmon caviar to sea urchin spines to green algae to knotweed—have been placed in complex antiaging formulations (Draelos, 2012). Nanoparticles are revolutionizing many areas of chemistry, physics, and possibly cosmetic formulation. The long-term effects of nanoparticles in the oceans of the world are not currently known. Yet, nanoparticles could be the next frontier in cosmetic dermatology (Sonneville–Aubrun et al., 2004). Nanoparticles have great potential to create topical cosmeceutical formulations that behave in ways that enable better penetration of active skin ingredients. Someday in the not-too–distant future we may be using nanoparticle therapy, nanoemulsions, polymeric nanoparticle spheres, and nanoliposomes to improve the appearance of the skin (Tadros et al., 2004). Nanotechnology may allow ingredients to exhibit new skin effects, improving cosmetics and skin care product efficacy. We will discuss a few of the specific and widely used enzymes, with the main focus on the nature of their activity (Table 12.1).

Superoxide Dismutase

Superoxide dismutase (SOD) belongs to the oxidoreductase class of enzymes and catalyzes oxidation and reduction reactions. SOD is one of the most effective and popular tropical enzymes so far being used in skin care products. After its discovery as a blue/

Table 12.1 Use of Enzymatic Activities in Cosmetic Products (Ugo Citernesi and Kathe Andersen; www.iralab.it/download/pubblicazioni/New_trends_in_drug.pdf)

Enzyme	Source	Cosmetic Use
Protease	Fungi	Peeling/antiaging/antiwrinkle
Lipases	Bacteria	Anticellulitis
Hyaluronidase	Bacteria	Moisturising agent
Tyrosinase	Yeast (recombinant) *Tenebrio molitor* Fungi	Tanning agent
Superoxide dismutase	Yeast (Recombinant)	Antifree radicals
Peroxidase	Horseradish, bacteria and yeast (recombinant)	Antifree radicals
Alkaline Phosphatase	Yeast and fungi (recombinant)	Antiwrinkle Energetic

green protein in 1938 by Mann and Leilin and its subsequent characterization as an enzyme and named as superoxide dismutase by McCord and Fridovitch in 1969, this enzyme has been frequently used in various fields for its effective role in catalyzing superoxide free radicals. Reactive oxygen species (ROS) are produced by cells during normal metabolic activities such as mitochondrial oxidative phosphorylation; however, levels of ROS vary with UV exposure and levels of antioxidant enzymes. Without inactivation, ROS damages macromolecules including lipid, proteins, and DNA (Zastrow et al., 2009). Numerous studies have tested the effects of solar radiation and oxidative stress on the skin (Lan et al., 2013), and oxidative stress has been linked to age-related loss of skin elasticity (Nylor et al., 2011), defective cellular signaling, and photoaging (Lee et al., 2012). Antioxidant enzymes mediate the removal of ROS, with different enzymes functioning in specific compartments thereby preventing ROS from reacting with DNA and other cell signal proteins, impairing their function (Fig. 12.1). Functionally, SODs are characterized as potential oxidizing and reducing agents, and many studies have demonstrated their applications in cosmetics and PCPs for younger looking skin. L'Oréal, a cosmeceutical company, was the first to obtain a European Union (EU) patent for this enzyme for its general use in cosmetics in the year 1973 (EU patent no. 2 287 889) and ever since the SOD from marine sources has been in use in developing PCPs. The main factor involved in utilizing enzymes as an active component of any PCP is its side effects and acceptability; many of the SODs derived from different sources were not effective or not considered suitable in the early days because they were reported to cause skin irritation. In 1987 Brooks Industries developed CuZn-SOD derived from yeast, which was formulated as a powder (Biocell SOD-Yeast CuZn-SOD) containing approximately 600 IU SOD activity. This form of yeast protein-bonded SOD is known to have excellent stability at 45°C in aqueous solution in comparison to pure forms of SOD.

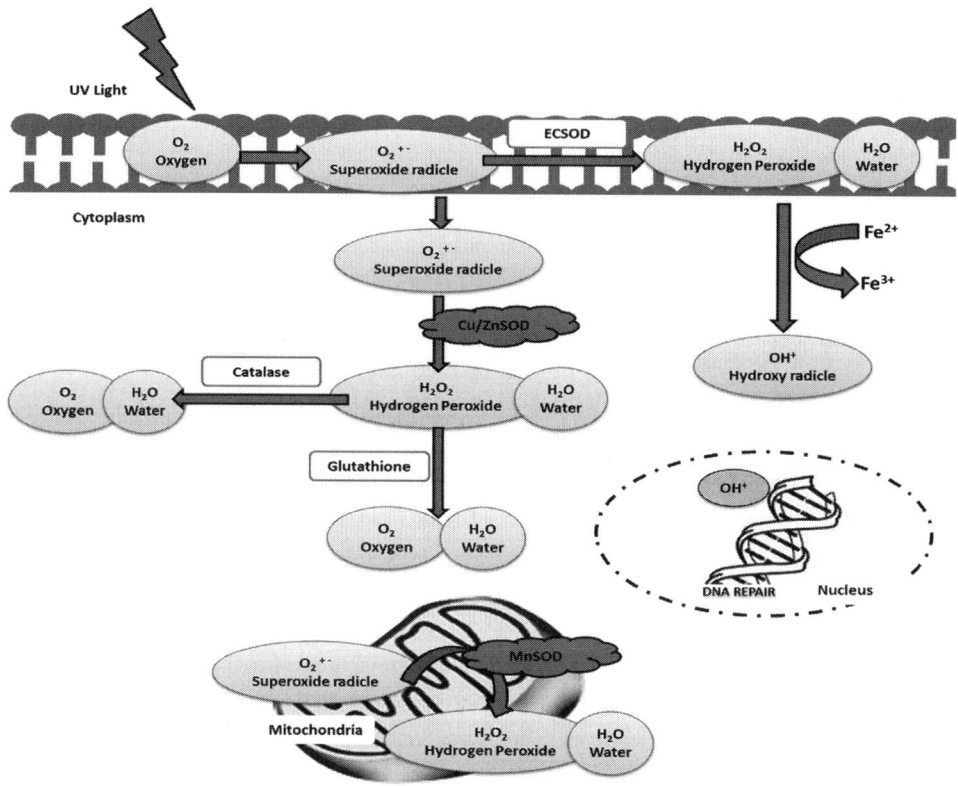

Figure 12.1 A schematic representation of different antioxidant enzyme functioning, in specific cell compartments for removal of reactive oxygen species. *ECSOD*, extracellular superoxide dismutase; *Cu/ Zn SOD*, copper/zinc superoxide dismutase; *MnSOD*, manganese superoxide dismutase. *Reproduced with input from Amaro-Ortiz, A., Betty Yan, B., D'Orazio, J.A., 2014. Ultraviolet radiation, aging and the skin: prevention of damage by topical cAMP manipulation. Molecules 19 (5), 6202–6219.*

This yeast SOD was also found to be nonirritating and nonsensitizing both in powder form and as an active 1% yeast CuZn-SOD liposome. Studies have also shown that these yeast-derived SODs are excellent antioxidants much better than commonly used antioxidants, like tocopherol and polyphenols. A few of the examples of products containing SOD as an active ingredient are presented in Table 12.2.

Peroxidase

There are two different types of hydroxyl free radical–scavenging enzymes, known as peroxidase and catalase belonging to the oxidoreductase class of enzymes. Plants are known to have heme-containing peroxidases, which are nonspecific peroxidases and are capable of acting on a variety of substrates including hydrogen peroxide. Similar nonspecific enzymes in animals are lactoperoxidase (thiocyanate ion oxidation), myeloperoxidase (phagocytosis), and thyroid peroxidase (iodine ion oxidation). However, the

Table 12.2 Commercial Products Containing Superoxide Dismutase as an Active Ingredient and Their Use (http://cosmetics.specialchem.com)

Products	Purpose	Manufacturer
Dismutin BT	Skin care:antiaging, antiinflammatory	DSN Nutritional Products, LLC
Dismutin BT 5000	Skin care- antiaging and antioxidant	DSN Nutritional Products, LLC
Liposystem complex	Skin care- Antioxidant Moisturizer	IRA Istituto Ricerche Applicate
Zymo Radical	Skin care- Smoothing Moistening Antiwrinkling	IRA Istituto Ricerche Applicate
Zymo Radical MD	Skin care- Antiwrinkling Moistening Smoothing	IRA Istituto Ricerche Applicate
Detox Duo	Skin care Protection Antioxidant	MD Skincare
Arch-Biocell SOD	Body lotion –Skin care antiinflammation Protective Antioxidant Soothing agent	LONZA
Brookosome SOD	Body lotion –Skin care antioxidant	LONZA
Chronosphere SOD	Body Lotion-skin care Antioxidant	LONZA
ProCircul8	Skin care Protective Antioxidant Antiinflammatory	LONZA

most studied one is the horseradish peroxidase obtained from the roots of horseradish. These free radical–scavenging enzymes are been extensively used in PCPs. For instance, fennel seed extracts containing peroxidase are being used in cosmetics because of their high-lipid peroxidation activities and low odor. The pale yellow/green liquid extract is also shown to have nonirritating and nonsensitizing activity and has shown much better protection activity than tocopherol. Lignin peroxidase, a novel skin-lightening active agent derived from a fungus is being studied with some interest for developing as an ingredient in products to treat pigmentation disorders. Some pigmentation disorders resulting from excessive sun exposure leading to solar lentigo are notoriously difficult to treat. Melanin is a very durable compound, and researchers have been largely unsuccessful in finding ways to break down melanin to reduce unwanted skin pigment. The existing topical treatments for skin lightening focus on the prevention of melanin formation by blocking tyrosinase and inhibiting its biosynthesis; by preventing the stimulation of melanocytes by UVA; or by blocking the transfer of melanosomes to keratinocytes via the PAR-2 receptor. The enzyme lignin peroxidase (LIP) was first identified by Gold et al. in 1984 and has been researched for many years as a potential agent to break down lignin to whiten wood pulp in paper production (Fig. 12.2). It was later found to break down eumelanin, which has a chemical structure similar to lignin. The development of lignin peroxidase as a skin-lightening agent resulted from these discoveries (US Patent and Trademark Office Patent Application 20060051305). This novel skin-lightening active ingredient is produced extracellularly during submerged fermentation of the fungus *Phanerochaete chrysosporium* (Woo et al., 2004) and then purified from the fermented liquid medium (Lonza of Switzerland). The LIP enzyme (trademarked as Melanozyme) identifies eumelanin in the epidermis and specifically breaks down the pigment without affecting melanin biosynthesis or blocking tyrosinase. Although there are other types of lignin peroxidase enzymes, at

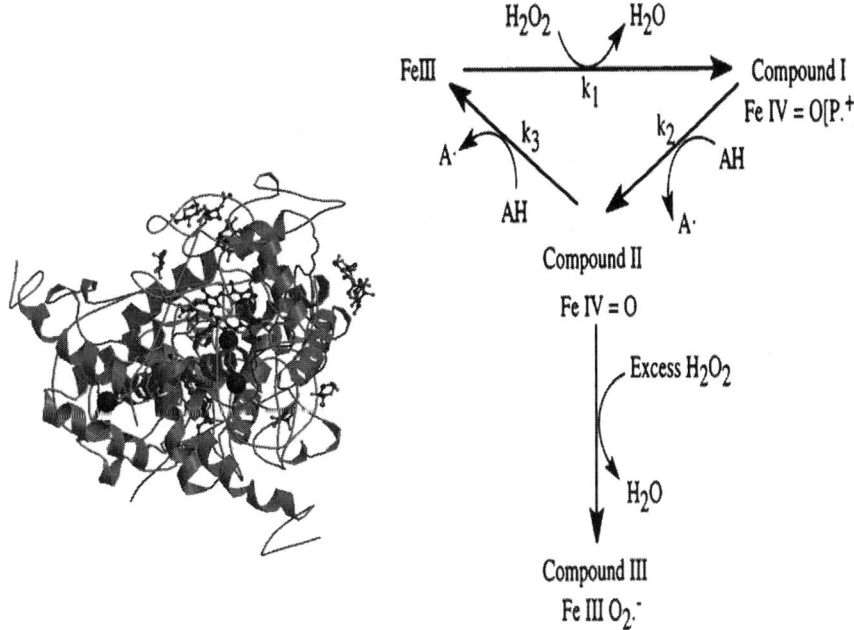

Figure 12.2 The crystal structure of lignin peroxidase at 1.70 Å resolution obtained from Protein Data Base and catalytic cycle of lignin peroxidase. *Adapted with permission from Gold, M.H., Wariishi, H., Valli, K., 1989. Extracellular peroxidases involved in lignin degradation by the white rot basidiomycete* Phanerochaete chrysosporium. *In: Whitaker, J., Sonnet, P. (Eds.), Biocatalysis in Agricultural Biotechnology. Toronto, Ontario, Canada: American Chemical Society, 127–140.*

this point, Melanozyme is the only one that has been developed and proved to be effective for skin lightening. Melanozyme is a glycoprotein active at pH 2–4.5. Melanozyme is currently proprietary and is available only in a new skin-lightening product known on the market as Elure. The safety of lignin peroxidase as a skin-lightening active ingredient has been demonstrated in preclinical studies with doses that are 17,000 times the recommended dose without prompting any side effects. LIP is non-mutagenic and nonirritating to eyes. The potential for skin irritation is very low, and in studies of 50 subjects each, there were no reports of skin irritation during acute sensitivity or cumulative sensitivity, or when used in sensitized skin. A few examples of products containing peroxidase as active ingredient are presented in Table 12.3.

Tyrosinase

Tyrosinase, an oxidase that is the rate-limiting enzyme for controlling the production of melanins is mainly involved in two distinct reactions of melanin synthesis: (1) the hydroxylation of a monophenol, and (2) the conversion of an o-diphenol to the corresponding o-quinone. o-quinone undergoes several reactions to eventually form melanin (Hideya et al., 2007; Kumar et al., 2011) (Fig. 12.3). Melanin synthesis in melanocytic cells is ultimately regulated by tyrosinase, a membrane-bound copper-containing glycoprotein, which is the critical rate-limiting enzyme. Tyrosinase is produced only by melanocytic cells, and following its synthesis and subsequent processing in the endoplasmic reticulum and Golgi, it is trafficked to specialized organelles, termed melanosomes, wherein the pigment is synthesized and deposited. In the skin and hair, the melanosomes are transferred from melanocytes to neighboring keratinocytes and are distributed in those tissues to produce visible color (Hideya et al., 2007). During the past years, the cosmetic industry increasingly worked with substances involved in natural melanin formation. The advantages here are obvious. Unlike the

Table 12.3 A Few of the Commercial Products Containing Peroxidase as an Active Ingredient and Their Uses (http://cosmetics.specialchem.com)

Products	Purpose	Manufacturer
Liposystem complex	Antioxidants Moisturizing agents Nourishing agents	IRA Istituto Ricerche Applicate
Zymo radical MD	Antiwrinkle agents Moisturizing agents Smoothening agent	IRA Istituto Ricerche Applicate
Zymo radical	Antiwrinkle agents Moisturizing agents Smoothening agent	IRA Istituto Ricerche Applicate
ABS fennel extract	Antiaging Antistress/Relaxing agents	Active Concepts

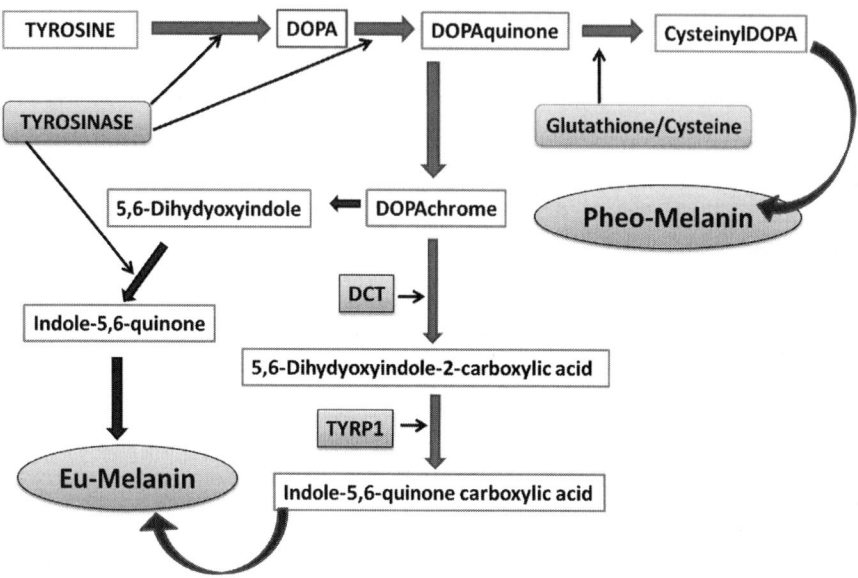

Figure 12.3 Schematic overview of melanin synthetic pathway and the involvement of melanogenic enzymes. Initial melanin synthesis is catalyzed by tyrosinase and is then divided into eumelanogenesis or pheomelanogenesis. The other melanogenic enzymes, that is, L-3,4-dihydroxyphenylalanine (DOPA) chrome tautomerase (DCT) and tyrosinase-related protein 1 (TYRP1), are involved in eumelanogenesis.

melanoidin process, a natural tan is induced and protection against UV radiation also is provided. It is a well-known fact that the enzyme tyrosinase transforms the amino acid tyrosine into dihydroxyphenylalanine (DOPA) and into its quinoid form, the DOPA quinone, which is the base for the formation of both the melanin types, eumelanin (dark brown) and pheomelanin (reddish yellow). The combination of both the types is responsible for the skin tone, which varies from skin to skin. The tyrosinase is controlled by UV radiation and induced by the α-melanocytes stimulating hormone (α-MSH). Further tyrosinase stimulators are the β-endorphins. Endorphin-related substances can be found in specific vegetable extracts as, eg, the chaste berry or chaste tree (vitex agnus castus), and together with synthetic acetyl tyrosine, a tyrosine pro-drug, they are able to induce the UV independent formation of melanin. Additional UV radiation will speed up and stimulate the melanin formation process after the product has been applied. New developments concentrate on additional tyrosinase activators and adequate transport systems to integrate the substances into the skin (Lautenschltens, 2007). Zymo-tan complex, a tanning activator, consists of tyrosine amino acids (precursors of melanine) and tyrosinase. Tyrosinase is an enzyme that catalyzes the reaction forming the melanin in the presence of solar radiation. The enzyme, present in several plants, has been also isolated from leucocytes, yeast, and milk. Some of the tyrosinase-based products are presented in Table 12.4.

Table 12.4 A Few of the Commercial Products Containing Tyrosinase as an Active Ingredient and Their Use (http://cosmetics.specialchem.com)

Products	Purpose	Manufacturer
Hydrosoluble zymo tan complex	Sunscreen agents, self-tanning agents	IRA Istituto Ricerche Applicate
Zymo tan complex PF	Sunscreen agents, self-tanning agents	IRA Istituto Ricerche Applicate
Zymo tan complex	Sunscreen agents, self-tanning agents	IRA Istituto Ricerche Applicate
Brookosome	Sunscreen agents, self-tanning agents	Lonza

Proteases

Proteases (also known as peptidases or proteinases), their substrates and inhibitors are of great relevance to biology, medicine, and biotechnology. Proteases are referred to as a group of enzymes that hydrolyze the protein bonds of amino acids (proteolysis). Proteases have evolved multiple times, and different classes of protease can perform the same reaction by completely different catalytic mechanisms (Gupta and Khare, 2007; Kalpana Devi et al., 2008). Proteases constitute the largest group of enzymes in bioindustry with an array of applications. They play an important role in industrial biotechnology, especially in detergents, foods, pharmaceuticals, and in PCPs. Proteolytic enzyme is essential for several physiological processes like digestion of food proteins, protein turnover, cell division, blood clotting cascade, signal transduction, processing of polypeptide hormones, etc. (Li et al., 2013). The vast variety of proteases, with specificity of their action and application, have attracted worldwide attention to exploit their physiological as well as biotechnological applications (Poldermans, 1990). They are considered eco-friendly because the appropriate producers of these enzymes for commercial exploitation are nontoxic and nonpathogenic and are designated as safe (Gupta et al., 2002). Proteases are used extensively in the pharmaceutical industry for preparation of medicines, such as ointments for debridement of wounds. They are also used in denture cleaners and as contact lens enzyme cleaners (Ogunbiyi et al., 1986). Proteases that are used in the food and detergent industries are prepared in bulk quantities and are used as crude preparations; whereas those used in medicine are produced in small amounts but require extensive purification before application (Bholay and Patil, 2012). The thermostability and their activity at high pH and the alleviation of pollution characteristics have made proteolytic enzymes an ideal candidate for laundry applications. Alkaline proteases are supplemented in different brands of detergents for use in home and commercial establishments. Enzymes have been added to laundry detergents for the last 50 years to facilitate the release of proteinaceous material in stains, such as those of milk and blood. The proteinaceous dirt coagulates on the fabric in the absence of proteinases as a result of washing conditions. The enzymes remove not only the stain, such as blood, but also other materials including proteins from body secretions and food, such as milk, egg, fish, and meat. An ideal detergent enzyme should be stable and active in the detergent solution and should have adequate temperature stability to be effective in a wide range of washing temperatures (Aurachalam and Saritha, 2009). A few examples of products containing different types of proteases as active ingredient are presented in Table 12.5.

Lipases

Lipases belong to hydrolases and exert their activity on the carboxyl ester bonds of triacylglycerols and other substrates. Their natural substrates are insoluble lipid compounds prone to aggregation in aqueous solution. Lipases are ubiquitous enzymes present in all types of living organisms. In eukaryotes they may be confined within an organelle (ie, the lysosome), or they can be found in the spaces outside cells and play roles in the metabolism, absorption, and transport of lipids. In lower eukaryotes and bacteria, lipases can be either intracellular or be secreted in order to degrade lipid substrates present in the environment, and in some pathogenic organisms (*Candida albicans*, *Staphylococcus* and *Pseudomonas* species, *Helicobacter pylori*) they can even act as virulence factors. Most bacterial lipases are sourced from *Pseudomonas, Burkholderia, Alcaligenes, Acinetobacter, Bacillus,* and *Chromobacterium* species; widely used fungal lipases are produced by *Candida, Humicola, Penicillium, Yarrowia, Mucor, Rhizopus,* and *Aspergillus* sp. Among the lipases from higher eukaryotes, porcine pancreatic lipase has been in use for several years as a technical enzyme (Lotti and Alberghina, 2007). Active lipases can mainly be found in cosmetics for surficial cleansing (anticellulite treatment) or overall body slimming, where they are responsible for the mild loosening and removal of dirt and/or small flakes of dead corneous skin (ie, peeling) and/or assist in breaking down fat deposits, often in combination with further enzymes, such as proteases. Further applications have been described for nose cleansing, makeup beauty masks, and hair care. Based on the broad variety of compounds derived from fats and carboxylic acids in cosmetic products, lipases and their hydrolytic, esterifying, and acylating activities show enormous potential for implementation in the production of cosmetic ingredients. In fact, a multitude of possible lipase-catalyzed syntheses have been described to date, and a variety of products have actually been commercialized. For classification, specialty esters, aroma compounds, and functional actives can essentially be distinguished (Marion et al., 2013). A few examples of products containing lipase as active ingredient are presented in Table 12.6.

Hyaluronidase

Hyaluronidase, enzymes that catalyze the hydrolysis (chemical decomposition involving the elements of water) of certain complex carbohydrates, such as hyaluronic acid (HA) and chondroitin sulfates, have been found in insects, leeches, snake venom, mammalian tissues (testis being the richest mammalian source), and in bacteria. HA has gained much importance in cosmetics for its popularity in cosmetic facial augmentation. HA is a naturally occurring glycosaminoglycan disaccharide present in skin, joint synovia,

Table 12.5 A Few of the Commercial Products Containing Protease as an Active Ingredient and Their Uses (http://cosmetics.specialchem.com)

Products	Purpose	Manufacturer
Protease	Antiaging agents Antiwrinkle agents Nourishing agents	Green Tech
Prozymex HBT LS 9142	Antiaging agents Lightening and whitening agents Exfoliants/peeling agents Smoothing agents	Laboratories Serobiologiques
Zymo acids	Conditioning agents	IRA Istituto Ricerche Applicate
Depil enzyme	Depilatory agents Antihair regrowth agents	IRA Istituto Ricerche Applicate
Zymo hair MD	Moisturizing agents	IRA Istituto Ricerche Applicate
Zymo lift MD	Antiwrinkle agents Moisturizing agents	IRA Istituto Ricerche Applicate
Okoumyrrhine	Antiaging agents Antiwrinkle agents Smoothing agents Antiinflammatory	Naturactiva
Dub karite	Antioxidants Antiaging agents Antiwrinkle agents Smoothing agents Antiinflammatory Moisturizing agents Healing agents Shining agents Regenerating agent Healing agent	Stearinerie dubois
PromaCare TA Prozymex HBT LS 9142	Lightening and whitening agents Antiaging agents Whitening agents Exfoliants/peeling agents Smoothing agents	Uniproma chemical Laboratoires Serobiologiques
Bromelain	Lightening and whitening agents Smoothing agents	Spec-chem Industry
BioNatural enzyme SK 320 P	Moisturizing agents Exfoliants/peeling agents Smoothing agents Antiwrinkle agents Conditioning agents	Bio-organic concepts
Eperuline PW LS 9627	Firming agents Toning agents Antiaging agents Antiinflammatory	BASF

Table 12.6 A Few of the Commercial Products Containing Lipase as an Active Ingredient and Their Uses (http://cosmetics.specialchem.com)

Products	Purpose	Manufacturer
CycloLipase	Slimming agents	Sederma Croda International Group
Zymo hair MD	Moisturizing agents	IRA Istituto Ricerche Applicate
Sopholiance	Antimicrobial and deodorants	Soliance
Zymo cell MD	Moisturizing agents Slimming agents	IRA Istituto Ricerche Applicate
Zymo clear MD	Moisturizing agents Antiacne agents	IRA Istituto Ricerche Applicate
Uncaryl	Antiaging agents Antiinflammatory Antiacne agents Slimming agents Antioxidants Sunscreen agents/UV filters	Cobiosa
Pheoslim	Slimming agents	Codif
Pheoslim G	Slimming agents	Codif
Lipocel-ErasePB	Slimming agents	PROTEOS Biotech
Lipocleansing-Erase HydraPB	Antiacne agents Peeling agents Smoothing agents	PROTEOS Biotech
Lipocleansing-SensitivePB	Antiacne agent Antiallergenic agent	PROTEOS Biotech
Spec-Chem-Climbazole	Antidandruff agents Antimicrobials	SpecChem Industry
Lipocel-Erase HYDRA PB	Slimming agents	PROTEOS Biotech
Facial cleaner	Antimicrobial Anticellulite Treatment/ Surfacial cleansing	JUJU Cosmetics
Revue SebumSoap	Antimicrobial Anticellulite Treatment/ Surfacial cleansing	Kanebo Cosmetics
Silhouette Sculptant Exfoliating Mousse 402	Anticellulite treatment	Maria Galland
Double Minceur Cible'e	Anticellulite treatment	Guinot
Bath additive with fat dissolving enzymes	Slimming agents	Ishizawa Laboratories

cartilage, and vitreous (Kablik et al., 2009). For its use as dermal filler, HA is chemically cross-linked to achieve the manufacturer's desired composition, which determines the filler's structure, longevity, and other properties. The properties of HA are adjusted in the manufacturing of different commercially available HA fillers, leading to their differing

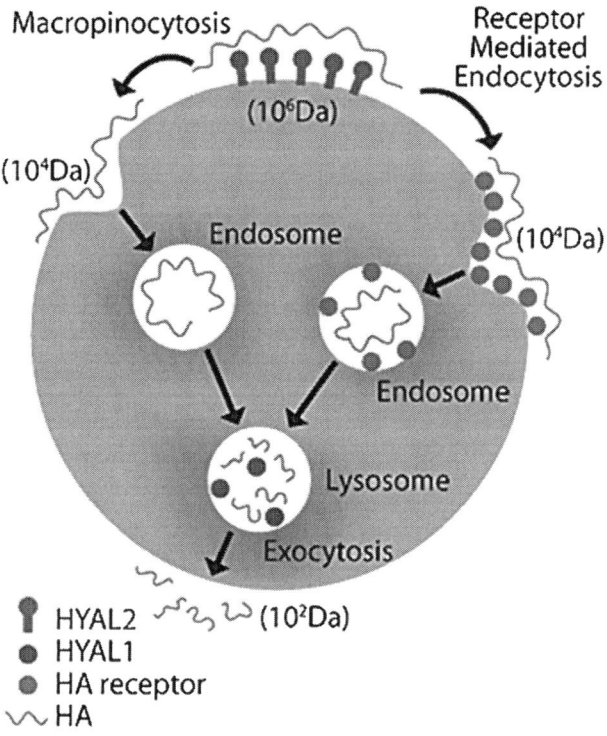

Figure 12.4 Schematic overview of hyaluronic acid (HA) endocytosis and processing. HMW- HA (~106 Da) is first degraded by hyaluronidase 2 (HYAL2) into smaller 104 -Da-sized fragments before it is taken up by a cell. The cell can either utilize surface HA receptors for receptor-mediated endocytosis or macropinocytosis. Once internalized the HA is degraded by hyaluronidase 1 (HYAL1) into small 102 Da fragments and then exocytosed. *Reproduced with input from Racine, R., Mummert, M.E., 2012. Hyaluronan endocytosis: mechanisms of uptake and biological functions. In: Brian, C. (Ed.), Biochemistry, Genetics and Molecular Biology – Molecular Regulation of Endocytosis. ISBN:978-953-51-0662-3.*

structural properties. These varying properties may inform clinicians as to which HA filler would be most appropriate for a specific clinical use. For example, a more highly cross-linked HA filler would likely be resilient in its ability to hold its form, making it suitable for the correction of deep wrinkles. In addition, a more monophasic filler might cleanly retain its form and clinically have a smoother appearance. Hyaluronidase is a naturally occurring enzyme capable of local degradation of HA (Fig. 12.4), thereby providing a means for correction or alteration of injected fillers. It is US Food and Drug Administration (FDA) approved as a temporary dispersion agent for injectable fluids, typically local anesthetics during retrobulbar blocks. It has been used clinically for over 60 years (Silverstein et al., 2012). In the event of complications with HA fillers, hyaluronidase has been used in attempt to reverse HA fillers (Lambros, 2004). Hyaluronidase hydrolyzes HA by splitting the bond between C1 of an N–acetylglucosamine moiety and C4 of a glucuronic acid moiety. It is FDA approved as an agent to increase tissue

Table 12.7 A Few Commercial Products Containing Hyaluronidase as an Active Ingredient and Their Use (http://cosmetics.specialchem.com)

Products	Purpose	Manufacturer
Hyaluronidase	Moisturizing agent	IRA Istituto Ricerche Applicate
Alpaflor Edelweiss B	Antiaging Antimicrobial Antioxidants	DSM
EcoCare Telmesteine	Antiinflammatory Antiaging	Ecochem
Phytessence Wakame	Anti-Inflammatory Antiaging Moisturizing agent Antioxidants	Croda
Hyaldrain PB	Slimming agents	PROTEOS Biotech
Hyalucorrect PB	Conditioning agents	PROTEOS Biotech

permeability to facilitate subcutaneous hydration, drug dispersion, and reabsorption of radiopaque dyes, so its use to reverse HA fillers is off-label. Different formulations of hyaluronidase are available, including a human recombinant agent and an ovine agent (Rao et al., 2014). A few examples of products containing lipase as active ingredient are presented in Table 12.7.

APPLICATION OF NANOPARTICLES FOR ENZYME IMMOBILIZATION

Nanotechnology plays a crucial role in developing elegant and effective cosmeceuticals by using smaller particles that are readily absorbed into the skin and repair damage easily and more efficiently (Singh et al., 2013a). Incorporation of nanotechnology in cosmeceuticals is aimed toward making incense of perfumes last longer, sunscreens to protect the skin, antiaging creams to fight back the years, and moisturizers to maintain the hydration of skin. Some of the innovations brought by nanotechnology intervention in the cosmeceutical arena are nanoemulsions (which are transparent and have unique tactile and texture properties), nanocapsules (which are used in skin care products), nanopigments (that are transparent and increase the efficiency of sunscreen products), liposome formulations (which contain small vesicles consisting of conventional cosmetic materials that protect oxygen or light sensitive cosmetic ingredients), niosomes, nanocrystals, solid lipid nanoparticles, carbon nanotubes, fullerenes, and dendrimers (Fig. 12.5). The primary advantages of using nanoparticles in cosmeceuticals include improvement in the stability of cosmetic ingredients (eg, vitamins, unsaturated fatty acids, and antioxidants) by encapsulating within the nanoparticles; efficient protection of the skin from harmful ultraviolet (UV) rays; aesthetically pleasing products (eg, in mineral sunscreens, using smaller particles of active mineral allows them to be applied without leaving a noticeable

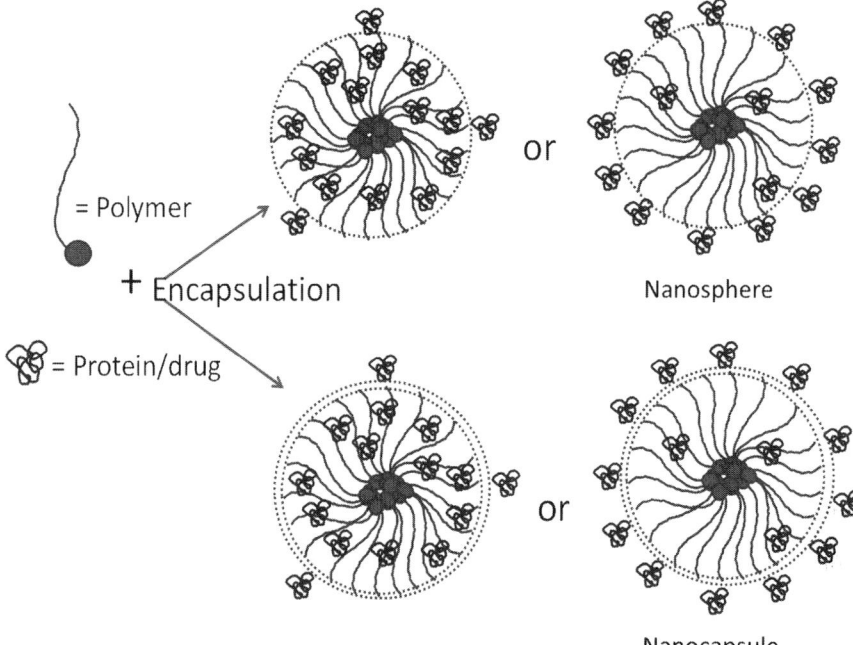

Figure 12.5 Schematic overview of different types of nanoencapsulations.

white cast); targeting of active ingredient to the desired site and controlled release of active ingredients for prolonged effect (Padamwar and Pokharkar, 2006; Mu and Sprando, 2010). LifePak Nano is a perfect example of a commercial product in which nanoencapsulation increases bioavailability of coenzyme Q10, protecting cells, tissues, and organs in the body against the ravages of aging (Lohani et al., 2014). There has been considerable interest in the development of enzyme immobilization techniques because immobilized enzymes have enhanced stability compared to soluble enzymes, and can easily be separated from the reaction. A few examples of nanoimmobilized enzymes are presented in Table 12.8. Approaches used for the design of immobilized enzymes have become increasingly more rational and are employed to generate improved catalysts for industrial applications. There are a variety of methods used to immobilize enzymes, the three of the most common being adsorption, entrapment, and cross-linking or covalently binding to a support. Currently, the major focus of enzyme immobilization has been in the development of robust enzymes that are not only active but also stable and selective in organic solvents. The ideal immobilization procedure for a given enzyme is one that permits a high-turnover rate of the enzyme while retaining high-catalytic activity over time. Proteins are immobilized either by physical adsorption to the surface of the nanoparticle or by covalent bonding to previously functionalized nanoparticles. Application of nanoparticles in formulations for PCPs has paved the way for utilizing these

Table 12.8 Examples of Nanoimmobilized Enzymes With Enhanced Activity (Singh et al., 2013b)

Enzyme	Applications	Kinetic Parameters	Nanoparticles Used
α–Chymotrypsin	Proteolysis (cleave peptide amide bonds)	Immobilized enzyme: $K_m = 31.7\,\mu M$, kcat $= 20.0\,s^{-1}$; Soluble enzyme: $K_m = 47.8\,\mu M$, kcat $= 17.8\,s^{-1}$	General
Glucose oxidase	Estimation of glucose level up to $300\,mg\,mL^{-1}$	Immobilized enzyme: $K_m = 3.74\,mM$, Soluble enzyme $= 5.85\,mM$	Gold NPs
Diastase	Starch hydrolysis	Immobilized enzyme: $K_m = 8414\,mM$, $V_{max} = 4.92\,\mu mol\,min^{-1}\,g{-1}$; Soluble enzyme: $K_m = 10,176\,mM$, $V_{max} = 2.71\,\mu mol\,min^{-1}\,mg^{-1}$	Fe impregnated silica NPs
Keratinase	Synthesis of keratin	Immobilized enzyme: Specific activity $= 129.0\,U\,mg^{-1}$; Soluble enzyme: Specific activity $= 37\,U\,mg^{-1}$	Iron oxide NPs
Horseradish peroxidase	catalyzes the conversion of chromogenic substrates (eg, TMB, DAB, ABTS) into colored products	Immobilized enzyme: $K_m = 0.8\,mM$, $V_{max} = 0.72\,\mu mol\,min^{-1}\,mg^{-1}$; Soluble enzyme: $K_m = 0.43\,mM$, $V_{max} = 0.35\,\mu mol\,min^{-1}\,mg^{-1}$	Nanoporous Cu NPs
Glucose oxidase	Estimation of glucose level	Immobilized enzyme: $K_m = 2.7\,mM$, $V_{max} = 28.6\,U\,\mu g^{-1}$; Soluble enzyme: $K_m = 9\,mM$, $V_{max} = 6.2\,\mu mol\,min^{-1}\,mg^{-1}$	Silver NPs
β–1,4-Glucosidase (*Agaricus arvensis*)	Lignocellulose hydrolysis	Immobilized enzyme: $K_m = 3.8\,mM$, $V_{max} = 3347\,\mu mol\,min^{-1}\,mg^{-1}$; Soluble enzyme: $K_m = 2.5\,mM$, $V_{max} = 3028\,\mu mol\,min^{-1}\,mg^{-1}$	Silica NPs
Diastase α-amylase	Hydrolyzing soluble starch	Immobilized enzyme: $K_m = 10.3\,mg\,mL^{-1}$; $V_{max} = 4.36\,\mu mol\,mL^{-1}\,min^{-1}$; Soluble enzyme: $K_m = 8.85\,mg\,mL^{-1}$; $V_{max} = 2.81\,\mu mol\,mL^{-1}\,min^{-1}$	Ag NP's doped gum acacia-gelatin-silica nanohybrid
Laccase	Bioremediation of environmental pollutants	Immobilized enzyme: K_m $(10^{-2}\,mM) = 10.7$, V_{max} $(10^{-2}\,mM\,min^{-1}) = 14.0$; Soluble enzyme: K_m $(10^{-2}\,mM) = 5.69$, V_{max} $(10^{-2}\,mM\,min^{-1}) = 7.7$	General
α-galactosidase (*Aspergillus terreus* gr.)	Animal feed	Immobilized enzyme: $K_m = 1.40\,mM$, $V_{max} = 20.16\,U\,mL^{-1}$; Soluble enzyme: $K_m = 4.2\,mM$, $V_{max} = 16.33\,U\,mL^{-1}$	Calcium alginate (Beads)

particles for developing these products with enhanced enzyme activity. For example, LifePak Nano, a nutritional antiaging program formulated to nourish and protect cells, tissues, and organs in the body with the specific purpose of guarding against the ravages of aging, has developed a product (Face Gel- Pharmanex/USA) that uses nanoparticles to enhance enzyme activity and in this product; nanoencapsulation increases bioavailability of coenzyme Q10 by 5–10 times (Lohani et al., 2014).

FUTURE PERSPECTIVES

The global personal care market, estimated at about $300 billion at the retail level, is a highly attractive segment of the consumer products space. The market has seen steady growth of 4.5% per annum in the last few years, from low-capital-intensive asset base, providing high return on capital to investors. Organics-based products is the fastest growing segment of the global personal care industry. Rising concerns for health safety, increasing go-green consciousness and growing consumer awareness toward hazards of synthetic chemicals have fueled the demand for organic PCPs, and increasing health awareness among consumers will continue to drive the growth of the organic personal care market during the forecast period. Among the organic and natural products, enzymes derived from various sources have found their specific utility in formulations of PCPs, which have been increasing rapidly. Application of enzymes in a given topical application depends on the nature of product as well as the limitations associated with the enzyme activity. Shorter shelf life of enzyme-based PCPs is a factor limiting consumer demand. Synthetic products are loaded with a large amount of preservatives in order to conserve their attributes. Enzyme-based PCP manufacturers have a hard time sourcing organic ingredients as an alternative to synthetic preservatives. The organic products containing enzymes as an active ingredient with natural preservatives have a short shelf life or need to be refrigerated. With the advent of nanotechnology, the effectiveness and durability of enzyme-based PCPs have been enhanced. Utilization of nanoparticles for enzyme stability and action is now being considered as an effective measure to address the problems that enzyme-based PCPs face. The enzyme groups that have proved to be quite useful are the oxidoreductases, proteases, and hydrolases, In addition, the search for new enzymes still continues. According to estimates by Kline & Co, a leading consultancy firm, the antiaging segment is the single largest product type in the personal care market and is the key growth engine. Skin care and hair care—the two largest segments of the market—are also the fastest growing, providing sizeable growth opportunities for suppliers. Cosmeceuticals, which are cosmetic products with drug-like benefits, has become the fastest-growing segment of the cosmetics and personal care industry. The global cosmeceuticals market offers huge potential among Asian countries, like Japan, China, and India, which are set to attract major players in the future. Though the market is at a nascent stage in developing countries, such as India and China, there remains a large untapped potential, with the desire to look young and fair.

LIST OF ABBREVIATIONS

Cu/Zn SOD Copper/zinc superoxide dismutase
DOPA Dihydroxyphenylalanine
ECSOD Extracellular superoxide dismutase
HA Hyaluronic acid
HYAL1 Hyaluronidase 1
HYAL2 Hyaluronidase 2
LIP Lignin peroxidase
MnSOD Manganese superoxide dismutase
NP Nanoparticle
PCP Personal care products
ROS Reactive oxygen species
SOD Superoxide dismutase
TYRP1 Tyrosinase-related protein 1
UV Ultraviolet

REFERENCES

Adrio, J.L., Demain, A.L., 2005. Microbial cells and enzymes—a century of progress. In: Barredo, J.L. (Ed.). Methods in Biotechnology. Microbial Enzymes and Biotransformations, vol. 17. Humana Press, Totowa, NJ, USA, pp. 1–27.

Adrio, J.L., Demain, A.L., 2014. Microbial enzymes: tools for biotechnological processes. Biomolecules 4, 117–139.

Amaro-Ortiz, A., Betty Yan, B., D'Orazio, J.A., 2014. Ultraviolet radiation, aging and the skin: prevention of damage by topical cAMP manipulation. Molecules 19 (5), 6202–6219.

Aurachalam, C., Saritha, K., 2009. Protease enzyme: an eco-friendly alternative for leather industry. Indian Journal of Science and Technology 2 (12), 29–32.

Barbalova, I., 2011. Global beauty and personal care: the year in review and winning strategies for the future. In-cosmetics. http://www.in-cosmetics.com.

Bholay, A.D., Patil, N., 2012. Bacterial extracellular alkaline proteases and its industrial applications. International Research Journal of Biological Sciences 1 (7), 1–5.

Draelos, Z.D., 2012. Cosmetics, diet, and the future. Dermatologic Therapy 25, 267–272.

Gold, M.H., Kuwahara, M., Chiu, A.A., Glenn, J.K., 1984. Purification and characterization of an extracellular H$_2$O$_2$-requiring diarylpropane oxygenase from the white rot basidiomycete, *Phanerochaete chrysosporium*. Archives of Biochemistry and Biophysics 234, 353–362.

Gold, M.H., Wariishi, H., Valli, K., 1989. Extracellular peroxidases involved in lignin degradation by the white rot basidiomycete *Phanerochaete chrysosporium*. In: Whitaker, J., Sonnet, P. (Eds.), Biocatalysis in Agricultural Biotechnology. American Chemical Society, Toronto, Ontario, Canada, pp. 127–140.

Gupta, A., Khare, S.K., 2007. Enhanced production and characterization of a solvent stable protease from solvent tolerant *Pseudomonas aeruginosa*. Enzyme and Microbial Technology 42, 11–16.

Gupta, R., Beg, Q.K., Chauhan, B., 2002. An overview on fermentation, downstream processing and properties of microbial proteases. Applied Microbiology and Biotechnology 60, 381–395.

Hideya, A., Hirofumi, K., Masamitsu, I., Vincent, H.J., 2007. Approaches to identify inhibitors of melanin biosynthesis via the quality control of tyrosinase. Journal of Investigative Dermatology 127, 751–761.

Johannes, T.W., Zhao, H., 2006. Directed evolution of enzymes and biosynthetic pathways. Current Opinion in Microbiology 9, 261–267.

Kablik, J., Monheit, G.D., Yu, L., Chang, G., Gershkovich, J., 2009. Comparative physical properties of hyaluronic acid dermal fillers. Dermatologic Surgery 35, 302–312.

Kalpana Devi, M., Rasheedha Banu, A., Gnanaprabhal, G.R., Pradeep, B.V., Palaniswamy, M., 2008. Purification, characterization of alkaline protease enzyme from native isolate *Aspergillusniger* and its compatibility with commercial detergents. Indian Journal of Science and Technology 1, 1–6.

Kumar, A., Singh, S., 2013. Directed evolution: tailoring biocatalysis for industrial application. Critical Review in Biotechnology 33, 365–378.

Kumar, C.M., Sathisha, U.V., Dharmesh, S., Rao, A.G., Singh, S.A., 2011. Interaction of sesamol (3,4-methylenedioxyphenol) with tyrosinase and its effect on melanin synthesis. Biochemistry 93 (3), 562–569.

Lambros, V., 2004. The use of hyaluronidase to reverse the effects of hyaluronic acid filler. Plastic and Reconstructive Surgery 114 (1), 260–277.

Lan, C.C., Wu, C.S., Yu, H.S., 2013. Solar-simulated radiation and heat treatment induced metalloproteinase-1 expression in cultured dermal fibroblasts via distinct pathways: implications on reduction of sun-associated aging. Journal of Dermatological Sciences 72, 290–295.

Lautenschläger, H., 2007. Self-tanning products – a beautiful sun-tan without sun. Kosmetische Praxis 2007 (6), 8–10.

Lee, C.W., Park, N.H., Kim, J.W., Um, B.H., Shpatov, A.V., et al., 2012. Study of skin anti-ageing and anti-inflammatory effects of dihydroquercetin, natural triterpenoids, and their synthetic derivatives. Bioorganicheskaia Khimiia 38, 374–381.

Ai, Q., Yi, L., Marek, P., Inverson, B.L., 2013. Commercial proteases: present and future. FEBS Letters 587, 1155–1163.

Lohani, A., Verma, A., Joshi, H., Yadav, N., Karki, N., 2014. Nanotechnology-based cosmeceuticals. ISRN Dermatology:843687.

Lotti, M., Alberghina, L., 2007. Lipases: molecular structure and function. In: Polaina, J., Maccabe, B.B. (Eds.), Industrial Enzymes. Springer, pp. 263–281.

Marion, B., Schumacher, A., Thum, O., 2013. Immobilised lipases in the cosmetics industry. Chemical Society Review 42, 6475–6490.

Mu, l, Sprando, R.L., 2010. Application of nanotechnology in cosmetics. Pharmaceutical Research 27 (8), 1746–1749.

Naylor, E.C., Watson, R.E., Sherratt, M.J., 2011. Molecular aspects of skin ageing. Maturitas 69, 249–256.

Ogunbiyi, L., Riedhammer, T.M., Smith, X., 1986. US Patent 4614549. Method for Enzymatic Cleaning and Disinfecting Contact Lenses.

Padamwar, M.N., Pokharkar, V.B., 2006. Development of vitamin loaded topical liposomal formulation using factorial design approach: drug deposition and stability. International Journal of Pharmaceutics 30 (1–2), 37–44.

Poldermans, B., 1990. Proteolytic enzymes. In: Gerhartz, W. (Ed.), Proteolytic Enzymes in Industry: Production and Applications. VCH Publishers, Weinheim, Germany, pp. 108–123.

Racine, R., Mummert, M.E., 2012. Hyaluronan endocytosis: mechanisms of uptake and biological functions. In: Brian, C. (Ed.), Biochemistry, Genetics and Molecular Biology – Molecular Regulation of Endocytosis. ISBN: 978-953-51-0662-3.

Rao, V., Chi, S., Woodward, J., 2014. Reversing facial fillers: interactions between hyaluronidase and commercially available hyaluronic-acid based fillers. Journal of Drug in Dermatology 13 (9), 1053–1056.

Silverstein, S.M., Greenbaum, S., Stern, R., 2012. Hyaluronidase in ophthalmology. Journal of Applied Research 12 (1), 1–13.

Singh, R.K., Tiwari, M.K., Singh, R., Lee, J.K., 2013a. From protein engineering to immobilization: promising strategies for the upgrade of industrial enzymes. International Journal of Molecular Science 14, 1232–1277.

Singh, R., Tiwari, S., Tawaniya, J., 2013b. Review on nanotechnology with several aspects. International Journal of Research in Computer Engineering and Electronics 2 (3), 1–8.

Sonneville-Aubrun, O., Simonnet, J.T., L'Alloret, F., 2004. Nanoemulsions: a new vehicle for skincare products. Advances in Colloid and Interface Science 108, 145–149.

Summer, J.B., 1926. The isolation and crystallization of the enzyme urease: preliminary paper. Journal of Biochemistry 69, 435–441.

Tadros, T., Izqulerdo, P., Esquena, J., Solans, C., 2004. Formation and stability of nano-emulsions. Advances in Colloid and Interface Science 108, 303–318.

Woo, S.H., Cho, J.S., Lee, B.S., Kim, E.K., 2004. Decolorization of melanin by lignin peroxidase from *Phanerochaete chrysosporium*. Biotechnology and Bioprocess Engineering 9, 256–260.

Zastrow, L., Groth, N., Klein, F., Kockott, D., Lademann, J., Renneberg, R., Ferrero, L., 2009. The missing link–light-induced (280–1600 nm) free radical formation in human skin. Skin Pharmacology and Physiology 22, 31–44.

CHAPTER 13

Strategies to Enhance Enzyme Activity for Industrial Processes in Managing Agro-Industrial Waste

M. Puri, R.E. Abraham
Deakin University, Waurn Ponds, VIC, Australia

INTRODUCTION

Worldwide a large magnitude of agricultural waste residues is being generated from current industrial processing practices (Goula and Lazarides, 2015). A wide range of agricultural/agro-industrial wastes, such as sugar cane bagasse, orange peels, rice straw, wheat straw, brewer's grain, distiller's grain, olive pomace, and their by-product residues are potentially suitable feedstocks for possible bioconversion into a range of value-added products of biotechnological interests (Mussatto et al., 2013; Molina-Calle et al., 2015; Kinab and Khoury, 2015). Often bioconversions of these agro-industrial residues can be facilitated either by whole cells (viable or nonviable), enzyme complexes, or enzymes, to produce many fine materials such as cellular proteins, organic acids, bioactives, prebiotic oligosaccharides, biologically important secondary metabolites, and nutraceuticals that have flourished in multibillion dollar markets (Nguyen et al., 2014).

There are many agro-industrial residues that can be enzymatically treated for producing valuables, however, it is not possible to describe more than a handful of these in this chapter. Various valuables, such as soluble dietary fibers, ie, pectin and naringin from citrus peel waste, biodiesel from soybean/olive pomace, and monomeric sugars from lignocellulose biomass, can be isolated by enzymatic pretreatment of respective agro-industrial waste.

Generally, enzymes, such as proteases, carbohydrases, lipases, and oxidoreductases are widely used in biotechnology industry due to their excellent catalytic activity (Cao, 2012). Nevertheless, the major challenges associated with the enzyme-catalyzed bioprocesses are high operational costs due to low stability and reusability of the enzymes when extended to large-scale industrial processes (Puri et al., 2013). The techniques that may improve enzyme stability and reusability are enzyme immobilization, enzyme modification, protein engineering, and solvent engineering. Among them, enzyme immobilization is the most frequently used to enhance enzyme characteristics. It has been commonly recognized that immobilization is able to stabilize enzymes against chemical and

Agro-Industrial Wastes as Feedstock for Enzyme Production
ISBN 978-0-12-802392-1
http://dx.doi.org/10.1016/B978-0-12-802392-1.00013-7

environmental attacks, and importantly the immobilized enzymes could be recovered and reused in a large-scale process.

In this chapter, we have described the role of enzymes in agro-biomass processing for producing interesting bioactives and coproducts. In addition, how immobilization of enzymes to various supports can bring stability and cost-effectivity to the process that may enhance its industrial acceptability is discussed. Some of the examples of whether the addition of enzymes can enhance yields of valuables from agro-industrial waste residue are also discussed.

ENZYME IMMOBILIZATION

Immobilization of enzyme refers to localizing of a biocatalyst in a certain region of defined space with retention of its catalytic activity so that immobilized biocatalysts can be used repeatedly and continuously (Kennedy and Cabral, 1987). Localization of the enzyme depends upon the nature of the enzyme, the type of carrier/support, and the mode of interaction followed for matrix and enzyme involved (Puri et al., 1996).

Immobilization can be broadly classified into various techniques such as adsorption, covalent binding, cross-linking, and encapsulation (Fig. 13.1). The enzyme gets adsorbed on the surface of the support in adsorption. In encapsulation, the enzyme molecules are entrapped within pores connected by small entrances (Santalla et al., 2011). The mode of interaction between enzyme and dynamic supports based on various immobilization methods is presented in Fig. 13.2. Covalent immobilization improves enzyme stability and leads to easy recovery (Puri et al., 2005; Verma et al., 2012). However, steric hindrance, diffusion, and structural changes during immobilization process are some of its drawbacks; whereas, cross-linking of an enzyme could overcome these limitations (Zhu et al., 2011).

Glutaraldehyde is commonly used for cross-linking as it generates an enzyme-polymer composite. Several immobilization studies are reported using glutaraldehyde as cross-linker (Singh et al., 2007; Pal and Khanum, 2011; Abraham et al., 2014). Xylanase from

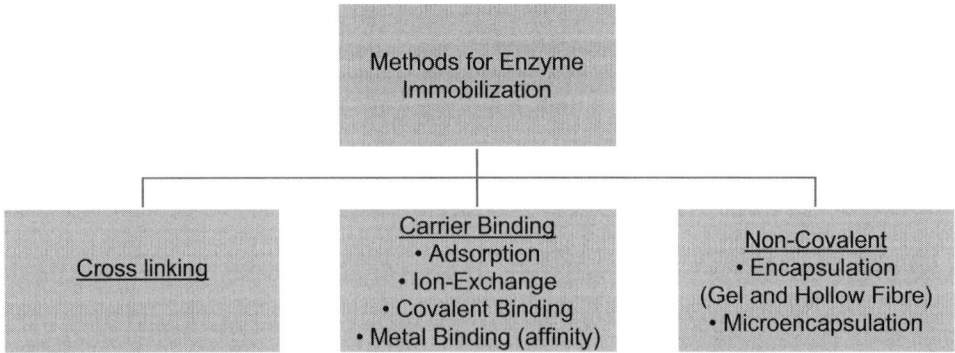

Figure 13.1 Various methods followed in the literature for enzyme immobilization.

Talaromyces thermophiles was immobilized using glutaraldehyde as cross-linker showing enhancement of pH stability from 7.0 to 8.0 (Maalej-Achouri et al., 2009; Cowan and Fernandez-Lafuente, 2011).

Immobilization enhances enzyme loading, its activity, and stability (Puri et al., 2010b; Verma et al., 2013). The enzyme tends to be more stable at a higher temperature, reusable, easier to handle and separate, and exhibits pH stability (Ansari and Husain, 2012). The reaction condition of immobilization largely affects the efficiency of the immobilized mixture. Some of the important parameters in biochemical characterization study include temperature, pH, enzyme loading, carrier/support loading, substrate concentration, agitation, and reaction time (Hanefeld et al., 2009; Brady and Jordaan, 2009). The enzyme gets denatured at an extreme temperature resulting in low or no activity; similarly, a moderate pH is required. A heavy loading of enzyme creates loose binding resulting in elution of more concentration of enzyme. The concentration of enzyme should be selected where minimal enzyme elution is observed in the reaction mixture with respect to time and agitation (Jordan et al., 2011). The experiment should be designed with the different ratio of enzyme and support material, which will allow a broad study to determine the best enzyme: support material immobilization combination in terms of binding efficiency. The percentage of binding efficiency is determined by the ratio of total amount of protein bound to the total amount of protein available (Abraham et al., 2014).

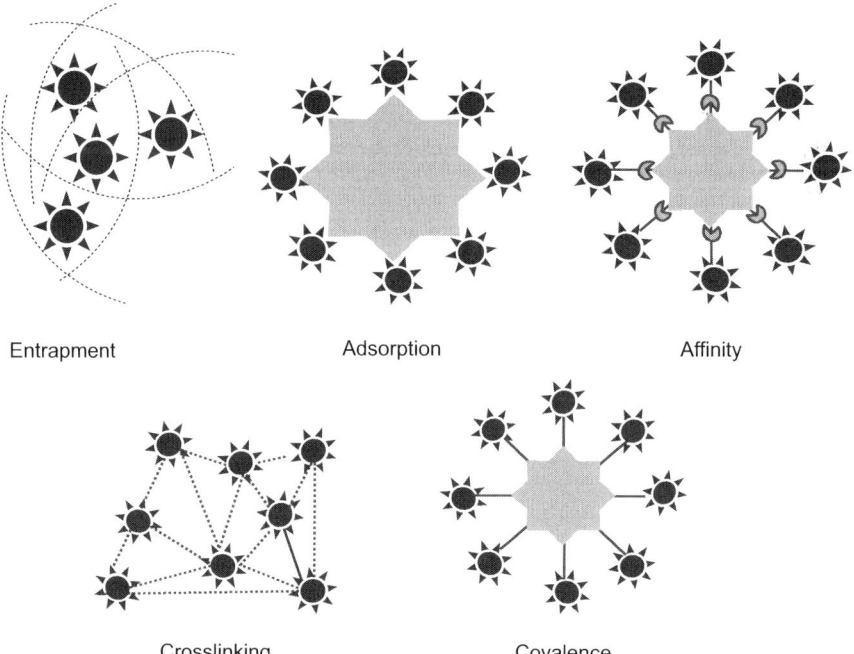

Figure 13.2 Mode of interaction between enzyme and support as result of immobilization.

There are inherent advantages and disadvantages associated with immobilization of enzyme and these are summarized in Table 13.1.

Different concentrations of substrate are used with immobilized enzyme to determine its activity and product formation. Immobilization offers various advantages over free enzyme (outlined in Table 13.1). During immobilization the properties of enzymes are enhanced. The enzyme becomes more stable at elevated conditions. Immobilized enzyme can work at higher temperature with good enzyme activity. This allows continuous usage of enzyme for a few cycles with easy separation from the reaction mixture. It further improves the storage stability of enzyme with prolonged life span and stable properties. There are various analytical techniques that are used to characterize the support, successful binding of enzyme and support. Some of them include Fourier transform infrared (FTIR), thermogravimetric analysis, Raman spectroscopy, transmission electron microscopy, scanning electron microscopy, and X-ray diffraction (Santalla et al., 2011; Deep et al., 2012).

DIFFERENT STRATEGIES FOR ENZYME IMMOBILIZATION ON VARIOUS SUPPORTS

Recently, various immobilization systems have been developed to use enzymes for the processing of agro-industrial waste in an efficient manner. Some of the commonly immobilized enzymes are cellulase, xylanase, and β-glucosidase, as well as other enzymes such as pectinase, lipase, laccase, inulinase, and rhamnosidase, which have been successfully immobilized on various support(s) (Table 13.2). These enzymes have improved their stability properties and appear to be efficient in hydrolyzing biomass or cellulose with good reusability (Jordan et al., 2011; Wang et al., 2011; Kobayashi and Fukuoka, 2013).

Many studies have been conducted on cellulase (EC 3.2.1.4) immobilization using different supports. Cellulase immobilized on acrylonitrile copolymer demonstrated enhanced pH stability (3.0–5.0); another study showed even broader pH stability (1.5–12.0) compared to free enzyme (Yuan et al., 1999; Hegedus et al., 2012). Few studies on cellulase immobilization on gamma-Fe_2O_3@Sio_2 core–shell particles have

Table 13.1 Advantages and Disadvantages of Immobilization of Enzymes

Advantages	Disadvantages
Mass transfer resistance	Cost of the fabrication process
Effective enzyme loading	Large-scale application
High surface area	Separation from the reaction medium
High mechanical strength	
Diffusional problems minimization	
Thermostability and reusability	
Storage stability	

Table 13.2 Various Supports Used for Enzyme Immobilization

Supports	Mode of Interaction	Enzyme	Application	References
Acrylamide	Covalent	Xylanase	Biomass hydrolysis	Sarbu et al. (2006)
Alginate	Entrapment, Metal binding	Pectinase, recombinant rhamnosidase	Clarification of juices, debittering of fruit juices	Puri et al. (1996, 2010b)
Alginate-glutaraldehyde	Covalent	Xylanase	Production of D-glutamate from L-glutamate	Pal and Khanum (2011)
Carbon nanotubes	Adsorption	Xylanase, Cellulase, Laccase	Industrial applications (food and agricultural industry)	Shah and Gupta (2008), Mubarak et al. (2014), Mukhopadhyay et al. (2015)
Coconut fiber	Covalent	Laccase	Clarification of apple juice	Bezerra et al. (2015)
Chitosan-graphene beads	Ionic	Lipase	Esterification	Lau et al. (2014)
Carrageenan	Covalent	β-glucosidase	Hydrolysis of macroalgae cellulosic residue	Tan and Lee (2015)
Duolite resin	Covalent	Inulinase	High-fructose syrup	Singh et al. (2007)
Ferric oxide pellets	Covalent	Xylanase		Soozanipour et al. (2015)
Glass beads	Covalent	Xylanases	Xylooligosaccharide	Maalej-Achouri et al. (2009)
Hollow fiber	Cross-link	Lipase	Biodiesel production	Okobira et al. (2015)
Mesoporous silica	Entrapment	Laccase	Industrial applications	Mansor et al. (2015)
Nanoparticles	Covalent	Galactosidase	Lactose hydrolysis	Verma et al. (2012)
Magnetic nanoparticle(s)	Covalent	Cellulase,	Biomass hydrolysis	Abraham et al. (2014)
Polyurethane foam	Adsorption	Pectinase	Hydrolysis of pectic oligosaccharides	Bustamante-Vargas et al. (2015)
Poly(styrene-methacrylic acid)	Covalent	Lipase	Biodiesel from soybean oil	Xie and Wang (2014)
Polyvinyl alcohol sponge	Entrapment	Pectinase	Orange juice clarification	Esawy et al. (2013)
Wood chips	Covalent	Naringinase	Flavonoid hydrolysis	Puri et al. (2005)

reported improvement in K_m and V_m value of reaction (Georgelin et al., 2010; Blanchette et al., 2012). Studies have demonstrated an increase of two-and-a-half- to threefold in K_m value due to immobilization of cellulase on various supports (Liang and Cao, 2012). An immobilization study on β-glucosidase from *Agaricus arvensis* on functionalized silicon oxide nanoparticle reported the alteration in substrate binding affinity of enzyme site due to immobilization (Singh et al., 2011). Cellulase enzyme complex produced on a nanoparticle demonstrated improved thermostability; it was found to be stable at 80°C for an hour after immobilization (Hegedus et al., 2012). Immobilized cellulases were found to successfully hydrolyze pretreated biomass. A study conducted on pretreated yellow poplar exhibited 45% substrate hydrolysis in 24 h (Wang et al., 2011). A similar study conducted on sodium hydroxide pretreated hemp hurd biomass presented a yield of 93% in comparison to 89% of hydrolysis by free cellulase (Abraham et al., 2014). This immobilized enzyme was prepared by using magnetic nanoparticle activated with glutaraldehyde. Alternatively, another report employing cellulase (*Aspergillus niger*) cross-linked with glutaraldehyde showed 52% of enzyme hydrolysis from rice hull (Cheng and Chang, 2013).

Recently cellulase enzyme has been immobilized on functionalized multiwalled carbon nanotubes through physical adsorption and reported an enzyme loading of 97% after immobilization. The enzyme retained about 52% of its activity after six cycles of reaction demonstrating an improved property of enzyme (Mubarak et al., 2014). Another study reported 77–80% of activity retained by cellulase after thermal stability analysis and six reusable cycles with minimum loss of activity (Mishra and Sardar, 2015). Supermagnetic nanoparticle was used to immobilize cellulase via physical adsorption and reported a binding efficiency of 95%, good stability at broader pH, long storage stability, and easy recovery (Khoshnevisan et al., 2011). Similarly, covalent immobilization of cellulase onto a magnetic particle reported hydrolysis of wheat straw with two reusable hydrolysis cycles (Alftrén and Hobley, 2014).

A xylanase (*endo*-1,4-β-xylanase; E.C.3.2.1.8) immobilized system was developed on a polysulfonate acrylate membrane. Covalent bond between amino acid of enzyme and acrylate group of the derivatized polysulfone was exhibited. The best immobilization reaction time was obtained in 24 h with storage stability of 38 days and was more stable at 55°C with improved pH stability. Maximum activity was observed at pH 6 (Cano et al., 2006). Immobilization of xylanase on acrylamide-grafted cellulose acetate flat membranes was successful in 7 h at 55°C. The confirmation of binding was achieved through FTIR analysis (Sarbu et al., 2006). Another study demonstrated the use of carbon nanotubes to immobilize xylanase. Hydrophobic surface of nanotube allowed binding, refolding, and immobilization of enzyme. Immobilization efficiency and the modification of α and β helix in the structure was confirmed using FTIR study (Shah and Gupta, 2008). In addition, a study conducted on xylanase using glutaraldehyde-alginate beads and entrapment method reported a maximum enzyme efficiency of 91%

with an increase in K_m (0.9–1.49%) and V_m (7092–8000 IU/mL) values. The temperature and pH stability were observed, and the improvement in thermostability was determined by increased half-life. The mixture demonstrated five successful reusable cycles (Pal and Khanum, 2011). Xylanase from *A. niger* was used for immobilizing Fe_3O_4-coated chitosan magnetic nanoparticle and demonstrated a good binding efficiency; it retained 87% of its activity after seven successive reusable cycles (Liu et al., 2014). Another study on xylanase used covalent immobilization attachment by response surface model on aluminium oxide pellets. They reported an immobilization yield of 83%, and the preparation retained 60% of its activity after 10 consecutive cycles. Functionalized magnetic nanoparticle used for immobilizing xylanase reported as high as 65% of initial activity after nine consecutive cycles of hydrolysis and exhibited a maximal activity at pH 6.5 and 60°C (Soozanipour et al., 2015). Xylanase immobilized on activated alginate beads using response surface methodology demonstrated as good as 85% of original activity after five continuous cycles with optimum conditions of pH 5.5 and temperature 40–45°C.

Pectinase (EC 3.2.1.15) from *A. niger* NRC1ami was evaluated for hydrolysis efficiency of dry untreated orange peels, HCl-treated orange peels, and NaOH-treated orange peels. The enzyme was entrapped in polyvinyl alcohol sponge and the optimum pH and temperature of the free and immobilized enzymes were shifted from 4, 40°C to 6, 50°C, respectively. The study of the pH stability of free and immobilized pectinase showed that the immobilization process protected the enzyme strongly from severe alkaline pHs. The immobilization process improved the enzyme thermal stability and could solve the orange juice haze problem completely. Immobilized enzyme was reused 12 times in orange juice clarification with 9% activity loss from the original activity (Esawy et al., 2013). Similarly, pectinase was successfully immobilized in rigid polyurethane foam (RFPU). The process achieved immobilization yield of 178.64%. The optimum reaction conditions for free and immobilized pectinase were pH 3.5 and 37°C and pH 4.5 and 55°C, respectively. The immobilization exerted a protective effect to the enzyme exposed to high temperatures, probably due to a result of the microenvironment created inside the RFPU. The kinetic parameters showed that the immobilization process did not change the affinity of the enzyme by the substrate. The immobilized pectinase in RFPU was reused six times in the hydrolysis of pectic oligosaccharides, keeping 35% of its initial activity (Bustamante-Vargas et al., 2015).

Another study reported the development of nanobiocatalytic system on functionalized carbon–based nanomaterial using lipase (EC 3.1.1.3) from *Candida rugosa*. A comparative study on physical adsorption and covalent attachment was conducted. The results demonstrated that 14 successive reusable cycles were obtained with covalent attachment with 60% initial activity, whereas the noncovalently immobilized enzyme lost 90% of its initial activity in nine uses (Pavlidis et al., 2012). Lipase from *C. rugosa* was immobilized onto chitosan/graphene oxide beads. Chitosan was incorporated with graphene oxide because the latter was able to enhance the physical strength of the chitosan beads by its superior mechanical integrity and low thermal conductivity. This study provided an enzyme-immobilizing carrier with excellent enzyme immobilization activity for an enzyme group requiring hydrophilicity on the immobilizing carrier. The overall esterification conversion of the prepared product was 78% at 60°C, and it attained conversions of 98% and 88% with commercially available lipozyme and novozyme, respectively, under similar experimental conditions (Lau et al., 2014).

Lipase immobilization was grafted onto a hollow fiber membrane by radiation-induced graft polymerization. This study controlled the length and density of the polymer brushes (amino ethanol functional groups) consisting of the glycidyl methacrylate (GMA) by changing the concentration of GMA monomer during radiation-induced graft polymerization. Immobilized lipase showed the highest activity on the grafted membrane when 5 wt% of glycidyl methacrylate as a monomer for the radiation-induced graft polymerization was used. High-efficiency esterification (approximately $1600 \, mmol \, h^{-1} \, g^{-1}$-membrane) was achieved in five-layer lipase on aminoethanol (AE) polymer brush than that in monolayer lipase on the polymer brush possessing only hydroxyl groups. Moreover, the polymer brush possessing AE functional groups for lipase immobilization maintained high activity during esterification of oleic acid to oleic acid methyl ester, and lipase deactivation was not observed over seven repeated cycles (Okobira et al., 2015).

Laccase (EC 1.10.3.2) enzyme finds enormous application in textile and dye industry and has also been used for degrading lignin. Recently an immobilized system was developed by immobilizing laccase with copper oxide nanoparticle and single-walled nanotube. The immobilized mixture was found to be stable at temperatures as low as 4°C and as high as 80°C (Mukhopadhyay et al., 2015). Recently, waste material from agro-industry has been used in the immobilization process to reduce costs. In this direction, a solid support for enzyme immobilization was prepared using the pretreated coconut fiber (CF) activated with glutaraldehyde and was used to immobilize laccase produced by *Trametes versicolor*. Immobilized enzyme retained up to $59 \pm 1\%$ of the initial activity and showed maximum immobilization profile of $98 \pm 1\%$. The thermal stability was higher when laccase was immobilized on alkaline-pretreated support with increments of 6.8-fold (laccase-glutaraldehyde-FC) of the soluble enzyme. The laccase-glutaraldehyde-CF achieved excellent results in the clarification of apple juice, reducing $61 \pm 1\%$ of the original juice color and $29 \pm 1\%$ of its turbidity, retaining up to 100% of the initial activity after a 10-times reuse assay (Bezerra et al., 2015).

Extracellular exoinulinase (EC 3.2.1.80) from *Kluyveromyces marxianus* YS-1, which hydrolyzes inulin into fructose, was immobilized on Duolite A568. Coupling of the enzyme to the phenol formaldehyde resin was done with glutaraldehyde. This weak ionic macroporus resin possessing amino group (eNH) was derivatized. The optimum temperature of immobilized enzyme was 55°C, which was 5°C higher than the free enzyme, and optimal pH was 5.5. Immobilized biocatalyst retained more than 90% of its original activity after incubation at 60°C for 3 h, whereas in the free form its activity was reduced to 10% under the same conditions, showing a significant improvement in the thermal stability of the biocatalyst after immobilization. Immobilized biocatalyst was effectively used in the batch preparation of high-fructose syrup from *Asparagus racemosus* raw inulin and pure inulin.

Alpha-L-rhamnosidase (EC 3.2.1.40) from *Bacillus* sp. is an enzyme that has been recognized as a member of glycoside hydrolase family 78 (Puri, 2012). The enzyme catalyzes the cleavage of terminal rhamnoside groups from naringin to prunin and rhamnose (Puri et al., 1996). It is widely distributed in fungi, yeast, and bacteria and plays an important role in the modification of the viscous property of gellan gum (Puri, 2012). This enzyme activity attracts growing biotechnological interest due to its role in rhamnose production (Puri and Banerjee, 2000), biotransformation of antibiotics and steroids (Perkins et al., 2015). Moreover, its hydrolyzed products can be used as starting material for the synthesis of substances applied in pharmaceuticals, cosmetics, and food technology (Perkins et al., 2015). Naringinase from *Penicillium* sp. was covalently immobilized to wood chips to improve its catalytic activity. The immobilization of naringinase on glutaraldehyde-coated woodchips through 1% glutaraldehyde cross-linking was optimized. The immobilization caused a marked increase in thermal stability of the enzyme. The immobilized naringinase was stable during storage at 4°C. No loss of activity was observed when the immobilized enzyme was used for seven consecutive cycles of operations. The efficiency of immobilization was 120%, while soluble naringinase afforded 82% efficacy for the hydrolysis of standard naringin under optimal conditions. Its applicability for debittering kinnow mandarin juice afforded 76% debittering efficiency (Puri et al., 2005, 2010b).

NANOMATERIALS AS DYNAMIC SUPPORTS FOR ENZYME IMMOBILIZATION

Use of nanomaterials to immobilize biocatalysts for better agro-industrial processing with respect to biodiesel production has recently been advocated (Verma et al., 2016). Nanomaterials offer many advantages due to their unique size and physical properties.

Immobilization of enzyme on a nanomaterial includes the nanoparticle, nanofibers, nanotubes, and nanocomposites (Puri et al., 2013). These nanomaterials provide high surface area, easier separation, and recovery, catalytic specificity, and binding capacity. Some of the commonly used nanomaterials are magnetic nanomaterials that include silicon, iron oxide, zinc oxide, chitosan, gold, and silver nanoparticles (Ansari and Husain, 2012). Aggregation, precipitation, thermodynamic stability, and monodispersity of nanomaterials are some of its limitations (Verma et al., 2013). Mesoporous material and solgels are also used for immobilization studies.

Studies have demonstrated the use of redox polymer microparticles to maintain activity for a longer time. Microparticles are required to have some of the essential factors such as transfer of electrons generated in an enzymatic reaction, maintain longer activity, and should be firmly attached to the surface (Lin et al., 2013). The modification of α-glucosidase with polyethylene glycol and aminoglucose improved the enzyme activity by fourfold; β-galactosidase covalently immobilized with porous ceramic support; β-galactosidase covalent attachment with Eupergit C (improved binding efficiency by 95%); and lipase from *Bacillus thermoleovorans* immobilized on porous polypropylene are some of the examples for improved stability and thermostability properties of enzymes (Cowan and Fernandez-Lafuente, 2011). Carbodiimide linkage is also used for covalent immobilization of lipases to nanomaterial (Dyal et al., 2003).

A magnetic composite poly (styrene-methacrylic acid) microsphere was prepared using oleic acid–coated magnetic nanoparticles as seeds by microemulsion copolymerization of styrene and methacrylic acid. The lipase from *C. rugosa* was then covalently bound to the magnetic polymer-coated microspheres by using 1-ethyl-3-(3-(dimethylamino) propyl) carbodiimide hydrochloride as an activation reagent. The immobilization of lipase enhanced the thermal and pH stability of immobilized enzyme when compared to free lipase. The immobilized enzyme showed high activity to soybean oil transesterification with methanol to produce biodiesel. It was found that the oil conversion of 86% was attained at a reaction temperature of 35°C for 24 h. The immobilized lipase was found to be stable with repeated use for four cycles without severe loss of its activity (Xie and Wang, 2014).

An immobilized enzyme can significantly contribute to the economic development of a conversion process. The stability properties of enzyme allow reusability, thus reducing the increased consumption of enzyme during conversion process (Ramani et al., 2010). The high consumption of enzyme during conversion of agricultural waste into by-product increases the production cost. At times the rigidity of biomass might cause a huge intake of an enzyme; in such instances, an immobilized enzyme would be helpful to use due to reusability property. Low activity of an enzyme, week attachment of enzyme and support, lack of substrate and enzyme interaction, and aggregation of immobilized preparation could be some of the technical limitations during the process (Miletić et al., 2012).

A strong attachment of enzyme to support is essential for its economic usage. Loose binding will separate enzyme from the support and will not be helpful during recycling of the process, resulting in an inefficient hydrolysis of substrate. Reports in the literature have suggested that enzymes retained more than 80% of their activity after six to seven consecutive reusable hydrolysis cycles (Liu et al., 2014, 2015; Mishra and Sardar, 2015). Currently, this technology is in a developing stage and usage of an immobilized enzyme to hydrolyze higher volumes of substrate requires more studies on its feasibility at advanced levels.

FUTURE DIRECTIONS

Recent trends in enzyme immobilization on nanomaterials have been found to be successful for improving the biochemical properties of enzymes for biomass hydrolysis. An ideal immobilization system that provides strong affinity to the nanosupport should be identified to improve enzyme stability and reusability. Development of a multienzyme system coated on a nanomaterial for hydrolyzing agro-industrial waste will significantly help the progress of recovering valuables. Further investigation is warranted to improve the current immobilization system by which enzyme hydrolysis reaction environment can be improved. The immobilized multienzyme system should suit different agro-industrial wastes based on good reusability.

ACKNOWLEDGMENTS

The authors thank the Centre for Chemistry and Biotechnology for supporting research in the area of waste to valuables.

REFERENCES

Abraham, R.E., Verma, M.L., Barrow, C.J., Puri, M., 2014. Suitability of magnetic nanoparticle immobilized cellulases in enhancing enzymatic saccharification of pretreated hemp biomass. Biotechnology for Biofuels 7, 90.

Alftrén, J., Hobley, T.J., 2014. Immobilization of cellulase mixtures on magnetic particles for hydrolysis of lignocellulose and ease of recycling. Biomass and Bioenergy 65, 72–78.

Ansari, S.A., Husain, Q., 2012. Potential applications of enzymes immobilized on/in nano materials: a review. Biotechnology Advances 30, 512–523.

Bezerra, T.M.D.S., Bassan, J.C., Santos, V.T.D.O., Monti, R., 2015. Covalent immobilization of laccase in green coconut fiber and use in clarification of apple juice. Process Biochemistry 50, 378–387.

Blanchette, C., Lacayo, C.I., Fischer, N.O., Hwang, M., Thelen, M.P., 2012. Enhanced cellulose degradation using cellulase-nanosphere complexes. PLoS One 7, E42116.

Brady, D., Jordaan, J., 2009. Advances in enzyme immobilization. Biotechnology Letters 31, 1639–1650.

Bustamante-Vargas, C.E., De Oliveira, D., Backes, G.T., Dallago, R.M., 2015. In situ immobilization of commercial pectinase in rigid polyurethane foam and application in the hydrolysis of pectic oligosaccharides. Journal of Molceular Catalysis B: Enzymatic 122, 35–43.

Cano, À., Minguillón, C., Palet, C., 2006. Immobilization of endo-1,4-β-xylanase on polysulfone acrylate membranes: synthesis and characterization. Journal of Membrane Science 280, 383–388.

Cao, M., 2012. Food related application of magnetic iron oxide nanoparticles: enzyme immobilization, protein purification, and food analysis. Trends in Food Science and Technology 27, 47–56.

Cheng, C., Chang, K.C., 2013. Development of immobilized cellulase through functionalized gold nanoparticles for glucose production by continuous hydrolysis of waste bamboo chopsticks. Enzyme and Microbial Technology 53, 444–451.

Cowan, D.A., Fernandez-Lafuente, R., 2011. Enhancing the functional properties of thermophilic enzymes by chemical modification and immobilization. Enzyme and Microbial Technology 49, 326–346.

Deep, A., Tiwari, U., Kumar, P., Mishra, V., Jain, S.C., Singh, N., Kapur, P., Bharadwaj, L.M., 2012. Immobilization of enzyme on long period grating fibers for sensitive glucose detection. Biosensors and Bioelectronics 33, 190–195.

Dyal, A., Loos, K., Noto, M., Chang, S.W., Spagnoli, C., Shafi, K.V.P.M., Ulman, A., Cowman, M., Gross, R.A., 2003. Activity of Candida rugosa lipase immobilized on Γ-Fe$_2$O$_3$ magnetic nanoparticles. Journal of the American Chemical Society 125, 1684–1685.

Esawy, M.A., Gamal, A.A., Abdel-Fattah, A.F., 2013. Evaluation of free and immobilized *Aspergillus niger* Nrc1ami pectinase applicable in industrial processes. Carbohydrate Polymer 92, 1463–1469.

Georgelin, T., Maurice, V., Malezieux, B., Siaugue, J.M., Cabuil, V., 2010. Design of multifunctionalized Γ-Fe₂O₃@Sio₂ core–shell nanoparticles for enzymes immobilization. Journal of Nanoparticle Research 12, 675–680.

Goula, A.M., Lazarides, H.N., 2015. Integrated processes can turn industrial food waste into valuable food by-products and/or ingredients: the cases of olive mill and pomegranate wastes. Journal of Food Engineering 167 (Part A), 45–50.

Hanefeld, U., Gardossi, L., Magner, E., 2009. Understanding enzyme immobilization. Chemical Society Reviews 38, 453–468.

Hegedus, I., Hancsok, J., Nagy, E., 2012. Stabilization of the cellulase enzyme complex as enzyme nanoparticle. Applied Biochemistry and Biotechnology 168, 1372–1383.

Jordan, J., Kumar, C.S.S.R., Theegala, C., 2011. Preparation and characterization of cellulase-bound magnetite nanoparticles. Journal of Molecular Catalysis B: Enzymatic 68, 139–146.

Kennedy, J.F., Cabral, J.M.S., 1987. Immobilization of enzymes on transition metal-activated support. Methods in Enzymology 135, 117–130.

Kinab, E., Khoury, G., 2015. Management of olive solid waste in Lebanon: from mill to stove. Renewable and Sustainable Energy Reviews 52, 209–216.

Kobayashi, H., Fukuoka, A., 2013. Synthesis and utilisation of sugar compounds derived from lignocellulosic biomass. Green Chemistry 15, 1740–1763.

Khoshnevisan, K., Bordbar, A.-K., Zare, D., Davoodi, D., Noruzi, M., Barkhi, M., Tabatabaei, M., 2011. Immobilization of cellulase enzyme on superparamagnetic nanoparticles and determination of its activity and stability. Chemical Engineering Journal 171 (2), 669–673.

Lau, S.C., Lim, H.N., Andou, Y., 2014. Enhanced biocatalytic esterification with lipase immobilized chitosan/graphen oxide beads. PLoS One 9, E104695.

Liang, W., Cao, X., 2012. Preparation of a Ph-sensitive polyacrylate amphiphilic copolymer and its application in cellulase immobilization. Bioresource Technology 116, 140–146.

Lin, X., Konno, T., Takai, M., Ishihara, K., 2013. Redox phospholipid polymer microparticles as doubly functional polymer support for immobilization of enzyme oxidase. Colloids and Surfaces B: Biointerfaces 102, 857–863.

Liu, M.-Q., Dai, X.-J., Guan, R.-F., Xu, X., 2014. Immobilization of *Aspergillus niger* xylanase a on Fe₃O₄-coated chitosan magnetic nanoparticles for xylooligosaccharide preparation. Catalysis Communications 55, 6–10.

Liu, M.-Q., Huo, W.-K., Xu, X., Jin, D.-F., 2015. An immobilized bifunctional xylanase on carbon-coated chitosan nanoparticles with a potential application in xylan-rich biomass bioconversion. Journal of Molecular Catalysis B: Enzymatic 120, 119–126.

Maalej-Achouri, I., Guerfali, M., Gargouri, A., Belghith, H., 2009. Production of xylo-oligosaccharides from agro-industrial residues using immobilized *Talaromyces thermophilus* xylanase. Journal of Molecular Catalysis B: Enzymatic 59, 145–152.

Mansor, A.F., Mohidem, N.A., Zawawi, W.N.I.W.M., Othman, N.S., Endud, S., 2015. The optimization of synthesis conditions for laccase entrapment in mesoporous silica microparticles by response surface methodology. Microporous and Mesoporous Materials 220, 308–314 Available online.

Miletić, N., Nastasović, A., Loos, K., 2012. Immobilization of biocatalysts for enzymatic polymerizations: possibilities, advantages, applications. Bioresource Technology 115, 126–135.

Mishra, A., Sardar, M., 2015. Cellulase assisted synthesis of nano-silver and gold: application as immobilization matrix for biocatalysis. International Journal of Biological Macromolecules 77, 105–113.

Molina-Calle, M., Priego-Capote, F., Luque De Castro, M.D., 2015. Development and application of a quantitative method for determination of flavonoids in orange peel: influence of sample pretreatment on composition. Talanta 144, 349–355.

Mubarak, N.M., Wong, J.R., Tan, K.W., Sahu, J.N., Abdullah, E.C., Jayakumar, N.S., Ganesan, P., 2014. Immobilization of cellulase enzyme on functionalized multiwall carbon nanotubes. Journal of Molecular Catalysis B: Enzymatic 107, 124–131.

Mukhopadhyay, A., Dasgupta, A.K., Chakrabarti, K., 2015. Enhanced functionality and stabilization of a cold active laccase using nanotechnology based activation-immobilization. Bioresource Technology 179, 573–584.

Mussatto, S.I., Moncada, J., Roberto, I.C., Cardona, C.A., 2013. Techno-economic analysis for Brewer's spent grains use on a biorefinery concept: the Brazilian case. Bioresource Technology 148, 302–310.

Nguyen, T.A.H., Ngo, H.H., Guo, W.S., Zhang, J., Liang, S., Lee, D.J., Nguyen, P.D., Bui, X.T., 2014. Modification of agricultural waste/by-products for enhanced phosphate removal and recovery: potential and obstacles. Bioresource Technology 169, 750–762.

Okobira, T., Matsuo, A., Uezu, K., 2015. Enhancement of immobilized lipase activity by design of polymer brushes on a hollow fibre membrane. Journal of Bioscience and Bioengineering 120, 257–262.

Pal, A., Khanum, F., 2011. Covalent immobilization of xylanase on glutaraldehyde activated alginate beads using response surface methodology: characterization of immobilized enzyme. Process Biochemistry 46, 1315–1322.

Pavlidis, I.V., Vorhaben, T., Tsoufis, T., Rudolf, P., Bornscheuer, U.T., Gournis, D., Stamatis, H., 2012. Development of effective nanobiocatalytic systems through the immobilization of hydrolases on functionalized carbon-based nanomaterials. Bioresource Technology 115, 164–171.

Perkins, S., Siddiqui, S., Puri, M., Demain, A.L., 2015. Biotechnological applications of microbial bioconversions. Critical Reviews in Biotechnology 18, 1–16.

Puri, M., Marwaha, S.S., Kothari, R.M., 1996. Studies on the applicability of alginate-entrapped naringinase for the debittering of kinnow fruit juice. Enzyme and Microbial Technology 18, 281–285.

Puri, M., Banerjee, U.C., 2000. Production, purification and characterisation of debittering enzyme naringinase. Biotechnology Advances 18, 207–217.

Puri, M., Kaur, H., Kennedy, J.F., 2005. Covalent immobilization of naringinase for the transformation of flavnoid. Journal of Chemical Technology and Biotechnology 80, 1160–1165.

Puri, M., Kaur, A., Singh, R.S., Schwarz, W.H., Kaur, A., 2010b. One-step purification and immobilization of his-tagged rhamnosidase for narigin hydrolusis. Process Biochemistry 45, 451–456.

Puri, M., 2012. Updates on nariginase: structural and biotechnological aspects. Applied Microbiology and Biotechnology 93, 49–60.

Puri, M., Barrow, C., Verma, M., 2013. Enzyme immobilization on nanomaterials for biofuel production. Trends in Biotechnology 31, 215–216.

Ramani, K., Boopathy, R., Vidya, C., Kennedy, L.J., Velan, M., Sekaran, G., 2010. Immobilization of *Pseudomonas gessardii* acidic lipase derived from beef tallow onto mesoporous activated carbon and its application on hydrolysis of olive oil. Process Biochemistry 45, 986–992.

Santalla, E., Serra, E., Mayoral, A., Losada, J., Blanco, R.M., Díaz, I., 2011. In-situ immobilization of enzymes in mesoporous silicas. Solid State Sciences 13, 691–697.

Sarbu, A., de Pinhob, M.N., do Rosário Freixoc, M., Goncalves, F., Udrea, I., 2006. New method for the covalent immobilization of a xylanase by radical grafting of acrylamide on cellulose acetate membranes. Enzyme and Microbial Technology 39, 125–130.

Shah, S., Gupta, M.N., 2008. Simultaneous refolding, purification and immobilization of xylanase with multi-walled carbon nanotubes. Biochimica et Biophysica Acta (BBA) – Proteins and Proteomics 1784, 363–367.

Singh, R.K., Zhang, Y.W., Nguyen, N.P.T., Jeya, M., Lee, J.K., 2011. Covalent immobilization of β-1,4-glucosidase from *Agaricus arvensis* onto functionalized silicon oxide nanoparticles. Applied Microbiology and Biotechnology 89, 337–344.

Singh, R.S., Dhaliwal, R., Puri, M., 2007. Production of high fructose syrup from Asparagus inulin using immobilized exoinulinase from *Kluyveromyces marxianus* Ys-1. Journal of Industrial Microbiology and Biotechnology 34, 649–655.

Soozanipour, A., Taheri-Kafrani, A., Isfahani, A.L., 2015. Covalent attachment of xylanase on functionalized magnetic nanoparticles and determination of its activity and stability. Chemical Engineering Journal 270, 235–243.

Tan, I.S., Lee, K.T., 2015. Immobilization of β-glucosidase from *Aspergillus niger* on κ-carrageenan hybrid matrix and its application on the production of reducing sugar from macroalgae cellulosic residue. Bioresource Technology 184, 386–394.

Verma, M., Barrow, C., Kennedy, J.F., Puri, M., 2012. Immobilization of galactosidase from *Kluyveromyces lactis* on funcationalised silicon dioxide nanoparticles: characterisation and lactose hydrolysis. International Journal of Biological Macromolecules 50, 432–437.

Verma, M., Barrow, C.J., Puri, M., 2013. Nanobiotechnology as a novel paradigm for enzyme immobilization and stabilisation with potential application in biodiesel production. Applied Microbiology and Biotechnology 97, 23–39.

Verma, M., Barrow, C.J., Puri, M., 2016. Recent trends in nanomaterial immobilized enzyme for biofuel production. Critical Reviews in Biotechnology 36, 108–119.

Wang, Y., Radosevich, M., Hayes, D., Labbé, N., 2011. Compatible ionic liquid-cellulases system for hydrolysis of lignocellulosic biomass. Biotechnology and Bioengineering 108, 1042–1048.

Xie, W., Wang, J., 2014. Enzymatic production of biodiesel from soybean oil by using immobilized lipase on magnetic microsophere as a biocatalysts. Energy and Fuels 28, 2624–2631.

Yuan, X., Shen, N., Sheng, J., Wei, X., 1999. Immobilization of cellulase using acrylamide grafted acrylonitrile copolymer membranes. Journal of Membrane Science 155, 101–106.

Zhu, X., Zhou, T., Wu, X., Cai, Y., Yao, D., Xie, C., Liu, D., 2011. Covalent immobilization of enzymes within micro-aqueous organic media. Journal of Molecular Catalysis B: Enzymatic 72, 145–149.

FURTHER READING

Puri, M., Gupta, S., Pahuja, P., Kanwar, J.R., Kennedy, J.F., 2010a. Cell disruption and covalent immobilization of galactosidase from *Kluyveromyces marxianus* Yw-1 for lactose hydrolysis in milk. Applied Biochemistry and Biotechnology 160, 98–108.

Teodor, S., Andrei, S., Damian, C.M., Patroi, D., Iordache, T.V., Budinova, T., Tsyntsarski, B., Ferhat Yardim, M., Sirkecioglu, A., 2015. Functionalized bicomponent polymer membranes as supports for covalent immobilization of enzymes. Reactive and Functional Polymers 96, 5–13.

CHAPTER 14

Biotechnological Production of Enzymes Using Agro-Industrial Wastes: Economic Considerations, Commercialization Potential, and Future Prospects

N. Gopalan, K.M. Nampoothiri
CSIR-National Institute for Interdisciplinary Science and Technology (NIIST), Trivandrum, India

INTRODUCTION

Industrial processes are purely driven by the economics of the given process. The viability of a product is governed mainly by the gap in demand and supply and the percentage profit that comes out of the whole process. Just like any other bioprocess, enzyme production also consists of upstream, fermentation/bioproduction and downstream processes. The whole economics of the process is governed by the cost involved in executing all three stages, as well as the demand and acceptability of the final product. Enzymes are considered as green alternatives to industrial catalysts. However, in the modern biotechnology era, enzymes are used in special niche areas, where enzymes are the most essential components to the core process, viz., stereo selective synthesis (aldolases, hydrolases, nitrilase), detergent formulations (lipase and proteases), juice liquefaction (pectinase, xylanase), therapeutics (asparaginase for cancer treatment), diagnostics (glucose oxidase for glucometers) (Kasturi et al., 1998; Wang, 2008; Clapés et al., 2010; Pal and Khanum, 2011). The final use of the enzyme dictates the purity of the enzyme required and hence the price of the enzyme preparation. Technical enzymes are employed in four major areas, viz., food and beverages, detergents, animal feed, biomass hydrolyzing, or biofuel along with other subsidiaries, such as leather industries, textile, paper–pulp and biopharmaceuticals (Kirk et al., 2002). While the largest industry segment that acts as consumers for enzymes still remains the food and beverage industry, there has been a shift of highest-selling enzymes from proteases until 2012 to carbohydrases in 2013–14. The increase in the sales of carbohydrase enzymes were accounted for by increasing adoption of enzymes in the food and beverage industry as well as active research and technological thrust toward the production of second-generation biofuels and allied products. The United Nations and World Bank led Sustainable Energy for All (SE4All) initiative and it

Agro-Industrial Wastes as Feedstock for Enzyme Production
ISBN 978-0-12-802392-1
http://dx.doi.org/10.1016/B978-0-12-802392-1.00014-9

313

was adopted by nearly 80 countries by 2013. A collective of \$50 billion toward its three objectives has been invested in total by different financial bodies, public and private sectors. The SE4All initiative has led to a spike in capital investment in second-generation bioethanol and allied fuel products, consequently heavy investments in enzymes have led to an economic spurt in the enzyme production sector. The major market share is held by European and American giants Novozymes, DSM, and DuPont, however, significant competition is raising its head from the Asian side, with many Chinese and Japanese players coming into the scenario (Hexa Research, n.d.; The Novozymes Report 2014, n.d.).

Global enzyme economics is very dynamic due to the different regulation policies as well as sociogeographical differences. However, there is an upward trend in the utilization, consumption of biocatalytic enzymes in day-to-day life of the common man, and in the industrial sector. The slowdown in the global economy has not curtailed the growth of the enzyme industry. With government policies shifting toward sustainable use of resources and focus on green technologies, enzyme-based economy has been displaying a healthy upward trend. It can, hence, be called an ancient technology paving way for the future.

FACTORS DECIDING THE ECONOMICS OF ENZYME PRODUCTION

Enzyme production is a very general idea, and the actual process of production of enzymes will differ considerably depending on various aspects, including raw material availability and cost, location of the production facility, energy requirements, the nature of the enzyme being produced, the microorganism (wild type or modified), the concentration of the product enzyme, the stability and other physical and physiological characteristics of the enzyme molecule/molecules in question, and the end user preferences (Castilho et al., 2000; Mukherjee et al., 2006; Barta et al., 2010).

The total cost of production (TCP) is governed by all the just-mentioned factors. The analysis for each part of enzyme production can be done separately, so as to understand areas where innovations for reducing costs must be carried out. Any bioprocess can be divided into three parts: upstream processes, fermentation process, and finally downstream processes for recovery and purification. The individual processes coming under these divisions as well as the scale of the whole process decides the economy of the whole process.

Upstream Processes

The primary raw material used for fermentative production of enzymes can come from any source of agricultural by-products obtained from cereal crops, spice crops, seasonal crops, oil seeds, fruits, vegetables, roots and tubers, postharvested plant materials and processing industries, etc. Traditional bioprocesses utilized refined sugars as carbon sources and salts containing nitrate or ammonium groups as nitrogen source, with certain processes using undefined nitrogen sources, such as yeast extract and peptone and beef extract. The ratio of the carbon to the nitrogen source plays a significant role in the yield of enzymes or any bioproduct, along with other factors present in the medium (Aiyer, 2005). Later, with the focus being shifted to make different bioprocesses economical, cheaper sources of carbon, such as cane molasses and lignocellulosic substrates or starch/starch hydrolysates, came to light (Furlan et al., 2000; Olsson et al., 2003). One such example is, cellulase production by *Aspergillus fumigatus* grown on mixed substrate of rice straw and wheat bran (Sherief et al., 2010). Slowly, bioprocesses using cheaper nitrogen sources, like corn steep liquor and urea, started to take center stage (Edwinoliver et al., 2009; Nascimento et al., 2009).

Agricultural/agro-industrial wastes may be rich in carbohydrates, proteins, or lipids and it may are available in solid or liquid forms. Any agricultural waste if used as a major component of the fermentation medium can reduce the total cost of production by decreasing the cost of the raw material used for the fermentation process. In biomass-based technology, costs associated with upstream processing of enzyme production involve the cost of the raw material, if there is any pretreatment (such as chemical or enzymatic) of raw material, transport of the raw material to production site, and associated labor and energy costs, etc. The nature of the raw material like starchy material or cellulosic material itself affects the fermentation/production process as a whole. Choosing appropriate, readily available, and cheaper substrates for the desired enzyme is a must. Preferably, the raw material could be a good inducer for the enzyme production as well. The best example is that of lignocellulosic materials serving as substrate and inducer for the production of cellulase enzymes (Lau et al., 2012). Raw material costs for a typical bioprocess range from 10% to 50% of the TCP, and hence using agro-residual wastes would be beneficial, as more capita can be allotted to other components of the process and recurring costs on raw materials could be reduced (Castilho et al., 2000).

Similarly, depending on the mode of fermentation, viz., solid state fermentation (SSF) or submerged fermentation (SmF), the infrastructure, and consequent maintenance will vary. Equipment used in the pretreatment of raw materials involve sterilization or pasteurization methods to decrease or eliminate the competing microbes already present on/in the fermentation medium/nutrient medium. Depending on the mode of fermentation involved for the process, the equipment and the scales for this purpose may vary. Generally for SSF, small-volume, high-temperature, and pressure-based sterilization equipment is used, which is economical. For SmF, the equipment generally is coupled with the fermentation vessel in the form of a temperature-exchange jacket, whose price is inclusive in the cost of the fermenter itself. Raw materials require additional pretreatment steps in certain cases, viz., defatting or destarching when the microorganism is required to produce enzymes free of lipases and amylases, eg, feruloyl esterases, require destarched wheat bran as a substrate or defatted oil cakes (Bonnin et al., 2002; Laszlo et al., 2006), or wet heat pretreatment, in which case additional equipment or associated facilities may be required (Palmqvist et al., 1997; Singh et al., 2011).

Depending on the course followed, there are differences in the economic ramifications of the process. Submerged fermentation involves the cultivation and subsequent production of the enzymes produced by the microorganism while suspended in a nutrient-rich fluid medium. The energy requirements to maintain the temperature of large quantities of liquids are high, along with increases in the cost of sterilizing or other pretreatment required for the process. Submerged fermentation processes have been used classically for major fermentative processes since the onset of industrial microbiology, starting with the fermentative production of antibiotics from fungi and actinomycetes during the era of World War II. Fermentative production of some carbohydrases and lipolytic enzymes through SmF of agro-wastes have been reported (Sharma and Satyanarayana, 2006; Teng and Xu, 2008; Singh et al., 2009; Vidyalakshmi et al., 2009; Nagar et al., 2010). Various reactor designs are available for enzyme production processes. However, the stirred tank reactor is generally preferred over other designs. Stirred tank reactors are closed reactors that are expensive and the costs incurred for setting these normally takes the top spot in the TCP (Castilho et al., 2000). The stirred tank reactor also causes an increase in the costs incurred to efficiently handle bulk liquid quantities. Overall, the equipment costs for SmF are very high compared to SSF. Labor required for running stirred tank reactor-based bioprocesses is generally confined to trained personnel in the field of bioprocess engineering (Max et al., 2010). Since the specific heat of water, which makes up for most of the mass of the fermentation medium, is very high, the heating and cooling cost for the operation of SmF processes would rise. If the batch time for fermentation is small, smaller volume fermenters may be employed. However, for processes with longer incubation periods, the fermenter volume should be increased to compensate for productivity, concurrently leading to increased cost of equipment used. Increasing the number of batches/fermenters running parallel may be desirable when downstream processes have a smaller batch time compared to the fermenter batch time.

Solid-state fermentation is the preferred method for production of various enzymes from agro-industrial substrates, especially when the substrates are insoluble in water. Solid-state fermentation is defined as a process where a microbe is grown on a substrate in the absence or near absence of free water. SSF is preferred over SmF because it was revealed that cultures grown in solid state do not show catabolite repression (Viniegra-González et al., 2003; Viniegra-González and Favela-Torres, 2006) that is normally a prevalent phenomenon in SmF, and as a result enzyme concentrations that are usually unattainable in SmF are encountered in SSF. Sensitivity to ionic contaminants in water is reduced in case of solid-state cultivation of microbes, reducing the dependency of processed water for the bioprocess (Shankaranand and Lonsane, 1994). SSF is carried out on simple tray fermenter, made of wood or aluminum (Fig. 14.1). Tray fermenters are cheap alternatives compared to submerged fermenters. The process of SSF involves wetting of the solid substrate with nutrient media to a certain initial

Figure 14.1 Filamentous fungi growing on agro-residual solid substrate (wheat bran)—a low-cost enzyme production strategy.

moisture level (50–90%) (Pandey, 2003). Since the microbe is grown on a solid substrate, normally the temperature and humidity is controlled for the whole unit where the SSF is carried out. Trays are essentially open fermenters, as lower moisture content and high amounts of inoculums ensures the absence of contamination. The labor involved with running an SSF does not require trained personnel; cheap labor force under the supervision of a moderately trained personnel suffices for smooth operation of batches. The costs incurred for cooling and other energy-based expenditures are lesser compared to SmF, as the specific heat capacity for water is very high and water makes up a very small percentage of the medium for SSF compared to SmF. SSF process is carried out with no control over the pH, while SmF processes require precise pH control using acids and alkalis. The process cannot be controlled as precisely as SmF, due to the heterogeneous nature of the fermentation medium, however, from the economic standpoint, since the process requires lesser control than SmF, and energy expenditure as well as lesser inputs for the raw material, SSF becomes a very viable option.

Upstream process requires the use of a robust microbial culture. Bacteria or fungi can be used for either SmF or SSF. Ideal industrial cultures are fairly immutable, produce either large amounts of enzyme, or produce in higher concentrations than relative wild-type varieties. When in relation with an enzyme as a product, the microbe involved varies with the final use of the enzyme. For example, amylase as an enzyme is used for hydrolyzing starchy substrates under benign conditions. However, different variants of the amylase enzyme produced by different organisms are used for different applications. Amylase used for saccharification of waste starchy biomass for further fermentation into ethanol uses ordinary amylase, while amylases used for baking may use amylase mixes that impart flavors to the bread, by virtue of other enzymes (mostly esterases) produced by the particular organism. Amylase treatment improves bread shelf life by modifying the starch structure (Martínez-Anaya, 1996). Therefore, procuring the right organism according to the target market is one of the most important investments that an enzyme producer has to carry out. Isolating an industrial strain from nature is not only a tedious process but also requires considerable capital investment. A procured microbe is a catalogued organism with most of the parameters known to the fermentation expert, hence guarantees tighter control over the process of fermentation. Table 14.1 lists some of the commercial strains for production of enzymes.

Downstream Processes

When a fermentation process is carried out, the enzyme/enzymes produced by the microorganisms are accompanied by contaminating molecules produced by the organism as well as molecules from the substrate. To make the enzyme as a commercial product, it must be concentrated to a usable level. Downstream processing involves the purification of the enzyme solution so that the activity of the said enzyme reaches to the highest. Downstream processes take up a significant portion of the TCP, whether the process is SmF or SSF. Downstream processes include processes like filtration, centrifugation, reverse osmosis, precipitation, flocculation, evaporative concentration, freeze-drying/spray-drying. For applications like pharmaceutical processes and medical treatment or diagnostics, the method for purification involves many final polishing technologies like affinity chromatography, crystallization depending on the enzyme as well as the level of purity required (Wheelwright, 1989). Most of the methods and technology coming under downstream processing are energy intensive, and the efficiency of these technologies can directly affect the cost and quality of the final product. Submerged and solid-state fermentation products have to be processed by filtration, centrifugation, or other cell biomass removal processes, and rotary drum filtration is one of the preferred methods to do it. The process of enzyme purification is carried out after the separation of the microbial cells from the liquid medium for SmF or the solid fermented extract in the case of SSF. The concentration of enzymes produced in SSF is much higher compared to that of SmF (Viniegra-González, 1998; Viniegra-González et al., 2003; Singhania et al., 2009), hence for the same amount of product, the volumes to be handled by equipment for downstream processing will be different, ie, volume of the SSF product will be lower, and hence lower scale of equipment

Table 14.1 List of Industrially Important Enzymes and Potent Producers

Enzyme	Organisms
Aminopeptidases	*Clostridium histolyticum, Vibrio proteolyticus, Aspergillus*
Alpha–amylase	*Bacillus amyloliquefaciens* subsp. *amyloliquefaciens*
Amylase	*Aspergillus foetidus, B. amyloliquefaciens* subsp. *amyloliquefaciens, Bacillus licheniformis, Endomyces fibuliger, Paenibacillus macerans, Paenibacillus polymyxa, Rhizomucor miehei, Rhizomucor pusillus, Thermoanaerobacter brockii* subsp. *finnii, Thermoanaerobacter ethanolicus*
Amylase, alkaline	*Bacillus halodurans, Bacillus pseudofirmus, Bacillus* sp.
β–amylase, alkaline	*Bacillus halodurans*
Amylase, halophilic	*Nesterenkonia halobia, Amylase, thermoacidophilic, Alicyclobacillus acidocaldarius* subsp. *acidocaldarius*
Amylase, thermostable	*Bacillus* sp., *Geobacillus stearothermophilus, Thermoanaerobacterium thermosaccharolyticum, Thermoanaerobacterium thermosulfurigenes*
L–asparaginase	*Cupriavidus necator, Escherichia coli, Wolinella succinogenes*
Cellulase	*Cellulomonas uda, Chaetomium globosum, Clostridium alkalicellulosi, Clostridium thermocellum, Phanerochaete chrysosporium, Thermoascus aurantiacus, Trichoderma reesei, Aspergillus niger, Aspergillus flavus, Aspergillus terreus*
Cellulase, alkaline	*Bacillus cellulosilyticus, Bacillus wakoensis*
Cellulase, enhanced	*Trichoderma reesei, Bacillus* sp. *Escherichia coli, Kluyveromyces marxianus, Sulfolobus solfataricus*
Invertase	*Clostridium pasteurianum, Rhizopus oryzae*
Laccase	*Heterobasidion annosum, Hypholoma fasciculare, Pleurotus cystidiosus, Pleurotus ostreatus, Spongipellis litschaueri, Trametes versicolor*
Lactase	*Kluyveromyces marxianus*
Nitrilase	*Pseudomonas brassicacearum* subsp. *brassicacearum, Pseudomonas putida, Variovorax* sp.
Pectinase	*Fusarium oxysporum, Pichia Canadensis, R. oryzae, Trichoderma reesei*
Peroxidase	*Geotrichum candidum, Inonotus weirii, Phanerochaete chrysosporium*
Protease	*Acremonium chrysogenum, Aneurinibacillus migulanus, Bacillus cereus, Bacillus circulans, Candida tropicalis, Coprinus cinereus, Coprinus radians, Dictyostelium discoideum, Lysobacter enzymogenes* subsp. *enzymogenes, Ruminobacter amylophilus, Streptomyces* sp.
Protease, acid	*Candida parapsilosis, Candida tropicalis*
Protease, alkaline	*Bacillus alcalophilus, Bacillus clausii, Bacillus licheniformis, Bacillus subtilis*
Xylanase	*Phanerochaete chrysosporium*

needs to be employed to process it. Depending on the batch size of the fermenter and batch time of the downstream process, optimum scale and number of equipment for fermentation and downstream processing must be fixed. For example, if the batch time for fermenter is about 24 h, while the downstream processing takes about 11 h, the ideal number of simultaneous fermenter runs per batch of downstream process is two (Castilho et al., 2000). However, if the capital required for setting up the fermenter exceeds the capital required for the downstream process equipment, the number of downstream processing units must be optimized with respect to the fermenters that can be deployed with the available capital. Other downstream costs include procurement of stabilizers (sorbitol, glycerol, mannitol, etc.) for increasing the shelf life of the final product.

Other Miscellaneous Factors

Fermentation processes have to evaluate for sustainability through periods of perturbation in the market price of the enzyme and perturbation in batch production capacity. Depending on the variance of the price from the past data, one can predict the probabilistic range of the price fluctuation of the enzyme in the market. Economic simulations to predict the viability of the process must be carried out. However, as a rule of thumb, when the TCP per unit enzyme is significantly lower than the market price, the product will be viable even with periods of large fluctuations in the market price. Depreciation/payback of finances is one of the major factors to be considered while discussing the economics of production of enzymes. With lower cost of the agro-residual wastes, the TCP per unit enzyme will be lesser than the market price, leading to higher profits. Higher profits will lead to a decrease in the depreciation period. Along with these factors, maintenance should also be performed to ensure the longevity of the setup. Since SmF utilizes expensive fermenters, with large volumes, maintenance of this equipment requires higher capital than the simple tray fermenters for SSF.

Waste management from either submerged or solid-state fermentation for enzyme production would require careful cost management, as it has to be ensured that the producer microbe/spent medium should not be released into the environment as a biohazard. Table 14.2 summarizes the most significant economic gain of using enzymes in industrial applications (adapted with modifications from Jegannathan and Nielsen, 2013).

COMMERCIALIZATION POTENTIAL OF ENZYMES

The global enzyme market was worth US$5.1 billion in 2014. With a compound annual growth rate of 6.3%, it is expected to grow to approximately US$7 billion (Industry Trends, n.d.) by 2017. The enzyme market is still in its growing phase, with developed countries being mature markets with less growth, but in a steady state and developing countries serving as an emerging market. Whether the enzyme is produced by costly raw materials or by making use of surplus raw materials like agro-residual wastes, the final acceptability of the enzyme preparation is the same, the only difference being that enzymes required for pharmaceutical processes and for the food industry must be relatively pure and must be approved as food additives or pharmaceutical process adducts by local governing bodies. The three largest manufacturers of industrial enzymes are Novozymes, DSM, and DuPont, with 48%, about 5%, and 21% (The Novozymes Report 2014, n.d., BCC market research). Enzymes provide considerable economic advantage over conventional processes; the major enzymes currently sold come under one of the following categories.

Table 14.2 Enzyme Application in Various Industries and the Economic Advantage

Industry	Application Process	Enzyme	Function of Enzyme	Economics Improvements Mainly Through Reduction in the Requirement of
Pulp and paper	Thermomechanical pulping	Cellulase	Softens wood chips	Energy
	Deinking	Cellulase	Acts on recycled fibers and facilitates ink loosening	Chemicals
	Bleaching	Laccase	Oxidizes lignin and enhances lignin removal	Bleaching chemicals, energy
		Xylanase	Hydrolyzes xylan and enhances lignin extraction	Bleaching chemicals
	Pitch control	Lipase	Hydrolyzes pitch	Cleaning agent, talc, energy
	Stickies control	Esterase	Hydrolyzes glue and controls stickies	Talc, solvent, energy
Leather	Beam house	Protease, lipase	Facilitates hair and fat removal from hides	Chemicals, energy
Textile	Scouring	Pectate–lyase	Degrades pectin and assists in removal of wax, etc. from raw cotton	Energy, water, chemicals, cotton
	Bleaching	Catalase	Hydrolyzes hydrogen peroxide to oxygen and water	Heat, electricity, water
	Bleaching	Aryl esterase,	Perhydrolyzes to form peracetic acid (bleaching activator)	Chemical, cotton, energy
Detergent	Laundry washing	Protease, lipase, amylase, cellulase	Removes stains from laundry	Chemicals, energy

Continued

Table 14.2 Enzyme Application in Various Industries and the Economic Advantage—cont'd

Industry	Application Process	Enzyme	Function of Enzyme	Economics Improvements Mainly Through Reduction in the Requirement of
Food and beverage	Degumming of soybean oil	Phospholipase	Hydrolyzes phospholipids	Oil, chemicals
	Hard stock production	Lipase	Interesterification of vegetable oil	Chemicals, energy
	Fruit juice production	Pectinase	Breaks down pectin in fruit	Fruits, Filtration
	White bread production	Amylase	Degrades starch and delays hardening of bread	Bread
	Steamed bread production	Amylase, lipase		
	Mozzarella cheese production	Phospholipase	Hydrolyzes phospholipids in milk	Milk
	100% barley beer production	Amylase, protease etc.	Converts starch to fermentable sugars	Barley, energy
		Amylase, protease, laccase	Laccase for flocculation and clarification of beer	Energy
Animal feed	Pig feed production	Xylanase	Depolymerizes xylans and enables better digestion	Feed
	Pig feed production	Phytase	Hydrolyzes phytate and releases phosphorus bound in feed	Inorganic phosphorus
	Poultry feed production	Phytase		
	Poultry feed production	Protease	Hydrolyzes protein in the feed	Feed protein
Fine chemicals	Aminobutanoic acid production	Lipase	Aminolysis in (S)-3-aminobutanoic acid production	Chemicals, waste
	α–naphthol production	Toluene orthomonooxygenase	Oxidizes naphthalene to alpha-naphthol	Chemicals, waste

Pharmaceuticals	(S)-2-3-Dihydro-1H-indole-2-carboxlic acid production	Phenylalanine ammonia lyase	Formation of C—N bond	Chemicals, energy
	γ-Aminobutyric acid production	Lipolase	Resolution of cyanodiester	Chemicals, energy, waste
	6-Aminopenicillanic acid production	Penicillin amidase	Deacylates penicillin molecule	Chemicals, energy
	7-Aminopenicillanic acid production	D-amino acid oxidase, glutaryl 7-ACA acylase	Oxidizes cephalosporin C salt and deacylates glutaryl 7-aminocephalosporic acid	Chemicals, energy
Cosmetics	Oleo chemical ester production	Lipase	Transesterification of vegetable oil	Chemicals, energy, raw material
Biodiesel	Methyl ester production	Lipase	Catalyzes the reaction of triglyceride and menthol to form methyl ester	Energy, chemicals, raw material

Carbohydrases

Carbohydrases were among the first commercial enzymes to be marketed to different industries. All enzymes that hydrolyze complex carbohydrate molecules to simpler monomeric sugars are called carbohydrases. Carbohydrases have been the largest segment of enzymes being commercialized, accounting for about 45% of sales in enzymes worldwide in 2013 (Enzymes Market Analysis by Product (Carbohydrase, Proteases, Lipases, Polymerases & Nucleases) and Segment Forecasts to 2020, n.d.). Enzymes coming under this banner include amylases, invertase, lactase, and pectinase used mostly by the food and beverage industry, and cellulase, xylanase, and β-glucosidase enzymes used by the biofuel, textile, and paper and pulp industries. Xylanase is a carbohydrase that takes a major share in animal feed industry among other carbohydrases (Sheppy, 2001), while amylases have a major share in the food industry. Food and beverages, animal feed are mature markets, however, the biofuels segment is an emerging market for carbohydrase and has led to carbohydrases climbing to the top. Various commercial uses of carbohydrases have been listed in Table 14.2.

Proteases

Proteases are enzymes that break down proteins into component amino acids and smaller peptides. Proteases are used heavily in the leather industry for dehairing of animal hides and softening leather. Alkaline proteases are used by detergent industries along with a range of other alkaline enzymes to reduce water requirements and to allow better cleansing of clothes. Neutral protease and acidic proteases are used in food industry to improve the nutritive quality of the food and in clarifying juices and malts and for tenderizing meats (Binod et al., 2013). Rennet was one of the first enzymes to be used on a commercial scale for the process of cheese making. Aminopeptidases are enzymes that act on terminal amino acids of peptides (amino or carboxyl) and have been used commercially for cheese ripening, with the aminopeptidases acting on hydrophobic peptides (with proline or leucine) to release free amino acids, ultimately reducing the bitterness of the product (Raksakulthai and Haard, 2003; Nampoothiri et al., 2008; Rahulan et al., 2012). Proteases were the leading segment among industrial enzymes, before carbohydrases, owing to heavy use in the detergent, dairy, and the food industries (Hexa Research, n.d.).

Esterases

Esterases are enzymes that catalyze the hydrolysis of an ester group from a variety of substrates so that the esterified acid is released. The major group of esterases that is used for industrial purposes is lipase. Lipases are enzymes that can catalyze deesterification or transesterification, a reaction that can be tailored to yield free acid or to yield another ester of the acid, normally an alkyl ester. Lipases are used as a part of the enzyme arsenal that is used in detergents; lipases are also used for degreasing of leather

(Binod et al., 2013). Novel applications include use of lipases as well as proteases in lens cleaning fluids for cleaning contact lenses (Binod et al., 2013). Immobilized lipases are also used for production of biodiesel from vegetable oils or nonedible oils like Jatropha by transesterification of triglycerides with methanol, to obtain methyl esters of long-chain fatty acids (Soumanou and Bornscheuer, 2003; Shah et al., 2004; Bajaj et al., 2010). Other esterases that are commercially used are acetyl xylan esterase, feruloyl esterases, which are components of blends of enzymes used for lignocellulosic saccharification, and pulp and paper bleaching enzymes.

Oxidoreductases

Oxidoreductases are diverse and are used for various purposes. Glucose oxidase is one of the commercial oxidoreductases used for production of gluconic acid and in biosensors for detection of glucose in biological fluids. Certain stereospecific oxidoreductases are used for reduction of aldehydes of a specific chirality to the corresponding alcohol (Table 14.2). Laccases are enzymes that are used for a variety of uses, from cross-linking phenolics to decolorizing dyes and clarification of beverages like beer (Dhillon et al., 2012a,b).

ENZYME TECHNOLOGIES WITH COMMERCIAL POTENTIAL

Some technologies that are under research have immense technological potential; some have been listed as follows. Bread wastes may be used for saccharification through enzymatic routes to produce sugars, which may be used as a raw material for various bioprocesses, including fuel production, and production of acids and polymers, where there is active research and the commercialization potential of such technologies is huge (Lam et al., 2014; Pleissner et al., 2014). Extraction of antioxidant compounds from agro-residual biomass has been studied for decades and holds immense potential as components of functional food additives. Enzymatic extraction of hydroxycinnamic acids from brewer's spent grain, wheat bran, and other cereal agro-residuals through enzymatic treatment of the same with enzymatic mixtures containing feruloyl esterases and other accessory enzymes (Faulds and Williamson, 1995; Li et al., 2006). Pigments from various vegetable and fruit sources that are available as wastes of food-processing industry like tomato, and carrots may be extracted with enzymatic cocktails containing pectinases and cellulases and other accessory enzymes so as to allow easy extractability of the pigments, which can be used as natural food colorants (Zuorro et al., 2013). Enzyme-assisted fiber extraction from various agro-residuals (Meyer et al., 2009) and production of tannase enzyme on tannin-rich substrates, with gallic acid as a side product (Treviño–Cueto et al., 2007), are all exploitable processes with untapped potential.

The high economic growth rate is accounted for by emerging markets, viz., South America, Africa, and Asia Pacific. Many countries are experiencing economic boosts, and with the rise in the per capita income of the average person, the capacity to purchase better quality goods increases, leading to demand-based markets for finer products made using marginally expensive procedures. No doubt enzymatic catalysis is a costlier affair as compared to inorganic catalysts in general. However, the quality of the final product is improved due to the specific action of the enzymes, and the process is normally benign on the environment. With stricter norms with regard to environmental protection, many industries have shifted to greener processes involving enzymes or whole-cell catalysis, and with bioprocess being the only option for certain biotransformations that are viable (stereoselective synthesis or stereoresolution), enzyme sales have been steadily growing. Manufacturers are looking at developing countries as emerging markets. Asia Pacific and Latin America account only for 17% and 11%, respectively, of the total sales, and have a huge potential for growth for Novozymes alone. The fast growth of the enzyme market has been synchronously stimulated by multiple factors along with the previously mentioned factors. The United Nations and World Bank led an initiative on sustainable energy for all (SE4All) that has led to a tremendous increase in investment toward renewable energy and making cellulosic ethanol a viable replacement for petroleum-based fuels. Many studies regarding the technoeconomic analysis for production of lignocellulosic ethanol have been carried out to optimize the process of on-site cellulase production and simultaneously produce ethanol from the hydrolyzed lignocellulosic feedstocks (Hamelinck et al., 2005; Gnansounou and Dauriat, 2010; Klein-Marcuschamer et al., 2012). Population growth and increased demand for food and proteins have only been a stimulus to the enzymes market (The Novozymes Report 2014, n.d.).

FUTURE PERSPECTIVES

Any bioprocess, because of the involvement of a biological entity, becomes difficult to control and predict unlike chemical processes, where yields and associated expenses can be estimated very accurately. When enzymes are produced by using agro-residual biomass, one more variable is thrown into the equation of the whole process. There will be batch-to-batch variation in the agro-residual biomass, with different conditions of growth of the same crop. However, even with such uncertainties, agro-residues are still lucrative as substrates for enzyme production due to the fact that they are unwanted wastes and will take up a very small percentage of total investment made per batch of enzymes produced. Process sustainability is decided by factors other than those related directly to product. The current scenario demands the perception of a bioprocess as a biorefinery-like process, where side products valorize the major product. The refinery concept applies to all of the bioprocesses, and even processes that are established may be further valorized by the specific activities of appropriate enzymes. Advances in the biorefinery system research basically demand value addition in the overall process. For example, cell surface expression of α-amylase was reported in *Corynebacterium glutamicum* for glutamate production, which could find use in whole crop biorefinery. Enzymatic extraction of value-added chemicals holds the key to sustainability in the future. Already established industries among the food sector may use enzymes for valorization of wastes produced as a result of large-scale food processing. The possibilities and the potential are virtually endless. With growth in population, waste generation and subsequent waste disposal will be a bigger problem than what it is today. Processes that target these wastes either as raw material from production of enzymes or as raw material for treatment with enzymes are more of a necessity than a side process that alleviates economic load.

The fruitful cerebration of the DNA sequencing technology developers and the corresponding reduction in cost and increase in the accuracy and speed of the sequencing procedure has boosted the development of engineered proteins/enzymes for better functionality. Directed evolution and high-throughput screening technologies have sped up discovery and improvement of enzymatic catalysts. Novozymes makes a hefty investment of about 14% total profits into research and development. With the increase in awareness of the average person about the role of enzymes in his day-to-day life and the advantages of enzyme-catalyzed processes (as an end user and as a global citizen), the demand for enzyme-based products is only going to increase. In summary, the global economy and the population needs to exploit opportunities that are presented through valorization of agro-residual biomass of different types so as to improve sustainability of not just the economy but also of the human community.

ACKNOWLEDGMENTS

The author NG thanks UGC-New Delhi for providing financial aid through fellowships, and the corresponding author thanks DST-CONACYT for providing financial aid, under the Indo-Mexican project for bilateral exchange to work on various biotechnological applications such as value-added chemicals from agro-residual substrates.

REFERENCES

Aiyer, P.D., 2005. Effect of C:N ratio on alpha amylase production by *Bacillus licheniformis* SPT 27. African Journal of Biotechnology 3 (10), 519–522.

Bajaj, A., Lohan, P., Jha, P.N., Mehrotra, R., 2010. Biodiesel production through lipase catalyzed transesterification: an overview. Journal of Molecular Catalysis B: Enzymatic 62 (1), 9–14.

Barta, Z., Kovacs, K., Reczey, K., Zacchi, G., 2010. Process design and economics of on-site cellulase production on various carbon sources in a softwood-based ethanol plant. Enzyme Research 2010.

Binod, P., Palkhiwala, P., Gaikaiwari, R., Nampoothiri, K.M., Duggal, A., Dey, K., Pandey, A., 2013. Industrial enzymes—present status and future perspectives for India. Journal of Scientific and Industrial Research 72, 271–286.

Bonnin, E., Saulnier, L., Brunel, M., Marot, C., Lesage-Meessen, L., Asther, M., Thibault, J.-F., 2002. Release of ferulic acid from agroindustrial by-products by the cell wall-degrading enzymes produced by *Aspergillus niger* I-1472. Enzyme and Microbial Technology 31 (7), 1000–1005.

Castilho, L.R., Polato, C.M., Baruque, E.A., Sant'Anna, G.L., Freire, D.M., 2000. Economic analysis of lipase production by *Penicillium restrictum* in solid-state and submerged fermentations. Biochemical Engineering Journal 4 (3), 239–247.

Clapés, P., Fessner, W.-D., Sprenger, G.A., Samland, A.K., 2010. Recent progress in stereoselective synthesis with aldolases. Current Opinion in Chemical Biology 14 (2), 154–167.

Dhillon, G.S., Kaur, S., Brar, S.K., 2012a. In-vitro decolorization of recalcitrant dyes through an ecofriendly approach using laccase from *Trametes versicolor* grown on brewer's spent grain. International Biodeterioration & Biodegradation 72, 67–75.

Dhillon, G.S., Kaur, S., Brar, S.K., Verma, M., 2012b. Flocculation and haze removal from crude beer using in-house produced laccase from *Trametes versicolor* cultured on brewer's spent grain. Journal of Agricultural and Food Chemistry 60 (32), 7895–7904.

Edwinoliver, N., Thirunavukarasu, K., Purushothaman, S., Rose, C., Gowthaman, M., Kamini, N., 2009. Corn steep liquor as a nutrition adjunct for the production of *Aspergillus niger* lipase and hydrolysis of oils thereof. Journal of Agricultural and Food Chemistry 57 (22), 10658–10663.

Enzymes Market Analysis by Product (Carbohydrase, Proteases, Lipases, Polymerases & Nucleases) and Segment Forecasts to 2020, n.d. Retrieved March 4, 2015, from: http://www.grandviewresearch.com/industry-analysis/enzymes-industry.

Faulds, C., Williamson, G., 1995. Release of ferulic acid from wheat bran by a ferulic acid esterase (FAE-III) from *Aspergillus niger*. Applied Microbiology and Biotechnology 43 (6), 1082–1087.

Furlan, S.A., Schneider, A.L., Merkle, R., de Fátima Carvalho-Jonas, M., Jonas, R., 2000. Formulation of a lactose-free, low-cost culture medium for the production of β-D-galactosidase by *Kluyveromyces marxianus*. Biotechnology Letters 22 (7), 589–593.

Gnansounou, E., Dauriat, A., 2010. Techno-economic analysis of lignocellulosic ethanol: a review. Bioresource Technology 101 (13), 4980–4991.

Hamelinck, C.N., Van Hooijdonk, G., Faaij, A.P., 2005. Ethanol from lignocellulosic biomass: techno-economic performance in short-, middle-and long-term. Biomass and Bioenergy 28 (4), 384–410.

Hexa Research, n.d. Global Enzymes Market Size, Market Share, Application Analysis, Regional Outlook, Growth Trends, Competitive Scenario and Forecasts, 2012 to 2020. Retrieved March 3, 2015, from: http://www.hexaresearch.com/research-report/global-enzymes-market-size-market-share-application-analysis-regional-outlook-growth-trends-competitive-scenario-forecasts-2012-2020/.

Jegannathan, K.R., Nielsen, P.H., 2013. Environmental assessment of enzyme use in industrial production–a literature review. Journal of Cleaner Production 42, 228–240.

Kasturi, C., Baeck, A., Wolff, A.M., 1998. Detergent Compositions Containing Lipase and Protease (Google Patents).

Kirk, O., Borchert, T.V., Fuglsang, C.C., 2002. Industrial enzyme applications. Current Opinion in Biotechnology 13 (4), 345–351.

Klein-Marcuschamer, D., Oleskowicz-Popiel, P., Simmons, B.A., Blanch, H.W., 2012. The challenge of enzyme cost in the production of lignocellulosic biofuels. Biotechnology and Bioengineering 109 (4), 1083–1087.

Lam, K.F., Leung, C.C.J., Lei, H.M., Lin, C.S.K., 2014. Economic feasibility of a pilot-scale fermentative succinic acid production from bakery wastes. Food and Bioproducts Processing 92 (3), 282–290.

Laszlo, J.A., Compton, D.L., Li, X.-L., 2006. Feruloyl esterase hydrolysis and recovery of ferulic acid from jojoba meal. Industrial Crops and Products 23 (1), 46–53.

Lau, M.W., Bals, B.D., Chundawat, S.P., Jin, M., Gunawan, C., Balan, V., Dale, B.E., 2012. An integrated paradigm for cellulosic biorefineries: utilization of lignocellulosic biomass as self-sufficient feedstocks for fuel, food precursors and saccharolytic enzyme production. Energy & Environmental Science 5 (5), 7100–7110.

Li, B., Smith, B., Hossain, M.M., 2006. Extraction of phenolics from citrus peels: II. Enzyme-assisted extraction method. Separation and Purification Technology 48 (2), 189–196.

Martínez-Anaya, M.A., 1996. Enzymes and bread flavor. Journal of Agricultural and Food Chemistry 44 (9), 2469–2480.

Max, B., Salgado, J.M., Rodríguez, N., Cortés, S., Converti, A., Domínguez, J.M., 2010. Biotechnological production of citric acid. Brazilian Journal of Microbiology 41 (4), 862–875.

Meyer, A.S., Dam, B.P., Lærke, H.N., 2009. Enzymatic solubilization of a pectinaceous dietary fiber fraction from potato pulp: optimization of the fiber extraction process. Biochemical Engineering Journal 43 (1), 106–112.

Mukherjee, S., Das, P., Sen, R., 2006. Towards commercial production of microbial surfactants. Trends in Biotechnology 24 (11), 509–515.

Nagar, S., Gupta, V.K., Kumar, D., Kumar, L., Kuhad, R.C., 2010. Production and optimization of cellulase-free, alkali-stable xylanase by *Bacillus pumilus* SV-85S in submerged fermentation. Journal of Industrial Microbiology & Biotechnology 37 (1), 71–83.

Nampoothiri, K.M., Rahulan, R., Pandey, A., 2008. Microbial aminopeptidases and their application in food processing. In: Porta, R., Di Pierro, P., Moriniello, L. (Eds.), Recent Research Developments in Food Biotechnology. Enzymes as additives or processing aids. Research Signpost, Trivandrum, pp. 55–84.

Nascimento, R., Junior, N., Pereira Jr., N., Bon, E., Coelho, R., 2009. Brewer's spent grain and corn steep liquor as substrates for cellulolytic enzymes production by *Streptomyces malaysiensis*. Letters in Applied Microbiology 48 (5), 529–535.

Olsson, L., Christensen, T.M., Hansen, K.P., Palmqvist, E.A., 2003. Influence of the carbon source on production of cellulases, hemicellulases and pectinases by *Trichoderma reesei* Rut C-30. Enzyme and Microbial Technology 33 (5), 612–619.

Pal, A., Khanum, F., 2011. Efficacy of xylanase purified from *Aspergillus niger* DFR-5 alone and in combination with pectinase and cellulase to improve yield and clarity of pineapple juice. Journal of Food Science and Technology 48 (5), 560–568.

Palmqvist, E., Hahn-Hägerdal, B., Szengyel, Z., Zacchi, G., Rèczey, K., 1997. Simultaneous detoxification and enzyme production of hemicellulose hydrolysates obtained after steam pretreatment. Enzyme and Microbial Technology 20 (4), 286–293.

Pandey, A., 2003. Solid-state fermentation. Biochemical Engineering Journal 13 (2), 81–84.

Pleissner, D., Lam, W.C., Han, W., Lau, K.Y., Cheung, L.C., Lee, M.W., Sun, Z., 2014. Fermentative polyhydroxybutyrate production from a novel feedstock derived from bakery waste. BioMed Research International 2014.

Rahulan, R., Dhar, K.S., Nampoothiri, K.M., Pandey, A., 2012. Aminopeptidase from Streptomyces gedanensis as a useful tool for protein hydrolysate preparations with improved functional properties. Journal of food science 77 (7), C791–C797.

Raksakulthai, R., Haard, N.F., 2003. Exopeptidases and their application to reduce bitterness in food: a review. Critical Reviews in Food Science and Nutrition 43.

Shah, S., Sharma, S., Gupta, M., 2004. Biodiesel preparation by lipase-catalyzed transesterification of Jatropha oil. Energy & Fuels 18 (1), 154–159.

Shankaranand, V., Lonsane, B., 1994. Ability of *Aspergillus niger* to tolerate metal ions and minerals in a solid-state fermentation system for the production of citric acid. Process Biochemistry 29 (1), 29–37.

Sharma, D., Satyanarayana, T., 2006. A marked enhancement in the production of a highly alkaline and thermostable pectinase by *Bacillus pumilus* dcsr1 in submerged fermentation by using statistical methods. Bioresource Technology 97 (5), 727–733.

Sheppy, C., 2001. The current feed enzyme market and likely trends. Enzymes in Farm Animal Nutrition. CABI Publishing, Wallingford, UK, pp. 1–10.

Sherief, A., El-Tanash, A., Atia, N., 2010. Cellulase production by *Aspergillus fumigatus* grown on mixed substrate of rice straw and wheat bran. Research Journal of Microbiology 5.

Singh, A., Singh, N., Bishnoi, N.R., 2009. Production of cellulases by *Aspergillus heteromorphus* from wheat straw under submerged fermentation. International Journal of Civil and Environmental Engineering 1 (1), 23–26.

Singh, A., Tuteja, S., Singh, N., Bishnoi, N.R., 2011. Enhanced saccharification of rice straw and hull by microwave–alkali pretreatment and lignocellulolytic enzyme production. Bioresource Technology 102 (2), 1773–1782.

Singhania, R.R., Patel, A.K., Soccol, C.R., Pandey, A., 2009. Recent advances in solid-state fermentation. Biochemical Engineering Journal 44 (1), 13–18.

Soumanou, M.M., Bornscheuer, U.T., 2003. Improvement in lipase-catalyzed synthesis of fatty acid methyl esters from sunflower oil. Enzyme and Microbial Technology 33 (1), 97–103.

Teng, Y., Xu, Y., 2008. Culture condition improvement for whole-cell lipase production in submerged fermentation by *Rhizopus chinensis* using statistical method. Bioresource Technology 99 (9), 3900–3907.

The Novozymes Report 2014, n.d. Retrieved March 3, 2015, from: http://report2014.novozymes.com/menu/the-novozymes-report-2014/report/sales-and-markets.

Treviño-Cueto, B., Luis, M., Contreras-Esquivel, J., Rodríguez, R., Aguilera, A., Aguilar, C., 2007. Gallic acid and tannase accumulation during fungal solid state culture of a tannin-rich desert plant (*Larrea tridentata* Cov.). Bioresource Technology 98 (3), 721–724.

Vidyalakshmi, R., Paranthaman, R., Indhumathi, J., 2009. Amylase production on submerged fermentation by *Bacillus* spp. World Journal of Chemistry 4 (1), 89–91.

Viniegra-González, G., Favela-Torres, E., 2006. Why solid-state fermentation seems to be resistant to catabolite repression? Food Technology and Biotechnology 44 (3), 397.

Viniegra-González, G., Favela-Torres, E., Aguilar, C.N., de Jesus Rómero-Gomez, S., Díaz-Godínez, G., Augur, C., 2003. Advantages of fungal enzyme production in solid state over liquid fermentation systems. Biochemical Engineering Journal 13 (2), 157–167.

Viniegra-González, G., 1998. Strategies for the Selection of Mold Strains Geared to Produce Enzymes on Solid Substrates. Advances in Bioprocess EngineeringSpringer, pp. 123–136.

Wang, J., 2008. Electrochemical glucose biosensors. Chemical Reviews 108 (2), 814–825.

Wheelwright, S.M., 1989. The design of downstream processes for large-scale protein purification. Journal of Biotechnology 11 (2), 89–102.

World Enzymes – Industry Market Research, Market Share, Market Size, Sales, Demand Forecast, Market Leaders, Company Profiles, Industry Trends, n.d. Retrieved March 4, 2015, from: http://www.freedoniagroup.com/World-Enzymes.html.

Zuorro, A., Lavecchia, R., Medici, F., Piga, L., 2013. Enzyme-assisted production of tomato seed oil enriched with lycopene from tomato pomace. Food and Bioprocess Technology 6 (12), 3499–3509.

FURTHER READING

Bansal, S., Goel, G., 2015. Commercial application of rumen microbial enzymes. In: Rumen Microbiology: From Evolution to Revolution. Springer India, pp. 281–291.

Milmo, S., 2012. The enzyme makers. Chemistry & Industry 76 (11), 28–31.

INDEX

'*Note*: Page numbers followed by "f" indicate figures and "t" indicate tables.'

Edwards Brothers Malloy
Ann Arbor MI. USA
November 16, 2016